Global Virology IV: Viral Disease Diagnosis and Treatment Delivery in the 21st Century

Charurut Somboonwit • Paul Shapshak
Pandjassarame Kangueane • S. Balaji
John T. Sinnott • Lynette J. Menezes
Asa Oxner
Editors

Global Virology IV: Viral Disease Diagnosis and Treatment Delivery in the 21st Century

Springer

Editors
Charurut Somboonwit
Department of Internal Medicine
Division of Infectious Disease and
International Medicine
University of South Florida
Morsani College of Medicine
Tampa, FL, USA

Pandjassarame Kangueane
Irulan Sandy Annex
Biomedical Informatics
Pondicherry, Pondicherry, India

John T. Sinnott
Department of Internal Medicine
University of South Florida
Morsani College of Medicine
Tampa, FL, USA

Asa Oxner
University of South Florida
Division of General Internal Medicine
Morsani Center for Advanced Healthcare
Tampa, FL, USA

Paul Shapshak
Department of Internal Medicine
University of South Florida
Morsani College of Medicine
Tampa, FL, USA

S. Balaji
Department of Biotechnology
Manipal Institute of Technology
Manipal Academy of Higher Education
Manipal, Karnataka, India

Lynette J. Menezes
University of South Florida
Morsani College of Medicine
Tampa, FL, USA

ISBN 978-3-031-57368-2 ISBN 978-3-031-57369-9 (eBook)
https://doi.org/10.1007/978-3-031-57369-9

This Springer imprint is published by the registered company Springer Nature Switzerland AG
The registered company address is: Gewerbestrasse 11, 6330 Cham, Switzerland

Preface

The twenty-first century brings notable changes to clinical virology, mirrored by rapid advancements across science and technology. *Global Virology IV: Viral Disease Diagnosis and Treatment Delivery in the 21st Century* reflects this evolving landscape, offering insights into innovative methods enhancing our approach to viral diseases.

This book is intended for a broad audience, from medical students to experienced professionals. It provides an update on the latest advances in virology and sheds light on trends influencing clinical practice.

Recent global events, including pandemics and viral outbreaks, have highlighted the importance of innovation in our field, especially for the care of vulnerable populations. While we have yet to understand viruses like COVID-19 fully, this volume focuses on the underlying principles and advancements that are likely to have a significant impact.

This book presents a series of new concepts and methods. Respectful of its predecessors, this book builds upon the foundation laid by GVI through GVIII. We prioritize exploring new ideas and practices, ensuring fresh and relevant content. The chapters bridge the gap between current practice and future applications. This book is not a mere repetition of past volumes but a step to guide readers to the frontier of clinical virology.

With this volume, we invite you to explore the ongoing evolution of clinical virology, which remains as dynamic and challenging as ever. We hope to provide a helpful resource that sparks curiosity, informs practice, and contributes to improving global health.

Welcome to the state of virology today and the directions it may head in the future.

Tampa, FL, USA	Charurut Somboonwit
Tampa, FL, USA	Paul Shapshak
Pondicherry, Pondicherry, India	Pandjassarame Kangueane
Manipal, Karnataka, India	S. Balaji
Tampa, FL, USA	John T. Sinnott
Tampa, FL, USA	Lynette J. Menezes
Tampa, FL, USA	Asa Oxner

Introduction

Global Virology IV: Viral Disease Diagnosis and Treatment Delivery in the 21st Century explores our relentless battle against viral diseases. This volume will explore the historical context and then contemporary scientific advancements. Hopefully, a coherent narrative of our species' response to viral threats will be seen, and notable progress in treatment and prevention strategies will be presented.

We begin with a detailed analysis of emerging viruses, zeroing in on the genetic complexities of SARS-CoV-2 variants and their broader public health ramifications. The astute reader will note the focus shift to mankind's enduring struggle with perhaps our threat—the orthomyxoviruses. Always a dynamic challenge this pathogen continues to evolve.

Our microscope then explores Ebola's intricate virology, examines strategies for outbreak management, and discusses preventive measures. In the context of the Zika virus, the book highlights the recent leaps in genetic understanding and the diagnostic and therapeutic innovations triggered by this emerging virus.

Subsequent chapters delve into the omnipresent oncogenic human papillomavirus and its impact on genital and extragenital disease. Then, we focus on the problem of hepatitis C in specific populations and the most recent advances in treatment.

Our authors explore the intersection of emerging viral infections with health and social issues by further examining NeuroAIDS, the mental health implications of antiretroviral therapy, and even the effects of marijuana on people living with HIV.

The discussion addresses the challenges immunocompromised individuals face when dealing with viral infections, focusing on those undergoing hematopoietic cell transplants. It explores the repercussions of COVID-19 for cancer patients and stem cell transplant recipients and impacts of the Epstein-Barr Virus's infection pathways and prevention approaches, and considers the effects of Human Herpesviruses 6A, 6B, 7, and 8 on stem cell transplantation. The text discusses the management of infections such as Cytomegalovirus, Herpes Simplex, Varicella Zoster Virus, and Hemophagocytic Lymphohistiocytosis, and notes the progress in developing antiviral therapies.

The final chapters provide an update on the latest antiviral medications, from cidofovir to brincidofovir, and introduce the innovative domain of radioimmunotherapy, blending radiology with immunotherapy to create new treatments.

Finally, *Global Virology IV: Viral Disease Diagnosis and Treatment Delivery in the 21st Century* is an informative resource on virology, charting the scientific progress and ongoing efforts to tackle viral diseases.

Contents

The Genome Sequence Analysis of SARS-CoV-2 Variants of Concern

Adithi Somayaji and S. Balaji

Abstract The coronavirus pandemic has been at the forefront of the news, with many variants emerging with varied infectivity and transmissibility. These variants are the result of mutations in the virus. The evolutionary relationships of various SARS-CoV-2 variants were analyzed with a major focus on the Omicron variant. There were 61 sequence representatives obtained from GISAID and GenBank databases and were aligned and the phylogenetic trees were represented using MEGAX software. Evolutionary substitution models were utilized in conjunction with clustering approaches such as UPGMA, Neighbor-Joining (NJ), Minimum Evolution, and Maximum Composite Likelihood. The sequences of the omicron variant showed a close phylogenetic relationship with the Alpha variant UK (MZ376737.1), whereas the others showed significant distance from the Omicron variant.

Keywords SARS-CoV-2 · Genome analysis · Phylogenetics · Variants · Mutation · Omicron

1 Introduction

In the 1960s, scientists discovered that viruses found in some unhealthy animals had a similar bristly shape and spiky protein protrusions as a virus that causes the common cold in humans. These viruses resembled the solar corona under the electron microscope, prompting researchers to create the moniker coronaviruses in 1968. Coronaviruses were thought to only cause minor clinical manifestations in humans

A. Somayaji · S. Balaji (✉)
Department of Biotechnology, Manipal Institute of Technology, Manipal Academy of Higher Education, Manipal, Karnataka, India
e-mail: adithi.somayaji@learner.manipal.edu; s.balaji@manipal.edu

C. Somboonwit et al. (eds.), *Global Virology IV: Viral Disease Diagnosis and Treatment Delivery in the 21st Century*,
https://doi.org/10.1007/978-3-031-57369-9_1

until the 2003 outbreak of severe acute respiratory syndrome (SARS), which revealed how easily these adaptable viruses might kill people [1]. Now, SARS-CoV-2, a novel RNA virus from the *Coronaviridae* family, has resulted in a global outbreak, with the World Health Organization (WHO) declaring it a pandemic on March 11, 2020 [2].

The coronavirus genomes are the largest RNA viruses, with 30,000 genetic bases. SARS-CoV-2, unlike its close relatives, may infect human cells at various target sites, with the throat and lungs being the most common targets. Coronaviruses are also one of the few RNA viruses with a genetic proofreading system, which prevents the virus from accumulating mutations that would weaken it [1]. The receptor-binding domain (RBDs) of coronavirus spike proteins is critical for the virus's ability to penetrate human cells, with the SARS-CoV-2 binding domain being especially effective [3]. The spike proteins are the most distinct and important features of SARS-CoV-2 (and coronaviruses in general). These spike proteins were discovered to aid virus entry via the angiotensin-converting enzyme II (ACE2) receptor [4]. Additionally, proprotein convertase furin preactivates SARS-CoV-2 entry into the cells, lowering its reliance on target cell proteases. The SARS-CoV-2 can maintain efficient cell penetration while avoiding immunological detection due to the RBD's increased binding affinity to hACE2 and the furin preactivation of the spike protein [5]. The SARS-CoV-2 is reported to infect people within less than a fortnight of exposure, with an average incubation time of 4–5 days. Symptoms such as cough, cold, fever, weariness, and weakness are common. However, severe cases lead to lung damage, pneumonia, and mortality.

2 COVID-19 Variants and Implications

The coronavirus genome changes and evolves leading to variants. These variants are due to the accumulation of such changes or mutations in the viral genome. Many environmental mutagens modify the genomic composition of SARS-CoV-2 including metal ions, UV radiation, and endogenous elements of organisms [6]. The mutations in the RBD significantly improve the affinity of binding in the RBD-hACE2 complex and are connected to quick dissemination in human populations, and the term variant of concern is utilized [7]. At present, there are five variants of concern—Alpha, Beta, Gamma, Delta, and Omicron.

The alpha strain (B.1.1.7), first identified in the UK in September 2020 and drove the UK's second and third waves, is considered a variant of concern by the WHO [8]. The B.1.1.7 has gradually developed a foothold in the USA because of its high rate of transmission and lethality, which is 30–70% more than the original strain identified in Wuhan, China [9]. The efficacies of vaccines were reported in literature against this strain: 93.7% with the Pfizer-BioNTech vaccine, 100% with Moderna, 85.6% with Novavax vaccine, 71–91% Sinovac, and 74.5% Oxford-AstraZeneca vaccine [10].

The beta strain (B.1.351) was intially discovered in May 2020 in South Africa with higher infectivity in the younger age groups than earlier versions. Because of the E484K mutation, this strain is fatal and is the primary cause of the third wave in South Africa. Since then, the strain has been detected in 80 countries. This strain has been observed to evade several existing vaccines [8]. The beta variant showed complete resistance to therapeutically important monoclonal antibodies and neutralizing antibodies in the plasma of SARS-CoV-2 recovering patients in another in vitro research [9]. Early tests suggest that the Pfizer vaccine is slightly less effective against beta than the wild-type SARS-CoV-2 vaccine (about 75%), although both Moderna and Pfizer claim that their vaccines are still 95% effective against severe sickness and mortality [10].

The delta strain (B.1.617.2) originated in India (June 2021), had accounted for more than 80–90% of all cases in the UK and the USA. The devastating second wave of infections in India has been driven by the delta variant [8]. The infection rate of children and teens was five times more than older adults, and most of the infections were among unvaccinated individuals. It is also possible that some vaccines are less effective at neutralizing the delta version. According to a study, the delta variant is 60% more transmissible and it is more infective in the airways due to the increased amount of virus in the infected person (1260 times more than people infected with wild-type SARS-CoV-2), i.e., they can expel an increased amount of virus into the air [10]. The effectiveness of vaccines varies from 67% with Oxford-AstraZeneca, 88% with Pfizer-BioNTech, as well as 90% according to Sputnik V manufacturer [10].

The gamma strain (P.1) emerged in Manaus, Brazil (November 2020), was responsible for two Coronavirus outbreaks. Based on the patients' data, it was suggested that this strain is twice as transmissible as earlier strains. Only 54–79% protection was reported in this variation, resulting in a lethal second wave [9]. On November 26, 2021, WHO identified the variant Omicron (B.1.1.529) as a variant of concern (VOC) [11]. There were more than 30 mutations in the Omicron variant that have resulted in amino acid alterations in the spike protein sequences, 15 of which are situated in the RBD, which is critical for viral-cell contact mediated by the ACE-2 receptor. The Omicron spike gene sequence has been used to make inferences about the transmission rate. These findings revealed a group of alterations near the S1-S2 furin cleavage region that could improve viral infectivity. Furthermore, docking experiments revealed that a combination of mutations in the RBD would result in this variant's high binding affinity for human ACE2. This needs to be confirmed via in vitro studies [12].

There are variants categorized as "variants of interest" and "variants under monitoring." For instance, Lambda and Mu are designated as variants of interest as of February 2, 2022, while GR (B.1.1.318), GR (C.1.2), and GH/490R are classified as variants under monitoring. A genetic change linked to modifications in receptor binding, lower neutralization by antibodies established against a previous infection or vaccination, a projected increase in transmissibility or disease severity, reduced treatment efficacy, or potential diagnostic impact are all factors that contribute to make a variant, a variant of interest [13] (Table 1).

Table 1 Variant-specific data [14–16]

Variant	Pango Lineage	Notable mutations	Amino acid number	Amino acid change	Mutation location	Comments	Announced/estimated appearance on	Country of origin
Alpha	B.1.1.7	S:N501Y	501	Asparagine to Tyrosine	Spike protein	May increase ACE2 binding Common to Beta B.1.351, Gamma P.1, Mu B.1.621, Omicron	14 Dec 2020	UK
		S:H69-	69	Histidine to deleted	Spike protein	Might affect antibody recognition, perhaps affecting some antibody-therapy treatments or immunity Common to Eta B.1.525, S:126A B.1.620, Omicron		
		N:D3L	3	Aspartic acid to leucine	Nucleocapsid	The N:D3L and N:S235F mutations in the nucleocapsid appear in the unstructured portions of the N-terminal domain and linker sections, respectively, outside the dimer and RNA interaction interfaces However, mutation research predicted that N:S235F confers a stabilizing impact on the nucleocapsid		
		N:S235F	235	Serine to Phenylalanine				

Variant	Pango Lineage	Notable mutations	Amino acid number	Amino acid change	Mutation location	Comments	Announced/ estimated appearance on	Country of origin
Beta	B.1.351	S:E484K	484	Glutamic acid to Lysine	Spike protein	S:E484 mutations may diminish convalescent serum neutralization substantially Common to Kappa B.1.617.1, Gamma, V3 P.1, Eta B.1.525, Iota B.1.526, Mu B.1.621, S:126A B.1.620, Omicron	Dec 2020	South Africa
		N:T205I	205	Threonine to Isoleucine	Nucleocapsid			
Gamma	P.1	N:P80R	80	Proline to Arginine	Nucleocapsid		Jan 2021	Brazil
Delta	B.1.617	S:L452R	452	Leucine to Arginine	Spike protein	14 of 35 RBD monoclonal antibodies had their neutralizing activity lowered or eliminated by S:L452R Common in Kappa B.1.617.1, Epsilon B.1.427/9	Late 2020	India
		S:P681H	681	Proline to Arginine	Spike protein	Might diminish the recognition by class 3 antibodies Located near the furin cleavage site, which could be crucial for immunological recognition Common in Kappa B.1.617.1, Omicron		
		S:G142D	142	Glycine to Aspartic acid	Spike protein	Found in the spike protein's N-terminal domain, is an antibody escape mutant that has been found in viruses cultured in the presence of a monoclonal antibody		

(continued)

Table 1 (continued)

Variant	Pango Lineage	Notable mutations	Amino acid number	Amino acid change	Mutation location	Comments	Announced/ estimated appearance on	Country of origin
Omicron	B.1.1.529	S:H655Y S:N679K S:P681H	655 679 681	Histidine to tyrosine Asparagine to Lysine Proline to histidine	Spike protein	A cluster of mutations associated with an increased rate of transmission at the S1-S2 Furin cleavage site	Nov 2021	Multiple countries
		S:Q498R S:N501Y	498 501	Glutamine to arginine Asparagine to tyrosine	Spike protein	Associated with increased binding affinity to ACE-2 receptor		
Lambda	C.37	S:L452Q	452	Leucine to Glutamine	Spike protein	Similar to S:L452R	Late 2020	South America
		S:R246-toS:G252-	246 252	Arginine and Glycine respectively to deleted	Spike	Is a characteristic 7 amino acid deletion		

Variant	Pango Lineage	Notable mutations	Amino acid number	Amino acid change	Mutation location	Comments	Announced/ estimated appearance on	Country of origin
Mu	B.1.621	S:Y144T and S:Y145S	144 145	Tyrosine to Threonine and Serine respectively	Spike protein	Spike protein has a 3 nucleotide/1 amino acid insertion with an insertion "N" at position 146	Jan 2021	Colombia
		S:D950N	950	Glutamic acid to Asparagine	Spike protein			
Iota	B.1.526	S:A701V	701	Alanine to Valine	Spike protein	S:A701V is close to the furin cleavage site, while S:D253G is in an N-terminal domain region that is a target for neutralizing antibodies S:A701V is common to Beta B.1.351 S:D253G is common to Lambda C.37	Late 2020	USA
		S:D253G	253	Aspartic acid to Glycine	Spike protein	S:A701V is close to the furin cleavage site, while S:D253G is in an N-terminal domain region that is a target for neutralizing antibodies. S:A701V is common to Beta B.1.351 S:D253G is common to Lambda C.37		
Kappa	B.1.617.1	S:L452R, S:E484Q, and S:P681R	452 484 681	Leucine to Arginine Glutamic acid to Glutamine Proline to Arginine	Spike protein	Moderately lower sensitivity to vaccine-elicited antibodies and may have decreased cell-entry efficiency	Oct 2020	India
Eta	B.1.525	S:Y144-	144	Tyrosine to deleted	Spike protein	Associated with certain antibody escape Common to Alpha B.1.1.7, Mu B.1.621, S:126A B.1.620, Omicron	Dec 2020	Multiple countries
		S:H69-	69	Histidine to deleted	Spike protein	Might affect antibody recognition, perhaps affecting some antibody-therapy treatments or immunity Common to Alpha B.1.1.7, S:126A B.1.620, Omicron		
		S:V70-	70	Valine to deleted	Spike protein	Common to Alpha B.1.1.7, S:126A B.1.620, Omicron		

3 Materials and Methods

3.1 Sequence Retrieval

The genome sequences of COVID-19 variants were retrieved from the GenBank and GSIAD database. One full sequence of each variant from at least one country was collected from the database as of February 2, 2022 (Table 2), with NC_045512 being the Wuhan reference sequence. The sequences were kept to a minimum of 61 genome sequences to avoid handling errors during analysis.

3.2 Sequence Alignment

The 60 representative genomes, which included the Wuhan reference sequence, were stored in FASTA format. The sequences were aligned in MEGA X software using ClustalX 2.1 option (default parameters) [17] and as well as Clustal Omega (online).

3.3 Codon Usage, Frequency Analysis, and Heatmaps

The data on codon usage (standard), nucleotide, and amino acid frequencies were obtained using MEGAX. The tab-separated value data sets (.csv files) of nucleotide and amino acid frequencies were used as input for ClustVis (https://biit.cs.ut.ee/clustvis/), an online program for creating heatmaps [18]. The unit variance method of row scaling and the singular value decomposition method of imputation was set as default options.

3.4 Phylogenetic Tree Construction

The MEGAX standalone package was used to obtain the phylogenetic tree using the Neighbor-Joining, UPGMA, minimum evolution, and maximum likelihood using composite maximum likelihood model and p-distances. The sequence file of the 60 representative genomes was uploaded to MEGAX to obtain the overall phylogenetic trees. The bootstrapping was enabled, and resampling was set to 100. The pairwise distance matrix was obtained using MEGAX. The .csv file of the matrix was used as input for ClustVis with default options.

Table 2 The analyzed dataset is displayed, along with accession numbers, sampled geographic location, variant type, year of collection, sequence length, and GC%

Sl. No	Accession No.	Location	Variant	Year (collection)	Length	GC%
1	NC_045512	Wuhan	Reference sequence	2019	29,903	37.9728
2	EPI_ISL_6756515	Japan	Alpha variant	2021	29,763	37.9800
3	MZ413979.1	Bangladesh	Alpha variant	2021	29,766	37.9567
4	MW509795.1	Egypt	Alpha variant	2021	29,793	37.9500
5	MW822594.1	Germany	Alpha variant	2020	29,851	37.9567
6	MT991088.1	Hong Kong	Alpha variant	2020	29,793	37.9669
7	MZ268634	India	Alpha variant	2020	29,790	37.9837
8	MZ376737.1	UK	Alpha variant	2020	29,840	37.9500
9	OK319445.1	USA	Alpha variant	2021	29,817	37.9500
10	EPI_ISL_5416540	Japan	Beta variant	2021	29,764	37.9721
11	MZ562483.1	Bangladesh	Beta variant (B.1.351.3)	2021	29,835	37.9433
12	MW580244.1	France	Beta variant (B.1.351.2)	2021	29,812	37.9770
13	MZ314998.1	Germany	Beta variant	2021	29,827	37.9601
14	MZ317890.1	India	Beta variant	2021	29,784	37.9635
15	MZ413998.1	Pakistan	Beta variant	2021	29,885	37.9365
16	OU471655.1	Slovakia	Beta variant	2021	29,885	38.0411
17	MZ202314.1	South Africa	Beta variant	2020	29,783	37.9584
18	MZ277392.1	Taiwan	Beta variant (B.1.351.3)	2021	29,880	37.9433
19	OU212934.1	UK	Beta variant	2021	29,885	37.9500
20	MW617734.1	USA	Beta variant	2021	29,812	37.9466
21	MZ212516.1	USA	Beta variant (B.1.351.3)	2021	29,881	37.9365
22	MZ213758.1	USA	Beta variant (B.1.351.2)	2021	29,880	37.9466
23	EPI_ISL_6832166	Japan	Delta variant	2021	29,769	37.9556
24	OK104648.1	Egypt	Delta variant (B.1.617.2)	2021	29,903	37.9331
25	MZ340535.1	India	Delta variant (B.1.617.2)	2021	29,802	37.9601
26	LC643049.1	Japan	Delta variant (B.1.617.2)	2021	29,882	37.9196
27	OU659423.1	UK	Delta variant (B.1.617.2)	2021	29,890	37.9669
28	MZ436591.1	USA	Delta variant (AY.2)	2021	29,844	37.9365
29	MZ491704.1	USA	Delta variant (AY.3)	2021	29,779	37.9331

(continued)

Table 2 (continued)

Sl. No	Accession No.	Location	Variant	Year (collection)	Length	GC%
30	MZ892623.1	Uzbekistan	Delta variant (B.1.617.2)	2021	29,825	37.9534
31	OD990459	UK	Epsilon variant (B.1.427)	2021	29,903	37.9095
32	MW795884	USA	Epsilon variant (B.1.427)	2021	29,904	37.9196
33	MZ970734	Djibouti	Eta variant	2021	29,709	37.9635
34	MW598416	Ghana	Eta variant	2021	29,888	37.9669
35	OD976033	UK	Eta variant	2021	29,830	37.9635
36	MW974897	USA	Eta variant	2021	29,709	37.9669
37	EPI_ISL_6228367	Japan	Gamma variant	2021	29,768	39.3140
38	MZ020420.1	Bangladesh	Gamma variant	2021	29,853	37.9348
39	MZ477753.1	Brazil	Gamma variant	2021	29,735	37.9601
40	MW642248	Italy	Gamma variant	2021	29,853	37.9298
41	OU385891.1	UK	Gamma variant	2021	29,898	38.0985
42	MW963199.1	USA	Gamma	2021	29,777	37.9534
43	MW967498.1	USA	Gamma (P.1.2)	2021	29,897	37.9196
44	MZ611957.1	Venezuela	Gamma	2021	29,810	37.9365
45	EPI_ISL_6887009	France	GH490R	2021	29,724	37.9559
46	EPI_ISL_6910522	USA	hCoV	2021	29,807	37.9609
47	MW984875	USA	Iota variant (B.1.526)	2021	29,704	37.9837
48	MZ315140	Germany	Kappa variant	2021	29,781	37.9736
49	MZ314697	India	Kappa variant	2021	29,801	37.9871
50	MZ571142	Morocco	Kappa variant	2021	29,818	37.9635
51	OU391429	UK	Kappa variant	2021	29,902	38.3144
52	MZ257515	USA	Kappa variant	2021	29,876	37.9736
53	MZ275289.1	Peru	Lambda variant (C.37)	2021	29,781	37.9365
54	EPI_ISL_4470504	Japan	Mu GH	2021	29,781	37.9537
55	OU476023	Slovakia	Mu variant(B.1.621)	2021	29,899	37.9196
56	MZ348460	USA	Mu variant(B.1.621)	2021	29,839	37.9888
57	MZ710933	USA	Mu variant(B.1.621)	2021	29,881	37.9365
58	OK124281	USA	Mu variant (B.1.621)	2021	29,693	37.9466
59	OL677199.1	Canada	Omicron variant	2021	29,684	37.9497
60	EPI_ISL_6640916	Botswana	Omicron variant	2021	29,684	37.9813

4 Results and Discussion

4.1 Codon Usage, Frequency Analysis, and Heatmaps

For 60 genomes, the average frequency (in percentage) of A, C, G, and T was 32.182, 18.3362, 29.865, and 19.6152, respectively. The average computations did not include unresolved bases ("N"), which can occur owing to weak signal quality, base call errors, and other factors which still need to be eliminated while trimming [19].

The GC% of the entire dataset was calculated to be 37.98%, while the GT% was computed to be 51.79% due to the higher frequency of Thymine bases. The amino acid frequency and nucleotide frequency of the 60 sequences are depicted as heatmaps in Fig. 1a, b respectively. The nucleotide heat map (Fig. 1a) likewise shows that T/U is the most common, followed by A. The most common codon was UUU (354) for phenylalanine, while the least common was CCG (25.7) for alanine, according to the codon usage frequency with UGU (314.9) also observed under the high-frequency bracket. Several codons come under a low-frequency bracket, including CGA, CGC, etc.

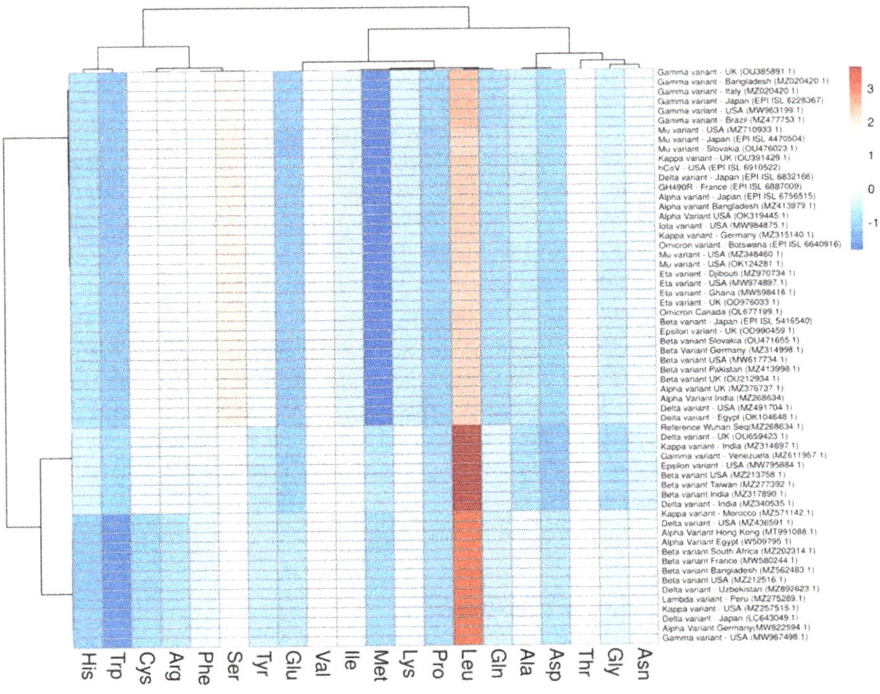

Fig. 1 (**a**) Amino acid frequency. (**b**) Nucleotide frequency

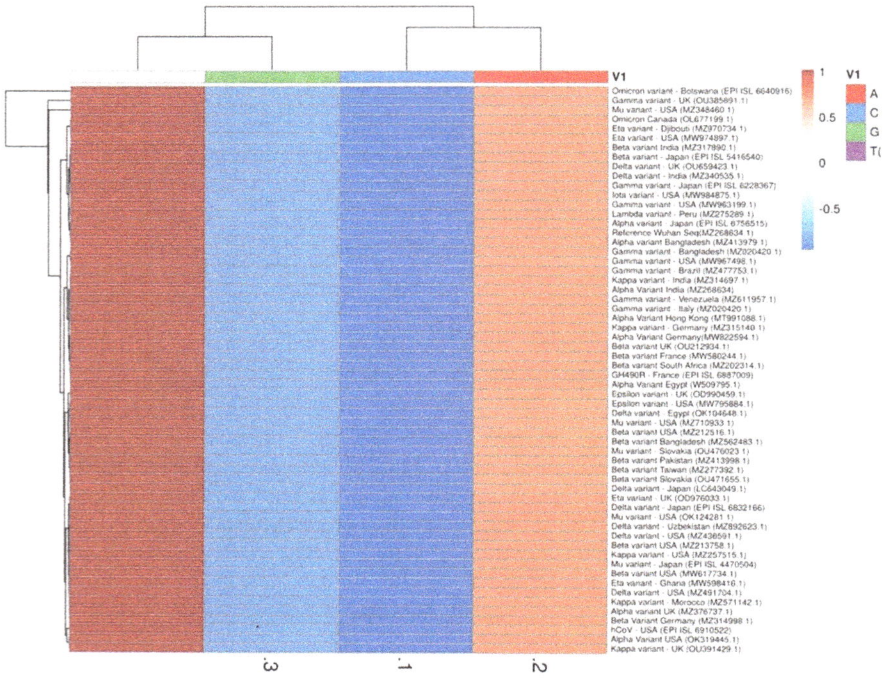

Fig. 1.1 (continued)

The amino acid heatmap (Fig. 1b) indicates a relatively leucine-rich pattern, followed by serine and threonine. The low-frequency bracket color of dark blue confirms the lack of ambiguous bases in the sequences. The least abundant amino acid was tryptophan (2.2963%), and the most abundant amino acid was computed to be leucine (11.673%), as shown in the figure. In both the heatmaps, three different clusters were observed with varied locations and variants. In the first cluster, all the variants of concern including omicron and at least one of the other variants were observed. Most of the sequences isolated from the USA were present in this cluster. The second cluster comprises the delta, gamma, beta, epsilon, and kappa from varied geographic locations. The reference sequence was also observed in the second cluster. In the third cluster, all the variants of concern, except omicron and Kappa and lambda variants, were observed. The geographical patterns were only readily observed in some of the clusters.

4.2 *Phylogenetic Tree Construction*

Two omicron variants were used to conduct the phylogenetic analysis: one from Botswana (the first sequenced omicron genome) and the other from Canada. They were aligned with alpha, beta, gamma, delta, mu, lambda, GH490R, kappa, eta, and epsilon variant genome sequences sequenced at various locations as well as SARS-CoV-2 USA isolates using online clustal omega website and MEGAX software. The phylogenetic analysis of the omicron variants was done using UPGMA/Maximum composite likelihood (MCL), UPGMA/p-distance, Neighbor-joining (NJ)/MCL, Maximum likelihood (ML)/MCL, and Minimum evolution/p-distance using MEGAX software as shown in Figs. 2, 3, 4, 5, and 6. The analysis conducted in this paper is like the study by Kandeel et al. 2021 [20].

All three phylogenetic trees showed a close relationship with the Alpha variant UK (MZ376737.1) and the Omicron variants. When developing the tree, the UPGMA model assumes all lineages advance at the same rate, and the mutation rate is ignored. The pairwise distance is used to build the tree. During tree construction, the NJ, on the other hand, examines the pace of evolution. The aggregate of linked log-likelihoods is referred to as a composite likelihood. Because the evolutionary links among the sequences cause all pairwise distances in a distance matrix to have correlations, the sum of their log-likelihoods is a composite likelihood. This distance is just the proportion (p) of nucleotide sites that differ between the two sequences being compared. As the study [20] suggests, a single evolution model cannot be used due to complex evolutionary methods of viruses and single-gene variations.

The distance matrix also shows the minimum distance between omicron variants and the alpha UK variant as shown in the heatmap (Fig. 7). Using NCBI nucleotide Blast, the identity percentage between Alpha variant and Omicron sequences was calculated, and Omicron Botswana-Alpha UK showed 99.69% identity and Omicron_Canada-Alpha_UK showed 99.97% identity. The reference sequence and the omicron variants showed significant distance (although not the highest), implying that the omicron variant is highly mutated. The Omicron variants are phylogenetically distant from the other variants, as shown by all four methods. However, in the UPGMA/MCL and UPGMA/p-distance methods, the two omicron sequences were more closely related than in the NJ/MCL and ML/MCL methods, where the omicron_Canada and alpha_UK were more closely related. This could be due to UPGMA models assuming the same evolutionary speed of all lineages and ignoring the mutation rates.

The delta and kappa variants were almost always clustered together, which could be due to their Indian origins. The Mu variant USA (MZ710993) and the GH490R_France variant showed a significant phylogenetic distance from other sequences. The beta variant India sequence showed the closest relationship with the reference sequence in all the phylogenetic methods.

Contrary to the study [20] from where we used the same sequences (GISAID obtained sequences), we observed that the Alpha variant Japan was not closely related to the omicron variants and instead had a closer relationship with some of the beta variant sequences (Slovakia, Japan, Pakistan, UK). This could be due to the higher number of sequences used, the different methods of phylogenetic tree construction, or changes in the models used to calculate distances.

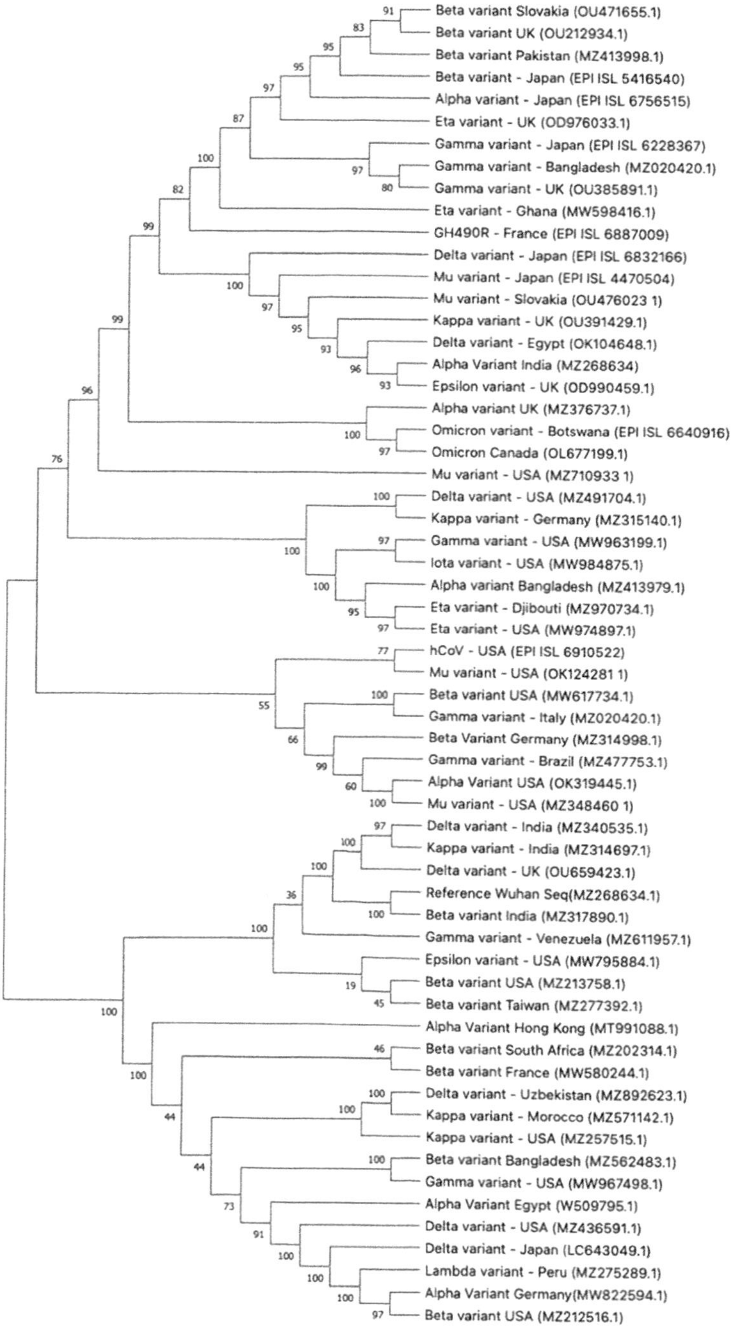

Fig. 2 Phylogenetic tree developed using UPGMA/Maximum composite likelihood model. Software used: MEGAX

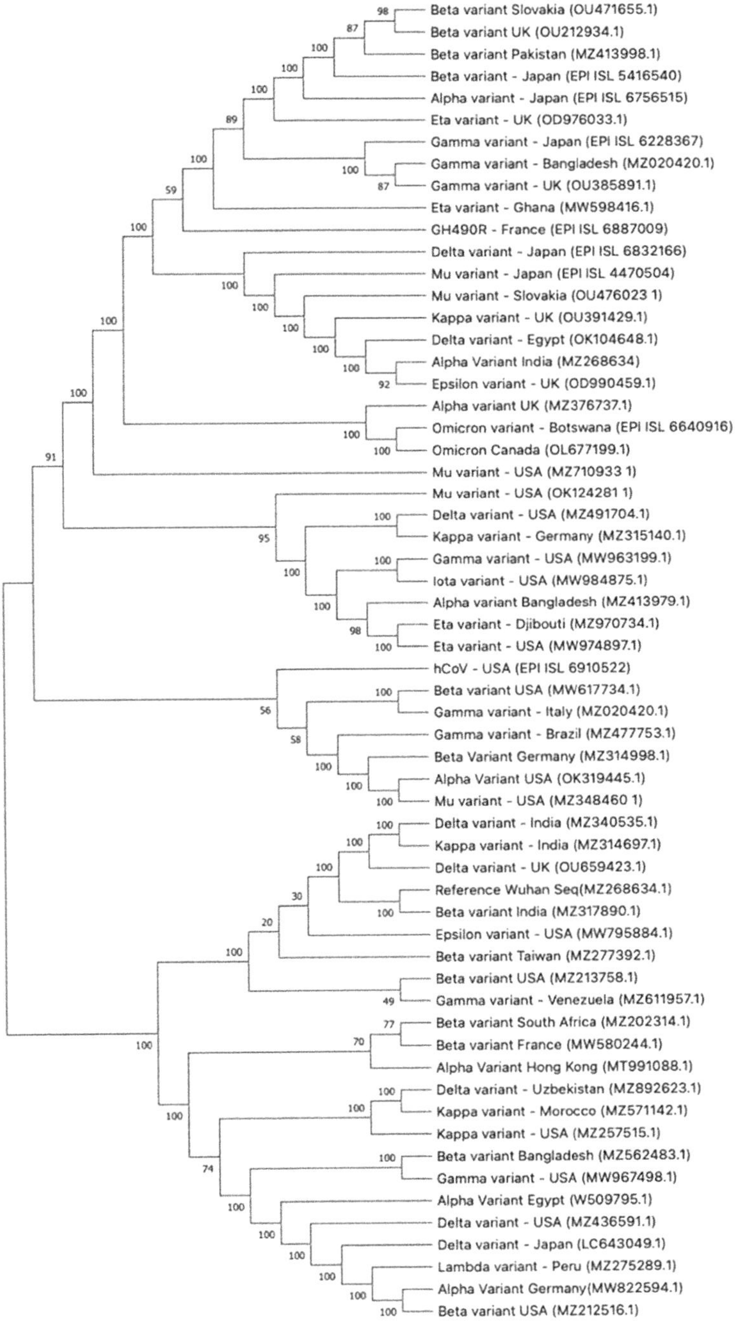

Fig. 3 Phylogenetic tree developed using UPGMA/p-distance model. Software used: MEGAX

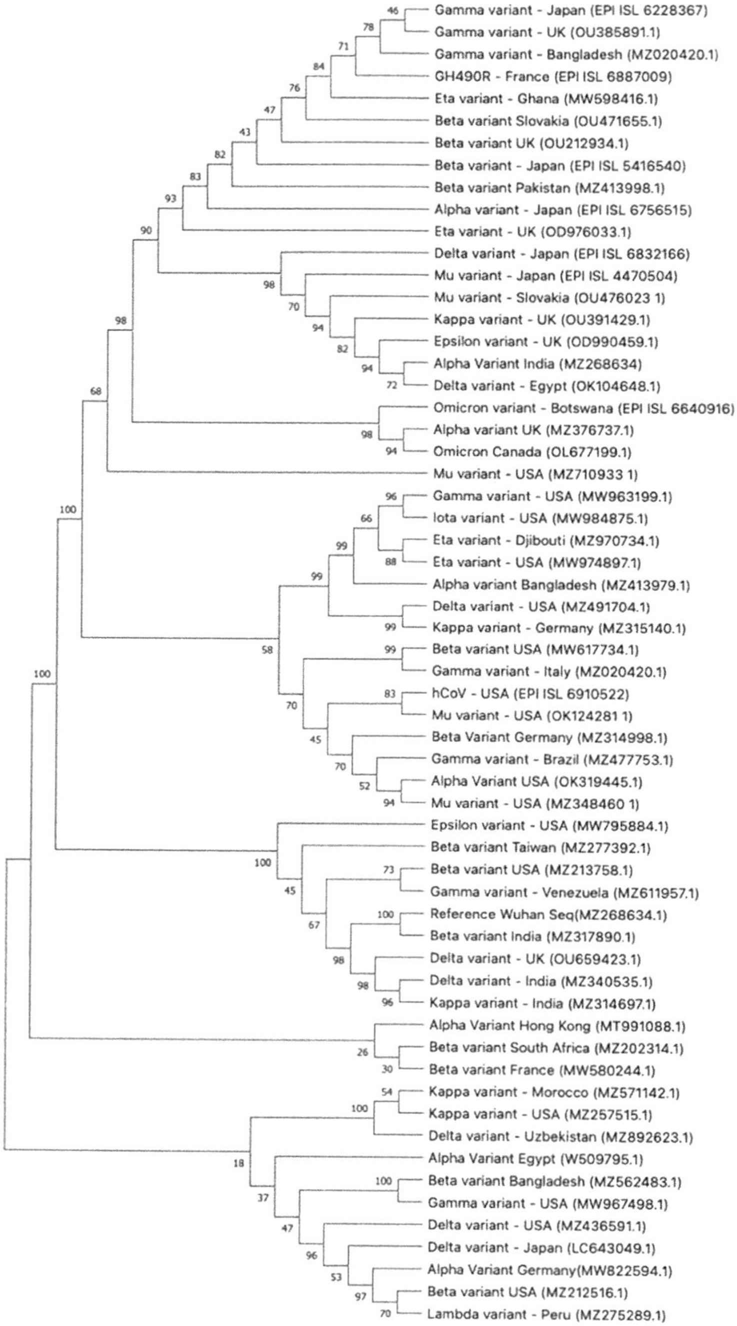

Fig. 4 Phylogenetic tree developed using Neighbor-Joining/Maximum composite likelihood model. Software used: MEGAX

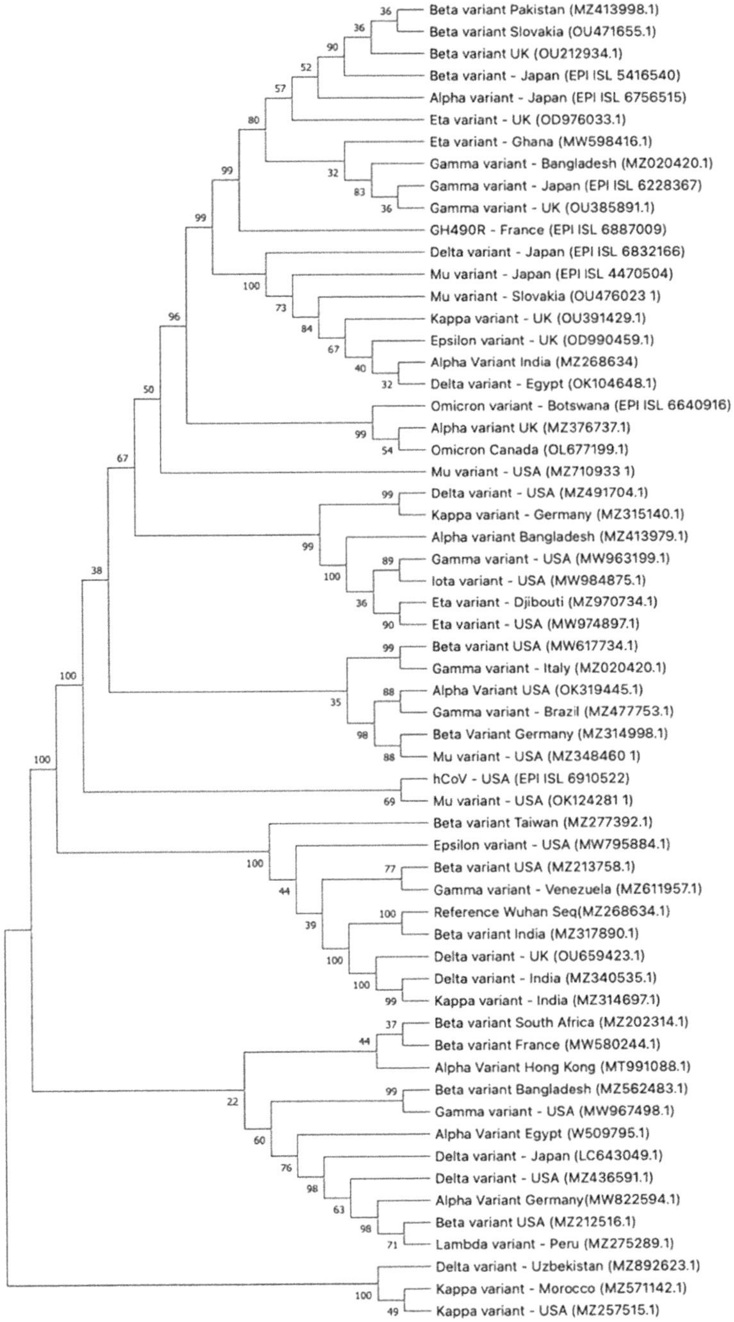

Fig. 5 Phylogenetic tree developed using Maximum Likelihood/Maximum composite likelihood model. Software used: MEGAX

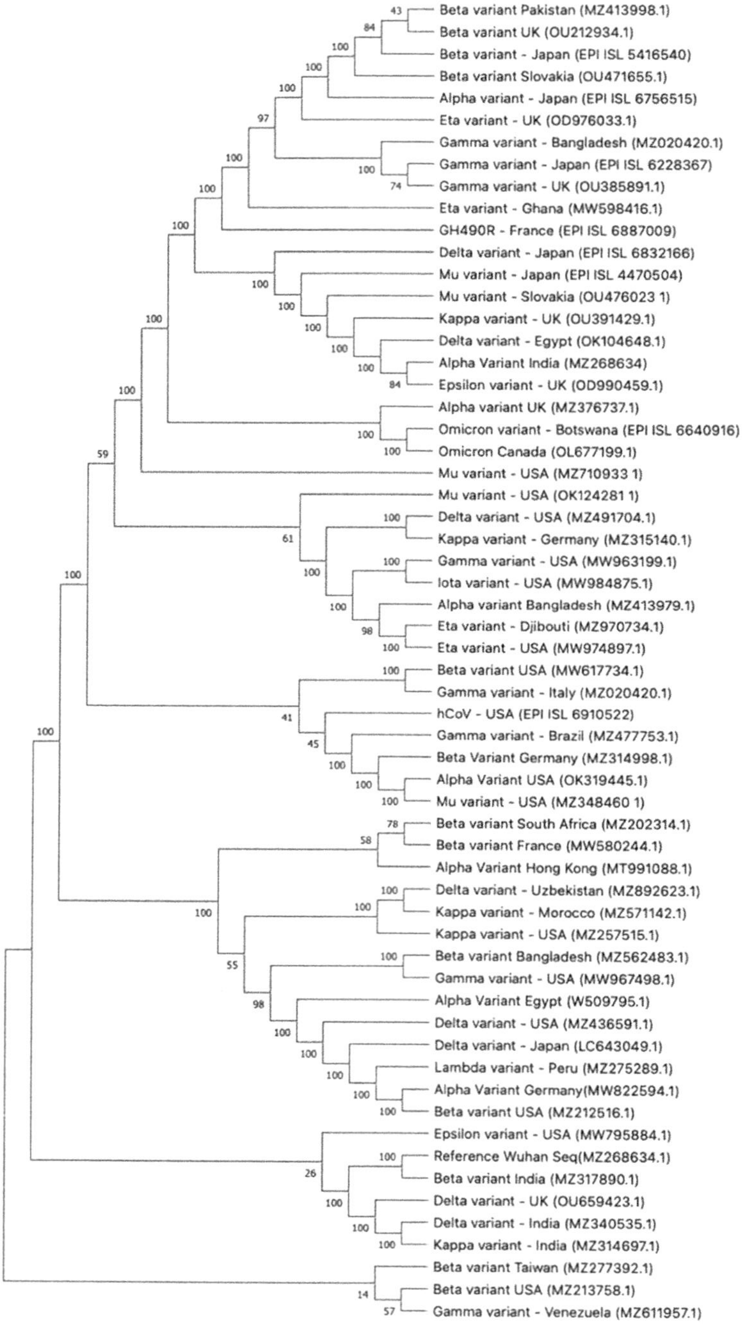

Fig. 6 Phylogenetic tree developed using Minimum evolution/p-distance model. Software used: MEGAX

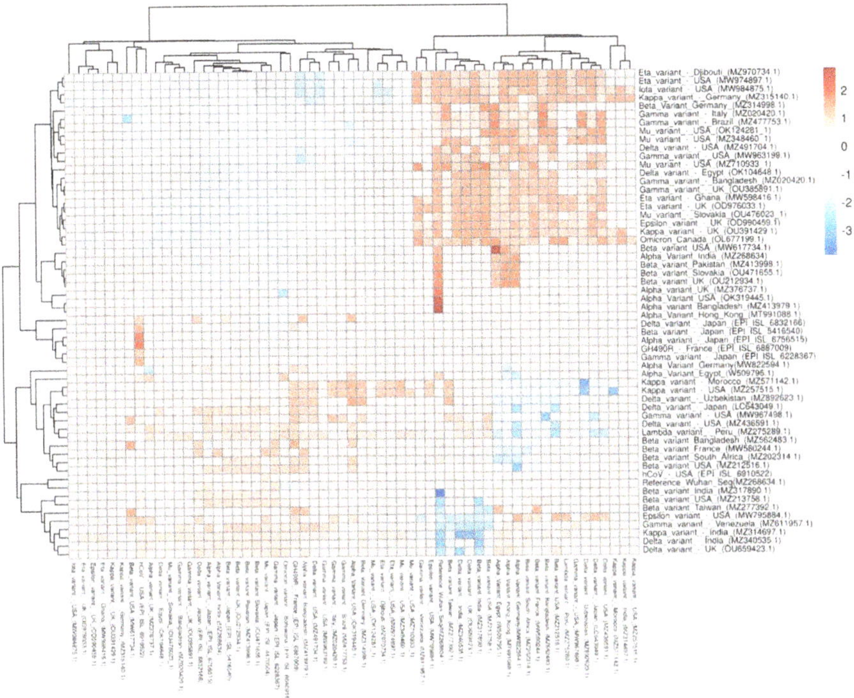

Fig. 7 Pairwise distance matrix heatmap. Software used: ClustVis (Online)

Moreover, similar to the phylogenetic results of the study by Yamin et al., the phylogenetic trees did not have any intermediate branches of evolution from the omicron variant (exception of Alpha variant UK), implying that the variant evolved differently from all the variants [21]. In addition, similar to their conclusion regarding the relationship between the delta variant and omicron, this study revealed a weak evolutionary relationship. In contrast to previous findings regarding gamma and omicron variants, this study did not find any relationship between gamma and omicron variants, which may be due to the differences in the analysis methods. Moreover, the covariant data demonstrates that the omicron variant has a very distinct mutation profile compared to other variants [14].

Acknowledgments The authors would like to acknowledge the support received from Manipal Institute of Technology and Manipal Academy of Higher Education.

Declaration of Competing Interests The authors declare no conflict of interest.

References

1. Cyranoski D (2020) Profile of a killer: the complex biology powering the coronavirus pandemic. In: Nature. https://www.nature.com/articles/d41586-020-01315-7. Accessed 20 Jun 2021
2. Behera P, Mazumder A, Arora M, et al (2020) SARS-CoV-2 epidemic in India: Epidemiological features and in silico analysis of the effect of interventions. F1000Res. https://doi.org/10.12688/f1000research.23496.1
3. Zhang T, Wu Q, Zhang Z (2020) Probable Pangolin Origin of SARS-CoV-2 Associated with the COVID-19 Outbreak. Current Biology 30:1346–1351.e2
4. Zhou H, Chen X, Hu T, et al (2020) A Novel Bat Coronavirus Closely Related to SARS-CoV-2 Contains Natural Insertions at the S1/S2 Cleavage Site of the Spike Protein. Current Biology 30:2196–2203.e3
5. Wang MY, Zhao R, Gao LJ, Gao XF, Wang DP, Cao JM (2020) SARS-CoV-2: Structure, Biology, and Structure-Based Therapeutics Development. Frontiers in Cellular and Infection Microbiology. https://doi.org/10.3389/fcimb.2020.587269
6. Sanjuán R, Domingo-Calap P (2016) Mechanisms of viral mutation. Cellular and Molecular Life Sciences 73:4433–4448
7. (2022) Variant of concern. In: Wikipedia. https://en.wikipedia.org/wiki/Variant_of_concern. Accessed 20 Jun 2021
8. Roy B, Dhillon J, Habib N, Pugazhandhi B (2021) Global variants of COVID-19: Current understanding. Journal of Biomedical Sciences 8:8–11
9. Robishaw JD, Alter SM, Solano JJ, Shih RD, DeMets DL, Maki DG, Hennekens CH (2021) Genomic surveillance to combat COVID-19: challenges and opportunities. The Lancet Microbe 2:e481–e484
10. Mahase E (2021) Covid-19: How many variants are there, and what do we know about them?: Video 1. BMJ n1971
11. Classification of omicron (B.1.1.529): SARS-CoV-2 variant of concern. https://www.who.int/news/item/26-11-2021-classification-of-omicron-(b.1.1.529)-sars-cov-2-variant-of-concern. Accessed 10 Jan 2022
12. Ferré VM, Peiffer-Smadja N, Visseaux B, Descamps D, Ghosn J, Charpentier C (2022) Omicron SARS-CoV-2 variant: What we know and what we don't. Anaesthesia Critical Care and Pain Medicine. https://doi.org/10.1016/j.accpm.2021.100998
13. (2021) SARS-CoV-2 Variant Classifications and Definitions. In: Centres for disease control and prevention. https://www.cdc.gov/coronavirus/2019-ncov/variants/variant-classifications.html?CDC_AA_refVal=https%3A%2F%2Fwww.cdc.gov%2Fcoronavirus%2F2019-ncov%2Fvariants%2Fvariant-info.html. Accessed 1 Jan 2022
14. Overview of Variants/Mutations. In: GISAID. https://covariants.org/variants. Accessed 1 Jan 2022
15. Outbreak.info. https://outbreak.info. Accessed 7 Sep 2021
16. Detect and analyze variants of SARS-CoV-2. https://cov-spectrum.org/explore/Switzerland/AllSamples/AllTimes. Accessed 2 Sep 2021
17. Stecher G, Tamura K, Kumar S (2020) Molecular evolutionary genetics analysis (MEGA) for macOS. Molecular Biology and Evolution 37:1237–1239
18. Metsalu T, Vilo J (2015) ClustVis: A web tool for visualizing clustering of multivariate data using Principal Component Analysis and heatmap. Nucleic Acids Research 43:W566–W570
19. Seetharaman B, Ramachandran A, Nandy K, Paul S (2017) Sequence accuracy in primary databases: A case study on HIV-1B. In: Global Virology II - HIV and NeuroAIDS. Springer New York, pp 779–822
20. Kandeel M, Mohamed MEM, Abd El-Lateef HM, Venugopala KN, El-Beltagi HS (2021) Omicron variant genome evolution and phylogenetics. Journal of Medical Virology. https://doi.org/10.1002/jmv.27515
21. Sun Y, Lin W, Dong W, Xu J (2022) Origin and evolutionary analysis of the SARS-CoV-2 Omicron variant. Journal of Biosafety and Biosecurity 4:33–37

Influenza: Clinical Challenges in the Twenty-First Century

Mindy M. Sampson, Christopher M. Polk, Anupama Neelakanta, and Catherine L. Passaretti

Abstract Influenza has had a significant impact on society and human health and may have had a similar impact in earlier historical times, if post-hoc diagnoses are correct. Significant progress has been made within diagnostics, vaccines, treatment, and infection control practices for influenza, which has resulted in improved control of this virus at both the seasonal epidemic and pandemic levels. Despite advances in knowledge of influenza, it is important that we continue to improve access to healthcare and address disparities between many populations. Poor outcomes occur when people infected with influenza have less access to healthcare or refuse it. With improved understanding of viral evolution, it is also critical for public health agencies to plan for continual future influenza epidemics and pandemics.

Keywords Influenza · Pandemic · Epidemics · H1N1 · Vaccine · Age · Race · Sex · Diagnostic testing · Baloxivir · Oseltamavir · Infection control · Healthcare · Disparities · Minorities

---

M. M. Sampson (✉)
Department of Internal Medicine, Stanford University, Palo Alto, CA, USA
e-mail: mindysam@stanford.edu

C. M. Polk
Department of Internal Medicine, Wake Forest School of Medicine, Winston-Salem, NC, USA
e-mail: Christopher.Polk@atriumhealth.org

A. Neelakanta · C. L. Passaretti
Department of Internal Medicine, Wake Forest School of Medicine, Winston-Salem, NC, USA

Department of Internal Medicine, Atrium Health, Charlotte, NC, USA
e-mail: Anupama.Neelakanta@atriumhealth.org; Catherine.Passaretti@atriumhealth.org

C. Somboonwit et al. (eds.), *Global Virology IV: Viral Disease Diagnosis and Treatment Delivery in the 21st Century*,
https://doi.org/10.1007/978-3-031-57369-9_2

1 Key Concepts

1. Small genomic variations of influenza viral proteins, antigen drift, can lead to changes in seasonal influenza subtypes, leading to epidemics. However, larger genomic mutations can lead to antigenic shifts, which potentially cause pandemics.
2. Disparities in healthcare access related to influenza infection continue to exist and need to be addressed—women and people of color are known to have more severe illness.
3. The gold standard for influenza testing utilizes polymerase chain reaction (PCR)-based methods. However, test rapidity is also an important factor given the need for early treatment and infection control interventions.
4. Neuraminidase inhibitors, which act by preventing viral particle release, are the mainstay of influenza treatment. Additionally, Baloxavir is a newer antiviral which inhibits viral replication by viral polymerase blockade.
5. Infection prevention practices are key to controlling the spread of influenza. This comprises a multifactorial approach that includes vaccination, masking, hand hygiene, surveillance, and chemoprophylaxis.

2 Influenza Background and Evolution

Influenza is an RNA virus, a member of the *Orthomyxoviridae* family. While cases of influenza-like illness (ILI) are thought to date back as far as 412 BC, the virus was first discovered in the 1930s [1]. Classically, influenza infection results in fever, chills, myalgias, headache, cough, and occasionally in more severe cases, pneumonia and respiratory failure [2, 3]. Individuals at extremes of age or with weakened immune systems are most prone to severe illness and complications associated with influenza infection [3]. Influenza symptoms typically develop 1–4 days post-exposure. Among exposed persons, approximately two thirds will develop clinical symptoms [4]. Some individuals may be asymptomatic and still shed virus [4]. Individuals can spread infection as early as 1 day post-exposure [4]. Most individuals are symptomatically improved and no longer shed virus within 5–10 days. However, children and highly immunocompromised individuals may shed virus for much longer periods [4, 5].

Influenza is classified into three types: influenza A, influenza B, and influenza C [6]. Influenza A and B are further subtyped based on their hemagglutinin (HA) and neuraminidase (NA) proteins [2, 3]. Influenza B, historically, undergoes less subtype genetic variation than influenza A, with only two distinct lineages in circulation since 1983, called Victoria and Yamagata strains [7]. However, the Yamagata lineage hasn't been isolated since 2020 and could be extinct [8]. There are several influenza C variants, which co-circulate, typically without seasonality nor severe disease association [9].

The evolution of influenza A, in particular, has had a major clinical impact on people. The consequence of genomic mutations in HA and NA of influenza A, is exposure of an unfamiliar virus to susceptible human hosts. Small genomic changes, called antigenic drift, occur regularly during viral replication and accumulate over the course of a flu season often leading to seasonal epidemics, when the virus is able to escape protective antibodies that result from prior infections [2, 3].

A more significant type of mutation, antigenic shift, occurs when new HA or NA are introduced into human populations [2, 3]. There are 18 different HA (H1–H18) and 11 different NA (N1–N11) proteins which can reassort to produce different subtype combinations of influenza in both humans and animals, including birds and mammals [10]. In recent years, H1N1 and H3N2 have been the predominantly circulating subtypes; however, many other subtypes occur in animal reservoirs [11]. Animal reservoirs are often asymptomatic, therefore allowing long-term infection or colonization [12]. Arguably, the most important animal reservoir leading to the persistence and evolution of influenza A is the avian species, specifically aquatic birds. Other known reservoirs include swine, horses, minks, and bats [13]. Influenza subtypes, novel to humans, can develop either through transmission directly from an animal reservoir or through introduction of the HA or NA from reassortment among human and animal strains [13]. This novel mixing of HA and NA proteins can result in viruses, which never previously caused infection in humans and to which humans have no preexisting immunity. This antigenic shift can result in worldwide pandemics if novel viruses develop the ability to spread effectively among humans [12]. There is a concern that, along with globalization, changes in contemporary societies, including the industrialization of livestock farming, will result in more frequent interactions between animals and humans. Consequently, genomic reassortment of influenza virus will occur more frequently, thereby increasing the risk for future influenza pandemics [13].

Influenza pandemics cause significant morbidity, mortality, economic loss, and social disruption. While pandemic-like events can be traced back to the 1500s, the 1918 Influenza pandemic is the most recognized and deadly of influenza pandemics in modern times. It left little of the world untouched with spread exacerbated by World War I and an unprepared public healthcare system [14]. The 1918 Pandemic was caused by the H1N1 subtype and resulted in over 50 million deaths globally over the course of the pandemic [15]. The 1918 Pandemic was particularly lethal in young adults with nearly half of the influenza-related deaths occurring in those between 20 years and 40 years of age [16]. As we have seen in pandemics since 1918, peaks of infection occurred in multiple waves over several years [17].

Since 1918, there have been pandemics of H2N2 in 1957 and H3N2 in 1968, both of which caused a death toll of over a million people worldwide [18, 19]. The first influenza pandemic of the twenty-first century occurred in 2009, due to an H1N1 subtype that was first discovered in circulation in Mexico and the USA, originating from recombination that occurred in North American swine [20]. Morbidity and mortality impacts were notable particularly in children, young adults, and pregnant women; though, overall death estimates were lower than prior pandemics. The 2009 pandemic response for the first time combined relatively rapid vaccine development and antiviral use to help mitigate the impact of viral spread setting the precedent for future pandemic response and public health preparedness [21].

The evolution and epidemiology of influenza is affected not only by human and animal behavior but also by the impact of other circulating respiratory viruses. It is impossible to discuss the current epidemiology of influenza without addressing the impact of the COVID-19 pandemic on influenza rates. Co-circulation of these two major viruses will occur for the foreseeable future. During the beginning of the COVID-19 pandemic, influenza rates were significantly reduced from March of 2020 into the spring of 2021 with only 0.2% of specimens testing positive for influenza. Influenza-related hospitalizations also decreased to a rate of 0.8 per 100,000 compared to prior seasons, which ranged from 62 to 102.9 per 100,000 [22]. Despite that period of lower rates, cases did increase again during the 2021–2022 influenza season but remained below what is seen during a typical flu season [23]. Public health policies and mitigation efforts during the COVID-19 pandemic such as social distancing, masking, and decrease in travel had a major role on the decrease in influenza and other respiratory viruses [24]. There are data to suggest that "viral interference," meaning respiratory viruses can have a positive and negative impact on the case rates of each other, may have played a role [25, 26]. In the summer of 2022, there have been unusual trends in influenza A within the USA and a high number of cases in the southern hemisphere, indicating the impacts of the COVID-19 pandemic on influenza rates still needs close watching [27]. Moving forward, we will need to consider both influenza and SARS-CoV-2 as public health mitigation strategies are implemented and clinical treatments evolve.

3 Demographics and Disparities

3.1 Age

Age has played a strong role in the likelihood of infection and severity of illness. Seasonal influenza results in a bimodal mortality distribution with higher incidence and death rate among the young and old [28]. During flu pandemics, however, this mortality distribution typically changes. During the 1918 pandemic the peak mortality rates were seen at age 28, perhaps due to residual immunologic memory response in older cohorts to an earlier Russian influenza pandemic of 1889–1890 [29]. In the 2009 H1N1 pandemic, older individuals who were born prior to 1957 had lower rates of infection [30]. This was likely due to prior exposure to previously circulating H1N1 variants, given younger generations were not exposed to those earlier similar seasonal variants [31]. This caused a higher case rate in the young which resulted in more deaths occurring in young adults. The typical U-shaped mortality curve representing higher death rates in the very young and old was replaced by a W-shaped distribution curve which was due to the higher death rate in young adults (see Fig. 1) [32].

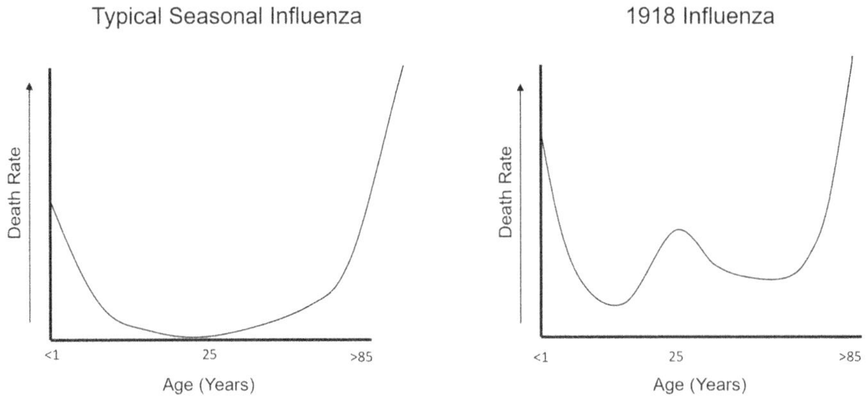

Fig. 1 Comparison of the classic U-shaped death rate curve which occurs in a typical influenza season to the W-shaped curve which was associated with the 1918 influenza pandemic

3.2 Race and Ethnicity

Centers for Disease Control and Prevention (CDC) data from the USA shows that communities of color have a higher risk of severe illness from influenza. Blacks are almost two times more likely to be hospitalized and require ICU admission compared to Whites [33]. American Indian, Alaska First Nation, and Hispanic people are also more likely to be hospitalized than Whites [33]. Underlying issues related to disparities in healthcare and education access, neighborhood socio-economic levels, comorbid medical conditions, and detrimental health behaviors exacerbate these disparities [34].

Vaccination for influenza is recommended by the CDC and is one of the most important public health interventions available to reduce cases and severity of illness. Therefore, it is important to consider differences in vaccine acceptance rates when comparing infection rates in marginalized and underserved groups given that lack of vaccination could put individuals at higher risk for severe disease outcomes. In the 2019–2020 flu season, less than half of the US population received their influenza vaccination [35]. Hispanic and Black adults had lower vaccination acceptance rates compared to White adults [35]. Reduced vaccination rates in communities of color and minorities has been persistent over time, which emphasizes the long-term inequities in healthcare and structural racism that play a role in perpetuating these differences [36, 37].

3.3 Sex

Historically young adult females have had worse disease outcomes including higher rates of hospitalizations and mortality compared to men [38–40]. This increased risk of severe disease is present regardless of pregnancy status, but pregnant women are also known to have further increased risks for severe diseases compared to the general population [41]. While pregnancy does not increase mortality in high-income nations, pregnant women with influenza, particularly those in the third trimester, are at increased risk of ICU admission [42–44]. This elevated risk of severe disease extends through the early post-partum period as well [43]. It is not well understood why these differences in outcomes based on sex occur but prior data and animal modeling suggests there are differences in the pro-inflammatory and anti-inflammatory immune responses where females tend to have increased pro-inflammatory immune dysregulation which can result in tissue damage and severe illness [45, 46].

Again, when considering infection rates in a population it is important to consider vaccine coverage. In recent flu seasons, women have been more likely to receive influenza vaccination compared to men [35, 47]. Interestingly, women have been shown to have an increased antibody response to influenza vaccination and therefore vaccination may be more efficacious in preventing infection in women [48].

4 Diagnostic Testing

Influenza can be a clinical diagnosis based on the history and exam in context of local influenza activity. According to the Infectious Disease Society of America (IDSA) guidelines, testing should be pursued in outpatients who present with ILI or acute respiratory symptoms when they are high risk for complications, such as the immunocompromised population, or in individuals who have chronic medical conditions such as asthma, where treatment may change based on testing results [49]. For hospitalized patients, it is recommended that anyone who presents with an acute respiratory illness, including pneumonia, and those with worsening of chronic cardiopulmonary conditions be tested for influenza [49]. Most recent guidelines do not recommend testing for influenza if clinical management would not change based on test results [49]. However, this approach became more complex during the COVID-19 pandemic given the differences in infection prevention interventions and treatment options based on test results; therefore, in most circumstances testing for both SARS-CoV-2 and influenza is recommended [50].

Appropriate specimens to send for influenza testing are nasal, midturbinate, nasopharyngeal, and lower respiratory tract specimens including tracheal aspirates or bronchial wash lavage. Of the upper respiratory specimens, the nasopharyngeal samples are the most sensitive, closely followed by midturbinate, and then a nasal specimen [51]. In patients with pneumonia, lower respiratory tract samples have greater sensitivity for diagnosis over nasopharyngeal, which can be falsely negative [52]. Optimal timing of a test is 1–4 days after symptom onset as maximal viral shedding occurs 48–72 hours after symptom onset [53].

The gold standard is polymerase chain reaction (PCR)-based tests; other tests include molecular assays using isothermal amplification and antigen-based tests [52, 54]. Antigen-based tests are often less costly, rapid, and more widely available but have limitations in utility due to reduced sensitivity and specificity. They are most useful early in the infection and when there are high levels of virus circulating in the community [49]. Fortunately, there are now rapid molecular tests available, which have been shown to be very sensitive and specific, allowing for both rapid and reliable testing [52].

During the COVID-19 pandemic, genomic sequencing played an incredibly important role in understanding the epidemiology, transmission, and clinical differences between viral variants as mutations in the genome resulted in rapid evolution of the SARS-CoV-2 virus [55]. An infrastructure of public health, academic, and private labs with the capability of performing this specialized testing was established during the pandemic, which increased public availability of sequencing data [56]. Prior to the COVID-19 pandemic, genomic sequencing of influenza specimens was common, but not anywhere near the frequency that has occurred for SARS-CoV-2; hopefully, this expansion of genomic sequencing technology will extend into other respiratory viruses, including influenza.

5 Treatment

Historically influenza treatment has been supportive; however, over the last 30 years, treatment options have progressed significantly after the development of oral antiviral medications. Once an influenza infection occurs in a human host, with or without prior vaccination, antiviral agents may be an option to mitigate impact and reduce severity. There are several different types of oral antiviral agents that have historically been used for the treatment of influenza. The first and oldest are the amantadanes, including amantidine and rimantidine, which work by preventing the release of the influenza viral RNA into the host cell and are effective only against influenza A. The amantadanes are not currently recommended for treatment of influenza as the current circulating influenza A strains, 2009 H1N1 and H3N2, have developed resistance to this class of antivirals [57]. Like antibacterial resistance, antiviral resistance can develop and should be monitored closely over time to ensure current medications continue to be effective. While amantadine resistance limits the current utility of that class of agents for influenza treatment, to date, rates of resistance to neuraminidase inhibitors and endonuclease inhibitors remains low [52, 58, 59].

Neuraminidase inhibitors act by preventing viral particle release through blocking cleavage of the neuraminidase glycoprotein on the surface of infected cells (see Fig. 2) [52]. This drug class is the mainstay of treatment and is available in oral (Oseltamivir), intravenous (Peramivir, Zanamivir), and inhalational (Zanamivir, Laninamivir in development) formulations [60]. Baloxavir is an endonuclease inhibitor which inhibits viral replication by blockade of viral polymerase; it is available as an oral capsule (see Fig. 2) [52]. All antivirals are best used early in the illness [61]. Early treatment has been associated with decreased duration of symptoms and risk of hospitalization, complications such as pneumonia and otitis, death

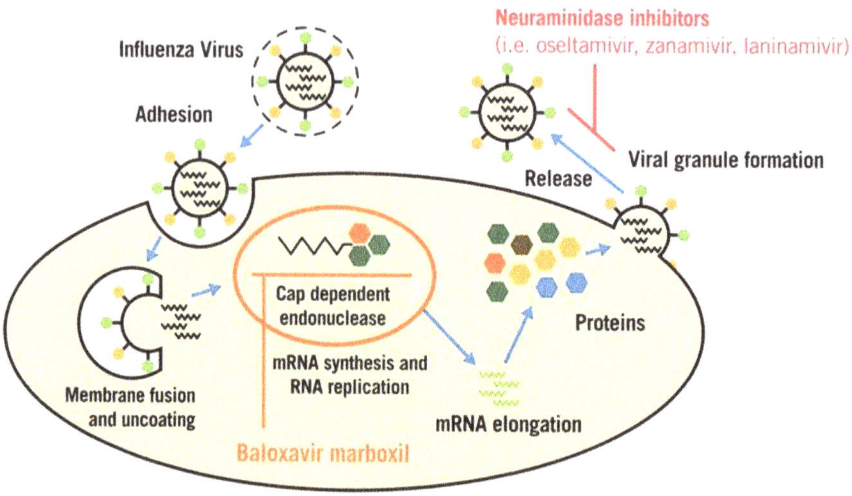

Fig. 2 The influenza life cycle and mechanism of action of antivirals. (*RNA* Ribonucleic acid, *mRNA* Messenger ribonucleic acid)

Table 1 When to consider treatment for individuals with influenza

Treatment recommended	1. Anyone hospitalized with influenza, regardless of duration of illness 2. Outpatients with severe, progressive illness, regardless of duration 3. Immunocompromised or other patients with chronic medical conditions at high risk of complications 4. Children <2 years and Adults >65 years 5. Pregnant women and women up to 2 weeks postpartum
Consider treatment	1. Outpatients with <2 days of symptoms 2. Outpatients who have household contacts who are immunocompromised or high risk of developing complications from influenza 3. Healthcare workers who care for patients who are immunocompromised or high risk of developing complications from influenza

[62–65]. Even in severe disease requiring hospitalization, benefit has been demonstrated with improved survival in patients who started treatment with Oseltamivir within 5 days of symptom onset [66]. Early antiviral therapy also reduces hospital and ICU length of stay and duration of mechanical ventilation [67, 68]. The IDSA guidelines for antiviral use recommend treatment regardless of vaccination status as soon as possible in individuals with an increased risk of severe disease (see Table 1) [49].

Choice of antiviral agent depends on individual patient factors such as access to medication, comorbidities, and severity of illness. Intravenous Peramivir is usually reserved for patients unable to take oral or inhalational antiviral agents [60]. Peramivir did not demonstrate benefit in the treatment of hospitalized patients; thus, other antivirals remain preferred [52]. Zanamivir retains activity against Oseltamivir-resistant viruses when that is a concern but is contraindicated in patients who are at risk for bronchospasm [52]. Neuraminidase inhibitors have been administered safely to children and in pregnancy, while Baloxivir is approved for children 12 years and up but does not have pregnancy data. Baloxavir has similar efficacy to Oseltamavir for uncomplicated influenza, but data in hospitalized patients with more severe complications of influenza are lacking. Baloxavir is currently only approved for outpatient therapy in patients with uncomplicated influenza [52, 59].

Oseltamivir and Zanamivir are recommended for a 5 day duration [60]. Baloxavir and Peramivir are administered as single doses in outpatient setting [60]. Longer courses of antivirals, particularly Oseltamivir, did not demonstrate increased efficacy, although clinicians can consider longer treatment durations in immunosuppressed patients with severe lower respiratory tract disease who may be slow to clear the virus [60]. Similarly, high-dose therapy with Oseltamivir has not been found to be of benefit [69]. Corticosteroids are not indicated as adjunctive therapy in influenza and may in fact lead to an increase in mortality [70]. Convalescent plasma with hyperimmune globulin from individuals with circulating antibodies against influenza has also been studied as a treatment modality but has not been found to have definitive benefit [52, 71].

6 Infection Prevention

6.1 Influenza Vaccine

The best prevention for influenza infection and severe viral disease is annual vaccination. Data analyzed across multiple years confirms persistent protection through vaccination against severe disease including hospitalization and death [72–74]. The most widely available and most widely used influenza vaccines are intramuscular quadrivalent inactivated viral vaccines composed of two subtypes of influenza A and two influenza B lineages [75]. In addition, recombinant intramuscular vaccines, live attenuated intranasal vaccines, and egg free formulations are available resulting in the panoply of influenza vaccinations currently available in the USA (see Table 2).

Each year influenza A subtypes are selected based on those predicted to be in circulation in the coming season in the Northern Hemisphere [75]. As such vaccine effectiveness can vary from year to year based on how well the vaccine composition is matched to the circulating viruses. Antigenic drift of the virus from strains included in the vaccine diminish vaccine efficacy, which has been observed particularly with late season influenza A infections [76]. Several different strategies have been deployed in an attempt to increase vaccine efficacy, particularly in the high-risk elderly or immunocompromised populations. These include adjuvanted vaccine and higher dose vaccination, with each of these approaches meeting with some success [77, 78]. Choice of a vaccine is dependent on patient factors such as egg allergy, age, immunosuppression, or a need to avoid live virus vaccines.

Table 2 Available influenza vaccines in 2021–2022 influenza season

	Trade name
Inactivated vaccines (administered IM)	
Egg-based	Fluarix Quadrivalent
	FluLaval Quadrivalent
	FluZone Quadrivalent
	Afluria Quadrivalent
	Fluzone high-dose Quadrivalent
	Fluad Quadrivalent (with MF59 adjuvant)
Cell culture-based	Flucelvax Quadrivalent
Recombinant HA (Administered IM)	Flubox Quadrivalent
Live Attenuated Egg-based (administered intranasally)	FluMist Quadrivalent

6.2 Masking and Hand Hygiene

While vaccines are the best and least onerous way to prevent influenza, other non-pharmaceutical interventions (NPI) can help prevent spread as well. NPIs are especially important in high-risk settings such as healthcare. In most cases, influenza is spread from person to person via large-particle respiratory droplets produced during activities like talking, coughing, or sneezing [79]. While most spread is thought to be due to large droplets that fall to the ground within 3–6 feet, there is some limited evidence for airborne transmission via smaller particles that can remain in the air longer and for further distances [79]. Source control by immediately masking individuals with respiratory symptoms and placing patients with suspected ILI is recommended in all healthcare settings by the CDC [80]. In addition, healthcare workers (HCW) should wear a well-fitted surgical mask during routine care of patients with confirmed or suspected influenza and a fit tested N95 or powered air purifying respirator when in the room during any aerosol generating procedure (intubation, bronchoscopy, etc.) of patients with confirmed or suspected influenza [80]. Eye protection, in addition, is recommended when there is reasonable anticipation of exposure to secretions or other bodily fluids [80].

While thought to be less common, transmission can also occur from contaminated surfaces to human mucosal surfaces of the eyes and mouth via hand transfer [79]. Influenza virus can survive for up to 48 hours on some hard surfaces such as stainless steel and plastic [81].

Hand hygiene is one of our most important defenses against infectious diseases and has been an important tool for prevention of several diseases including respiratory infections [82]. Good hand hygiene practices and avoiding touching eyes, nose, or mouth helps protect against acquiring influenza [83].

6.3 Impact of Presenteeism

For individuals infected with influenza or experiencing ILI, the CDC recommends staying home for at least 24 hours after a fever has resolved [84]. However, often people attend work despite being ill, which is called sickness (or contagious) presenteeism [85]. Working ill can lead to continued spread of infections [86, 87]. Many cultural and social factors have been linked to the occurrence of sickness presenteeism including sense of duty, economic pressures, and lack of coverage [85]. This behavior occurs across many job types and disciplines and has been a particular problem in HCWs where there is the potential to expose vulnerable populations [85, 88]. It is critical that employers take this into consideration in employee health policies by supporting and encouraging individuals to stay home when ill to reduce further spread to their workforce and the people they serve.

6.4 Antiviral Prophylaxis

Prophylaxis is used to prevent illness prior to infection occurring. This is most often recommended after an exposure to someone who has a known influenza infection, but occasionally it is recommended prior to a potential exposure in high-risk populations. Currently Oseltamivir, inhaled Zanamivir, and oral Baloxavir are used for prophylaxis [60]. Prophylaxis should be given within 48 hours of the exposure [49]. The IDSA guidelines recommend risk stratification to determine which individuals and populations may benefit the most from these medications [49]. In the community, it is only recommended to receive prophylaxis if individuals are at high risk for complications [49]. Certain highly immunocompromised populations may benefit from continuing antivirals for the duration of flu season, but for most individuals with high-risk conditions only a short course is recommended [49].

6.5 Prevention Strategies in Healthcare Settings

Preventing spread of influenza to individuals at increased risk for complications is critically important in our community and especially in high-risk healthcare or congregate living settings where concentration of patients at risk for severe disease is even higher. While prevention of the spread of influenza is largely similar between community and healthcare settings, there are a few nuances worth notating.

Likely the most important intervention in healthcare settings is the promotion of annual influenza vaccines [80]. Vaccination protects both patients and HCWs, high vaccine uptake in HCWs has been associated with lower mortality in patients and less disruption due to HCW absenteeism due to ILI [89, 90]. Ensuring HCWs do not work when ill is also an essential practice to prevent spread of infection to both co-workers and patients. Policies should instruct HCWs who have ILI to not attend work, and they should remain out of work for at least 24 hours after fever resolution [80]. Non-punitive and flexible policies assist with creating a culture where employees report illness and reduce presenteeism.

Surveillance in healthcare, particularly in long-term care or congregated living settings, is key for early detection and control of influenza cases. In long-term care settings, it is recommended that during influenza season, active surveillance occur to identify any residents, HCWs, or visitors who have ILI [91]. According to the CDC, an outbreak should be called if two or more cases of influenza are identified in a facility within 72 hours [91]. Implementation of droplet precautions for all confirmed or suspected cases is key to prevent continued spread [91]. When an outbreak occurs, it is recommended that anyone with ILI be tested for influenza [91]. In an outbreak setting, treatment with antivirals should be given promptly to anyone with confirmed or suspected influenza. In addition, patients or HCWs that may have been exposed during the outbreak should be given chemoprophylaxis to assist with preventing further cases and reduce likelihood of severe illness if an infection does occur [91]. Treatment with Oseltamivir is known to reduce viral shedding with a potential for reduction in transmission, and Zanamivir has demonstrated efficacy in reduced secondary attack rates in household contacts with treatment of the index case [92, 93].

7 Conclusion

Influenza will continue to circulate globally for the foreseeable future, with profound impacts on human health and society, particularly during inevitable pandemics. Fortunately, through use of infection control practices, including annual vaccinations, and an expanding array of antiviral medications, there is better preparation to mitigate these impacts than in the past. However, much progress is still needed to improve disease outcomes, particularly for the most vulnerable populations including the immunosuppressed and underserved communities, which experience disparities in healthcare.

Acknowledgements We would like to acknowledge Lewis Hughes for the design of Fig. 2.

Potential Conflicts of Interest C.P. reports serving on a paid Gilead advisory board for Remdesivir. At the time of the writing C.P. was employed by Atrium Health, he is now employed by ViiV Healthcare. All other authors have no potential conflicts of interest to report.

References

1. Ghendon Y. Introduction to pandemic influenza through history. Eur J Epidemiol. 1994;10(4):451–3.
2. Cox NJ, Fukuda K. Influenza. Infect Dis Clin North Am. 1998;12(1):27–38.
3. Treanor J. Influenza (Including Avian Influenza and Swine Influenza). In: Bennett JED, R; Blaser, MJ, editor. Mandell, Douglas, and Bennett's Principles and Practice of Infectious Diseases. Eigth Edition ed. Philadelphia, PA: Elsevier Inc.; 2015. p. 2000–24.
4. Carrat F, Vergu E, Ferguson NM, Lemaitre M, Cauchemez S, Leach S, et al. Time lines of infection and disease in human influenza: a review of volunteer challenge studies. Am J Epidemiol. 2008;167(7):775–85.
5. Ng S, Lopez R, Kuan G, Gresh L, Balmaseda A, Harris E, et al. The Timeline of Influenza Virus Shedding in Children and Adults in a Household Transmission Study of Influenza in Managua, Nicaragua. Pediatr Infect Dis J. 2016;35(5):583–6.
6. Hall E. Influenza. In: Hall EW, AP; Hamborsky, J; Morelli, V; et al., editor. Center for Disease Control and Prevention Epidemiology and Prevention of Vaccine-Preventable Diseases. 14th ed. ed: Public Health Foundation; 2021.
7. Xu X, Lindstrom SE, Shaw MW, Smith CB, Hall HE, Mungall BA, et al. Reassortment and evolution of current human influenza A and B viruses. Virus Res. 2004;103(1–2):55–60.
8. Koutsakos M, Wheatley AK, Laurie K, Kent SJ, Rockman S. Influenza lineage extinction during the COVID-19 pandemic? Nat Rev Microbiol. 2021;19(12):741–2.
9. Sederdahl BK, Williams JV. Epidemiology and Clinical Characteristics of Influenza C Virus. Viruses. 2020;12(1).
10. Types of Influenza Viruses: Centers for Disease Control and Prevention; 2021 [updated 11/2/2021. Available from: https://www.cdc.gov/flu/about/viruses/types.htm#:~:text=Influenza%20A%20viruses%20are%20classified,)%2C%20A(H3N2).
11. Influenza A Subtypes and Species Affected: Centers for Disease Control and Prevention; [updated 9/27/2018. Available from: https://www.cdc.gov/flu/other/animal-flu.html.
12. Cox NJ, Subbarao K. Global epidemiology of influenza: past and present. Annu Rev Med. 2000;51:407–21.
13. Kessler S, Harder TC, Schwemmle M, Ciminski K. Influenza A Viruses and Zoonotic Events-Are We Creating Our Own Reservoirs? Viruses. 2021;13(11).
14. Unwin RJ. The 1918 Influenza Pandemic: Back to the Future? Kidney Blood Press Res. 2021;46(5):639–46.
15. History of 1918 Flu Pandemic: Centers for Disease Control and Infection Prevention; [updated 3/21/2018. Available from: https://www.cdc.gov/flu/pandemic-resources/1918-commemoration/1918-pandemic-history.htm.
16. Simonsen L, Clarke MJ, Schonberger LB, Arden NH, Cox NJ, Fukuda K. Pandemic versus epidemic influenza mortality: a pattern of changing age distribution. J Infect Dis. 1998;178(1):53–60.
17. Saunders-Hastings PR, Krewski D. Reviewing the History of Pandemic Influenza: Understanding Patterns of Emergence and Transmission. Pathogens. 2016;5(4).
18. Viboud C, Simonsen L, Fuentes R, Flores J, Miller MA, Chowell G. Global Mortality Impact of the 1957–1959 Influenza Pandemic. J Infect Dis. 2016;213(5):738–45.
19. 1968 Pandemic (H3N2 virus): Centers for Disease Control and Prevention; [updated 1/2/2019. Available from: https://www.cdc.gov/flu/pandemic-resources/1968-pandemic.html.
20. Smith GJ, Vijaykrishna D, Bahl J, Lycett SJ, Worobey M, Pybus OG, et al. Origins and evolutionary genomics of the 2009 swine-origin H1N1 influenza A epidemic. Nature. 2009;459(7250):1122–5.
21. Blasio BF, Iversen BG, Tomba GS. Effect of vaccines and antivirals during the major 2009 A(H1N1) pandemic wave in Norway – and the influence of vaccination timing. PLoS One. 2012;7(1):e30018.

22. Olsen SJ, Winn AK, Budd AP, Prill MM, Steel J, Midgley CM, et al. Changes in Influenza and Other Respiratory Virus Activity During the COVID-19 Pandemic – United States, 2020–2021. MMWR Morb Mortal Wkly Rep. 2021;70(29):1013–9.
23. Weekly U.S. Influenza Surveillance Report. Center for Disease Control and Prevention 6/10/2022.
24. Noh JY, Seong H, Yoon JG, Song JY, Cheong HJ, Kim WJ. Social Distancing against COVID-19: Implication for the Control of Influenza. J Korean Med Sci. 2020;35(19):e182.
25. Chotpitayasunondh T, Fischer TK, Heraud JM, Hurt AC, Monto AS, Osterhaus A, et al. Influenza and COVID-19: What does co-existence mean? Influenza Other Respir Viruses. 2021;15(3):407–12.
26. Nickbakhsh S, Mair C, Matthews L, Reeve R, Johnson PCD, Thorburn F, et al. Virus-virus interactions impact the population dynamics of influenza and the common cold. Proc Natl Acad Sci U S A. 2019.
27. Australian Influenza Surveillance Report. Australian Government Department of Health; 2022 06/10/2022. Contract No.: 5.
28. Collaborators GBDI. Mortality, morbidity, and hospitalisations due to influenza lower respiratory tract infections, 2017: an analysis for the Global Burden of Disease Study 2017. Lancet Respir Med. 2019;7(1):69–89.
29. Gagnon A, Miller MS, Hallman SA, Bourbeau R, Herring DA, Earn DJ, et al. Age-specific mortality during the 1918 influenza pandemic: unravelling the mystery of high young adult mortality. PLoS One. 2013;8(8):e69586.
30. Fisman DN, Savage R, Gubbay J, Achonu C, Akwar H, Farrell DJ, et al. Older age and a reduced likelihood of 2009 H1N1 virus infection. N Engl J Med. 2009;361(20):2000–1.
31. Hancock K, Veguilla V, Lu X, Zhong W, Butler EN, Sun H, et al. Cross-reactive antibody responses to the 2009 pandemic H1N1 influenza virus. N Engl J Med. 2009;361(20):1945–52.
32. Greer AL, Tuite A, Fisman DN. Age, influenza pandemics and disease dynamics. Epidemiol Infect. 2010;138(11):1542–9.
33. O'Halloran AC, Holstein R, Cummings C, Daily Kirley P, Alden NB, Yousey-Hindes K, et al. Rates of Influenza-Associated Hospitalization, Intensive Care Unit Admission, and In-Hospital Death by Race and Ethnicity in the United States From 2009 to 2019. JAMA Netw Open. 2021;4(8):e2121880.
34. Cordoba E, Aiello AE. Social Determinants of Influenza Illness and Outbreaks in the United States. N C Med J. 2016;77(5):341–5.
35. Flu Vaccination Coverage, United States, 2019–20 Influenza Season: Centers for Disease Control and Prevention; 2020 [Available from: https://www.cdc.gov/flu/fluvaxview/coverage-1920estimates.htm.
36. Callahan AG, Strassberg ER, Rhoades CP, Varghese L, Schulkin J, Power ML. Pregnant Women's Opinions and Acceptance of Influenza and Tdap Vaccines. J Womens Health (Larchmt). 2022;31(5):656–64.
37. Callahan AG, Coleman-Cowger VH, Schulkin J, Power ML. Racial disparities in influenza immunization during pregnancy in the United States: A narrative review of the evidence for disparities and potential interventions. Vaccine. 2021;39(35):4938–48.
38. Klein SL, Passaretti C, Anker M, Olukoya P, Pekosz A. The impact of sex, gender and pregnancy on 2009 H1N1 disease. Biol Sex Differ. 2010;1(1):5.
39. Serfling RE, Sherman IL, Houseworth WJ. Excess pneumonia-influenza mortality by age and sex in three major influenza A2 epidemics, United States, 1957–58, 1960 and 1963. Am J Epidemiol. 1967;86(2):433–41.
40. Eshima N, Tokumaru O, Hara S, Bacal K, Korematsu S, Tabata M, et al. Sex- and age-related differences in morbidity rates of 2009 pandemic influenza A H1N1 virus of swine origin in Japan. PLoS One. 2011;6(4):e19409.
41. Jamieson DJ, Honein MA, Rasmussen SA, Williams JL, Swerdlow DL, Biggerstaff MS, et al. H1N1 2009 influenza virus infection during pregnancy in the USA. Lancet. 2009;374(9688):451–8.

42. Maravi-Poma E, Martin-Loeches I, Regidor E, Laplaza C, Cambra K, Aldunate S, et al. Severe 2009 A/H1N1v influenza in pregnant women in Spain. Crit Care Med. 2011;39(5):945–51.
43. Mertz D, Kim TH, Johnstone J, Lam PP, Science M, Kuster SP, et al. Populations at risk for severe or complicated influenza illness: systematic review and meta-analysis. BMJ. 2013;347:f5061.
44. Mertz D, Geraci J, Winkup J, Gessner BD, Ortiz JR, Loeb M. Pregnancy as a risk factor for severe outcomes from influenza virus infection: A systematic review and meta-analysis of observational studies. Vaccine. 2017;35(4):521–8.
45. Klein SL, Hodgson A, Robinson DP. Mechanisms of sex disparities in influenza pathogenesis. J Leukoc Biol. 2012;92(1):67–73.
46. Wang C, Lashua LP, Carter CE, Johnson SK, Wang M, Ross TM, et al. Sex disparities in influenza: A multiscale network analysis. iScience. 2022;25(5):104192.
47. Flu Vaccination Coverage, United States, 2020–21 Influenza Season: Centers for Disease Control and Prevention; 2021 [updated 10/7/2021. Available from: https://www.cdc.gov/flu/fluvaxview/coverage-2021estimates.htm.
48. Cook IF. Sexual dimorphism of humoral immunity with human vaccines. Vaccine. 2008;26(29–30):3551–5.
49. Uyeki TM, Bernstein HH, Bradley JS, Englund JA, File TM, Fry AM, et al. Clinical Practice Guidelines by the Infectious Diseases Society of America: 2018 Update on Diagnosis, Treatment, Chemoprophylaxis, and Institutional Outbreak Management of Seasonal Influenzaa. Clin Infect Dis. 2019;68(6):895–902.
50. Testing Guidance for Clinicians When SARS-CoV-2 and Influenza Viruses are Co-circulating: Centers for Disease control; [updated 2/9/2022.
51. Frazee BW, Rodriguez-Hoces de la Guardia A, Alter H, Chen CG, Fuentes EL, Holzer AK, et al. Accuracy and Discomfort of Different Types of Intranasal Specimen Collection Methods for Molecular Influenza Testing in Emergency Department Patients. Ann Emerg Med. 2018;71(4):509–17 e1.
52. O'Driscoll LS, Martin-Loeches I. Management of Severe Influenza. Semin Respir Crit Care Med. 2021;42(6):771–87.
53. Yang JR, Lo J, Ho YL, Wu HS, Liu MT. Pandemic H1N1 and seasonal H3N2 influenza infection in the human population show different distributions of viral loads, which substantially affect the performance of rapid influenza tests. Virus Res. 2011;155(1):163–7.
54. Mitamura K, Yamazaki M, Ichikawa M, Yasumi Y, Shiozaki K, Tokushima M, et al. Clinical usefulness of a rapid molecular assay, ID NOW influenza A & B 2, in adults. J Infect Chemother. 2021;27(3):450–4.
55. Chen X, Kang Y, Luo J, Pang K, Xu X, Wu J, et al. Next-Generation Sequencing Reveals the Progression of COVID-19. Front Cell Infect Microbiol. 2021;11:632490.
56. Chen Z, Azman AS, Chen X, Zou J, Tian Y, Sun R, et al. Global landscape of SARS-CoV-2 genomic surveillance and data sharing. Nat Genet. 2022;54(4):499–507.
57. Antiviral Drug Resistance among Influenza Viruses: Centers for Disease Control and Prevention; [Available from: https://www.cdc.gov/flu/professionals/antivirals/antiviral-drug-resistance.htm.
58. Behillil S, May F, Fourati S, Luyt CE, Chicheportiche T, Sonneville R, et al. Oseltamivir Resistance in Severe Influenza A(H1N1)pdm09 Pneumonia and Acute Respiratory Distress Syndrome: A French Multicenter Observational Cohort Study. Clin Infect Dis. 2020;71(4):1089–91.
59. Shirley M. Baloxavir Marboxil: A Review in Acute Uncomplicated Influenza. Drugs. 2020;80(11):1109–18.
60. For Clinicians: Antiviral Medication: Centers for Disease Control and Prevention; [updated 6/13/2022. Available from: https://www.cdc.gov/flu/professionals/antivirals/summary-clinicians.htm.
61. Palomba E, Castelli V, Renisi G, Bandera A, Lombardi A, Gori A. Antiviral Treatments for Influenza. Semin Respir Crit Care Med. 2021;42(6):859–72.

62. Hsu J, Santesso N, Mustafa R, Brozek J, Chen YL, Hopkins JP, et al. Antivirals for treatment of influenza: a systematic review and meta-analysis of observational studies. Ann Intern Med. 2012;156(7):512–24.
63. Jefferson T, Jones MA, Doshi P, Del Mar CB, Heneghan CJ, Hama R, et al. Neuraminidase inhibitors for preventing and treating influenza in healthy adults and children. Cochrane Database Syst Rev. 2012;1:CD008965.
64. Muthuri SG, Myles PR, Venkatesan S, Leonardi-Bee J, Nguyen-Van-Tam JS. Impact of neuraminidase inhibitor treatment on outcomes of public health importance during the 2009–2010 influenza A(H1N1) pandemic: a systematic review and meta-analysis in hospitalized patients. J Infect Dis. 2013;207(4):553–63.
65. Muthuri SG, Venkatesan S, Myles PR, Leonardi-Bee J, Al Khuwaitir TS, Al Mamun A, et al. Effectiveness of neuraminidase inhibitors in reducing mortality in patients admitted to hospital with influenza A H1N1pdm09 virus infection: a meta-analysis of individual participant data. Lancet Respir Med. 2014;2(5):395–404.
66. Kositpantawong N, Surasombatpattana S, Siripaitoon P, Kanchanasuwan S, Hortiwakul T, Charernmak B, et al. Outcomes of early oseltamivir treatment for hospitalized adult patients with community-acquired influenza pneumonia. PLoS One. 2021;16(12):e0261411.
67. Campbell AP, Tokars JI, Reynolds S, Garg S, Kirley PD, Miller L, et al. Influenza Antiviral Treatment and Length of Stay. Pediatrics. 2021;148(4).
68. Moreno G, Rodriguez A, Sole-Violan J, Martin-Loeches I, Diaz E, Bodi M, et al. Early oseltamivir treatment improves survival in critically ill patients with influenza pneumonia. ERJ Open Res. 2021;7(1).
69. Lee N, Hui DS, Zuo Z, Ngai KL, Lui GC, Wo SK, et al. A prospective intervention study on higher-dose oseltamivir treatment in adults hospitalized with influenza a and B infections. Clin Infect Dis. 2013;57(11):1511–9.
70. Nedel WL, Nora DG, Salluh JI, Lisboa T, Povoa P. Corticosteroids for severe influenza pneumonia: A critical appraisal. World J Crit Care Med. 2016;5(1):89–95.
71. Beigel JH, Aga E, Elie-Turenne MC, Cho J, Tebas P, Clark CL, et al. Anti-influenza immune plasma for the treatment of patients with severe influenza A: a randomised, double-blind, phase 3 trial. Lancet Respir Med. 2019;7(11):941–50.
72. Tenforde MW, Chung J, Smith ER, Talbot HK, Trabue CH, Zimmerman RK, et al. Influenza Vaccine Effectiveness in Inpatient and Outpatient Settings in the United States, 2015–2018. Clin Infect Dis. 2021;73(3):386–92.
73. Rolfes MA, Flannery B, Chung JR, O'Halloran A, Garg S, Belongia EA, et al. Effects of Influenza Vaccination in the United States During the 2017–2018 Influenza Season. Clin Infect Dis. 2019;69(11):1845–53.
74. Chung H, Buchan SA, Campigotto A, Campitelli MA, Crowcroft NS, Dubey V, et al. Influenza Vaccine Effectiveness Against All-Cause Mortality Following Laboratory-Confirmed Influenza in Older Adults, 2010–2011 to 2015–2016 Seasons in Ontario, Canada. Clin Infect Dis. 2021;73(5):e1191-e9.
75. Grohskopf LA, Alyanak E, Ferdinands JM, Broder KR, Blanton LH, Talbot HK, et al. Prevention and Control of Seasonal Influenza with Vaccines: Recommendations of the Advisory Committee on Immunization Practices, United States, 2021–22 Influenza Season. MMWR Recomm Rep. 2021;70(5):1–28.
76. Tenforde MW, Kondor RJG, Chung JR, Zimmerman RK, Nowalk MP, Jackson ML, et al. Effect of Antigenic Drift on Influenza Vaccine Effectiveness in the United States-2019–2020. Clin Infect Dis. 2021;73(11):e4244-e50.
77. McConeghy KW, Davidson HE, Canaday DH, Han L, Saade E, Mor V, et al. Cluster-randomized Trial of Adjuvanted Versus Nonadjuvanted Trivalent Influenza Vaccine in 823 US Nursing Homes. Clin Infect Dis. 2021;73(11):e4237-e43.
78. DiazGranados CA, Dunning AJ, Kimmel M, Kirby D, Treanor J, Collins A, et al. Efficacy of high-dose versus standard-dose influenza vaccine in older adults. N Engl J Med. 2014;371(7):635–45.

79. Bridges CB, Kuehnert MJ, Hall CB. Transmission of influenza: implications for control in health care settings. Clin Infect Dis. 2003;37(8):1094–101.
80. Prevention Strategies for Seasonal Influenza in Healthcare Settings: Centers for Disease Control and Prevention; [updated 5/13/217/6/22]. Available from: https://www.cdc.gov/flu/professionals/infectioncontrol/healthcaresettings.htm.
81. Bean B, Moore BM, Sterner B, Peterson LR, Gerding DN, Balfour HH, Jr. Survival of influenza viruses on environmental surfaces. J Infect Dis. 1982;146(1):47–51.
82. Ryan MA, Christian RS, Wohlrabe J. Handwashing and respiratory illness among young adults in military training. Am J Prev Med. 2001;21(2):79–83.
83. Liu M, Ou J, Zhang L, Shen X, Hong R, Ma H, et al. Protective Effect of Hand-Washing and Good Hygienic Habits Against Seasonal Influenza: A Case-Control Study. Medicine (Baltimore). 2016;95(11):e3046.
84. Stay Home When You Are Sick: Centers for Disease Control and Prevention; [updated 8/31/2021. Available from: https://www.cdc.gov/flu/business/stay-home-when-sick.htm#:~:text=CDC%20recommends%20that%20workers%20who%20have%20flu%20symptoms,%28at%20least%204-5%20days%20after%20flu%20symptoms%20started%29.
85. Kinman G. Sickness presenteeism at work: prevalence, costs and management. Br Med Bull. 2019;129(1):69–78.
86. Huttunen R, Syrjanen J. Healthcare workers as vectors of infectious diseases. Eur J Clin Microbiol Infect Dis. 2014;33(9):1477–88.
87. Widera E, Chang A, Chen HL. Presenteeism: a public health hazard. J Gen Intern Med. 2010;25(11):1244–7.
88. Chiu S, Black CL, Yue X, Greby SM, Laney AS, Campbell AP, et al. Working with influenza-like illness: Presenteeism among US health care personnel during the 2014–2015 influenza season. Am J Infect Control. 2017;45(11):1254–8.
89. Carrera M, Lawler EC, White C. Population Mortality and Laws Encouraging Influenza Vaccination for Hospital Workers. Ann Intern Med. 2021;174(4):444–52.
90. Frederick J, Brown AC, Cummings DA, Gaydos CA, Gibert CL, Gorse GJ, et al. Protecting Healthcare Personnel in Outpatient Settings: The Influence of Mandatory Versus Nonmandatory Influenza Vaccination Policies on Workplace Absenteeism During Multiple Respiratory Virus Seasons. Infect Control Hosp Epidemiol. 2018;39(4):452–61.
91. Guidance: Outbreak Management in Long-Term Care Facilities: Centers for Disease Control and Prevention; [updated 11/17/2020. Available from: https://www.cdc.gov/flu/professionals/infectioncontrol/ltc-facility-guidance.htm.
92. Yu H, Liao Q, Yuan Y, Zhou L, Xiang N, Huai Y, et al. Effectiveness of oseltamivir on disease progression and viral RNA shedding in patients with mild pandemic 2009 influenza A H1N1: opportunistic retrospective study of medical charts in China. BMJ. 2010;341:c4779.
93. Hayden FG, Gubareva LV, Monto AS, Klein TC, Elliot MJ, Hammond JM, et al. Inhaled zanamivir for the prevention of influenza in families. Zanamivir Family Study Group. N Engl J Med. 2000;343(18):1282–9.

Ebola: Virology, Clinical Considerations, and Outbreak Response and Prevention

Asa Oxner and Andrew Myers

Abstract Ebola viral disease (EVD) is globally known as a highly fatal and transmissible hemorrhagic fever that creates headlines with intimidating outbreaks. It is caused by infection with one of the ebolaviruses, harbored within a yet undiscovered reservoir in West, Central, and East Africa. Here we review Ebola's viral characteristics and pathogenesis; clinical considerations for diagnosis, treatment, and care of survivors; and finally, management and prevention of outbreaks.

Keywords Ebola · Hemorrhagic fever

1 Virus Characteristics, History, and Geography

Ebola virus disease (EVD) is caused by related viruses within the ebolavirus genus in the family of *Filoviridae* [1]. Ebolaviruses are single-stranded, negative-sense RNA viruses that have a filamentous shape [2]. Marburg is also a *Filoviridae* that causes hemorrhagic fevers in humans but is not discussed here. The ebolavirus RNA codes for ten proteins whose functions are partially understood. Some proteins enable attachment and entry into cells via highly conserved receptors that recognize cells from numerous species, including humans, nonhuman primates, pigs, and bats [3]. Other proteins code for reverse transcriptase, help the virus elude the host immune system, and cause infected cells to activate coagulation cascades causing disseminated intravascular coagulation [3–5].

There are four species of ebolaviruses that infect humans: Zaire ebolavirus, Sudan ebolavirus, Taï ebolavirus, and Bundibugyo ebolavirus [6]. There are two other ebolavirus species that are not known to cause illness in humans but do cause

A. Oxner (✉) · A. Myers
University of South Florida, Division of General Internal Medicine, Morsani Center for Advanced Healthcare, Tampa, FL, USA
e-mail: aoxner@usf.edu; awm@usf.edu

C. Somboonwit et al. (eds.), *Global Virology IV: Viral Disease Diagnosis and Treatment Delivery in the 21st Century*,
https://doi.org/10.1007/978-3-031-57369-9_3

illness in other animals (Reston in primates and pigs, Bombali in bats) [6]. Species names are based on their location of original identification and isolation, for example, Taï from Cote d'Ivoire and Bundibugyo from Uganda.

Zaire and Sudan ebolaviruses were both first described in 1976 when two outbreaks occurred in Central Africa [6]. The two villages impacted by these outbreaks were located about 500 kilometers apart. Although they were initially thought to be a single outbreak that traveled to a satellite location, ultimately, genetic sequences confirmed they were different strains of ebolaviruses [6]. The slightly earlier and more studied outbreak was in a village located near the Ebola river within the Democratic Republic of Congo (Zaire at that time), which scientists used to name the virus genus. Zaire ebolavirus has the most severe clinical presentation of EVD and the highest mortality rate, ranging from 50% to 90%. The largest outbreak to date was in 2014–2016 across West Africa—beginning with a child from Guinea, then his household and local healthcare workers, and ultimately spreading throughout Sierra Leone and Liberia [7]. There were even a few cases in Nigeria, Mali, and Senegal related to immigration [7]. This outbreak was the first and only event to involve widespread urban transmission, leading to 28,616 confirmed cases and 11,310 confirmed deaths in Guinea, Sierra Leone, and Liberia [7].

2 Life Cycle and Transmission

Ebola's lifecycle occurs entirely within a single infected cell, replicating RNA, assembling its virion, and then budding off to infect other cells. The reservoir host is believed to be African fruit bats however, conclusive evidence of this is yet to be established. Scientists cite population growth with subsequent encroachment into forested areas and interaction with wildlife as contributors to the emergence of EVD in humans since the 1970s. However, virologic studies date the ebolaviruses as much older [6]. Additionally, transportation and mobile phone infrastructure development have enabled migration and disease notification in the modern era compared to presumed self-isolating, unreported outbreaks previously.

Ebolaviruses are transmitted via contact with infected body fluids (blood, urine, saliva, sweat, feces, vomit, breast milk, amniotic fluid, semen) or fomites [6]. Common methods of transmission include during burial preparations and caring for sick persons. Viral load has been correlated to mortality; thus, deceased patients have exceedingly high viral loads in blood and many other tissues (over 10^6), making body preparation for burial highly infectious [3]. Ebolaviruses can survive in blood for several days at room temperature and on dry surfaces for several hours [2]. Because of its durability in the environment, the virus is highly infectious and manifests with outbreaks in communities and healthcare settings.

3 Clinical Presentation and Pathogenesis in Humans

EVD is a hemorrhagic fever (HF). Early symptoms include fever, malaise, anorexia, headaches, and other body pains. Because the geography of EVD overlaps with malaria endemicity, early stages of illness are commonly misdiagnosed as malaria, contributing to both failure in infection prevention and insufficient hydration to avoid progression to multiorgan failure. Two to five days later, patients typically develop secretory diarrhea, vomiting, and abdominal pains—the gastrointestinal fluid losses reach 10 liters per day [1]. Around 25–50% of infected patients develop hemorrhagic complications. The incubation period is 2–21 days, so infected and exposed persons should remain isolated for 21 days to avoid further transmission [6] (Fig. 1).

EVD pathogenesis is characterized by a) profoundly high inflammatory cytokine levels released from macrophages, which results in coagulation abnormalities and distributive septic shock from increased vascular permeability; and b) dysregulation of dendritic cells reducing the production of interferon and preventing interferon signaling, which contributes to triggered apoptosis of lymphocytes and failed adaptive/humoral immunity [1, 3–5]. The varied clinical presentation depends on the virus strain and host cell responses. In some patients, macrophages are activated to release tissue factor and coat infected cells with fibrin, both activating intravascular coagulation. Additionally, some patients manifest protein C deficiency, which is an anticoagulant, but the pathogenesis of how ebolaviruses cause this deficiency is not yet understood [3].

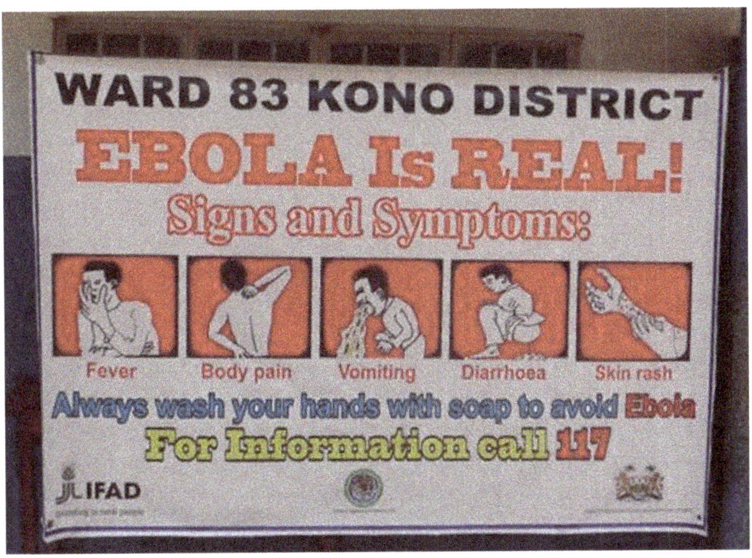

Fig. 1 Sign educating healthcare workers and community on the symptoms of Ebola and reducing transmission by hand washing and caring for sick persons in a facility. Seen posted at a healthcare facility in Kono, Sierra Leone during the 2014–2016 outbreak

4 Diagnostics

Polymerase chain reaction (PCR) is the gold standard diagnostic method for EVD. Medical diagnostic companies (notably Biofire, Altona, and Cepheid) have developed platforms for Ebola PCRs with acceptable sensitivity, specificity, and rapid turnaround time [2, 8, 9]. All have been produced considering the practicality of collecting and transporting infected blood and the infrastructure available for laboratories and supply chains in resource-limited settings.

Additionally, several bedside antigen platforms have been developed, but the sensitivity of these does not reach >90% compared to PCR, so they are not acceptable as the sole diagnostic for field use [10, 11]. However, these antigen tests have been used for postmortem surveillance during active outbreaks where high pretest probability improves the performance of their positive and negative predictive value [12]. In that setting, they were helpful to field teams in quickly returning results and engaging the families in safer burial and public health actions.

Ebolavirus typically reaches detectable levels for PCR 3 days after symptoms begin [2]. Thus, false-negative tests within the first 72 hours of symptoms are possible, and recommendations are for retesting before ending isolation in patients with high clinical suspicion of infection [1]. Test of clearance/discharge from isolation recommendations have been established during prior outbreaks using repeat PCR testing and cycle times. ELISA tests for Ebola antibodies are also available and can be used to confirm a person's prior infection by an ebolavirus [9].

5 Treatment

The mainstay of treatment for EVD is the management of septic shock and electrolyte depletion from secretory diarrhea via aggressive fluid and electrolyte resuscitation [1, 3, 6]. This can be accomplished orally with oral rehydration salts or intravenously (IV) if patients cannot maintain oral intake. Simultaneous use of supportive medications to lessen nausea, diarrhea, fever, and pain is recommended. Clinicians should have a low threshold to initiate testing or empiric treatment against concurrent malaria, aspiration pneumonia, or bacteremia from translocation across the gut [1, 3, 6] (Fig. 2).

In the 2018–2020 Ebola outbreak in the Democratic Republic of Congo, four medications specifically targeting Zaire ebolavirus underwent clinical trials: ZMapp and REGN-EB3 (both monoclonal antibody cocktails), Mab114 (a single monoclonal antibody), and remdesivir (an antiviral). At 28 days, death had occurred in 35% of the Mab114 group vs. 49.7% in the ZMapp group (p0.007) and 33.5% of the REGN-EB3 group vs. 51% in the ZMapp group (p0.002) [13]. The FDA ultimately approved both REGN-EB3 and Mab114 based on these data as a single-dose infusion for all persons aged 0–100 years [14]. The trial showed improved survival when the medications were given early in the course of symptoms [13]. REGN-EB3

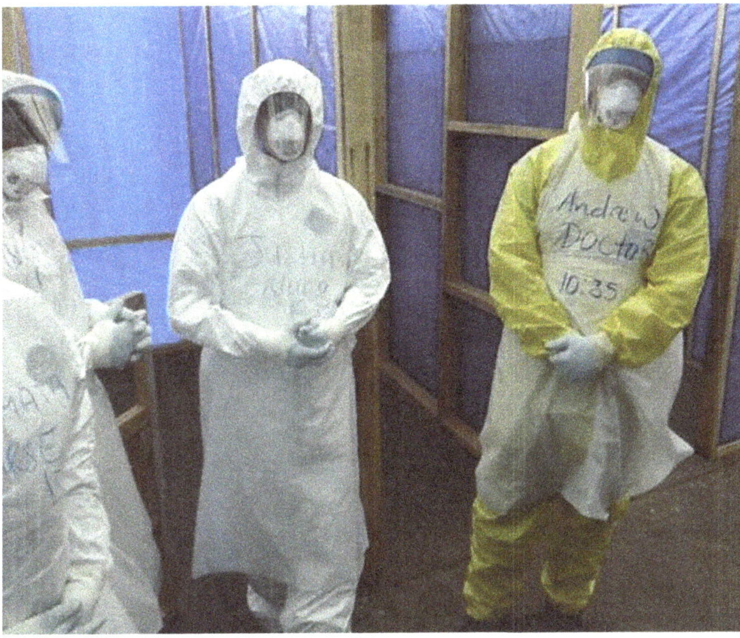

Fig. 2 Personal Protective Equipment used by staff in Sierra Leone (both national and international). Notice full body suit with hood, apron, double gloves (one under sleeves and one over), N95, face shield. Provider name, role, and time entering the ETU were written on apron

and Mab114 neutralize Zaire ebolavirus by binding to its glycoprotein, thus preventing the virus from entering human cells. The virus-monoclonal antibody complex is then broken down by the host's cellular immune system, preventing replication [6]. The currently approved medications for EVD are only protective against Zaire ebolavirus. Still, additional monoclonals are in the preclinical stages of development against Sudan and Bundibugyo strains and a mixed efficacy treatment against all strains [1].

6 Survivor Considerations

Ebolaviruses can remain in immunologically privileged tissues after recovery from EVD; PCR-positive fluids have been collected from the testes, interior of the eyes, placenta, and central nervous system up to 18 months after acute infection [1, 3, 6, 15]. There are reported case studies of sexual transmission from male survivors [16]. Survivors have antibodies against ebolavirus for up to 10 years, but the level of protective immunity they confer for the same or different strains is not fully known.

Ebola survivors suffer post-viral syndromes, including inflammatory arthritis, uveitis/iritis, encephalitis/brain fog, headache, galactorrhea, and increased dyspepsia/peptic ulcer disease [3, 15, 17]. Treatment of these should be with supportive medications for symptoms. Several of these symptoms are presumed to be sequelae of inflammation and cross-reacting antibodies [18], yet no specific recommendations exist for anti-autoimmune therapies to treat patients.

Depression and posttraumatic stress syndrome are also common in survivors and exacerbated by social isolation and stigma [15, 19]. It has not been studied whether these would respond to treatment with SSRIs and counseling, but we recommend those modalities. Further research is needed to help survivors and inform future outbreaks.

7 Prevention

The cornerstone of Ebola prevention is the quarantine of sick or exposed persons, personal protective equipment for caregivers and healthcare workers, sensitization of populations about the high likelihood of transmission during burial preparations, and need for hand hygiene [6]. Recently, scientists have added vaccination as a prevention strategy (Fig. 3).

In 2019, the US Food and Drug Administration (FDA) approved the Ebola vaccine rVSV-ZEBOV as a single-dose vaccine protective against Zaire ebolavirus. The vaccine is recommended for all exposed contacts (ages 0–100 years, pregnant women included) during an outbreak using a ring vaccination strategy [1, 6]. Additionally, the FDA and CDC recommend the Ebola vaccine for pre-exposure prophylaxis in adults aged 18 years and above who are at potential occupational risk of exposure to Zaire ebolavirus. Vaccine efficacy was 88% overall and improved to 97% when the vaccine had at least 10 days to mount an immune response before exposure [20]. Also, vaccinated patients who still developed EVD after vaccination in the 2018–2020 Democratic Republic of Congo outbreak had a higher survival rate than unvaccinated patients [1].

There have been some patients living with HIV who received the Ebola vaccine in West Africa, and it appears safe and immunogenic in this population, but further studies in immunocompromised individuals are needed. Similarly, open-label clinical trials in West Africa included pregnant and lactating women without statistical differences in pregnancy outcomes or congenital abnormalities. Because of this, the World Health Organization (WHO) recommends including pregnant and lactating women in ring vaccination campaigns during an active outbreak.

The approved Ebola vaccine is a live-attenuated vesicular stomatitis virus (VSV) recombined with a full-length ebolavirus glycoprotein. Side effects are typical febrile reactions and possible anaphylaxis if the individual has a history of allergy to rice proteins, as the vaccine is produced with rice-derived recombinant human albumin. The persistence of humoral immunity is still being studied to determine future recommendations for or against booster doses [1, 6, 21]. Vaccines against other ebolavirus strains are being developed, but none are currently approved.

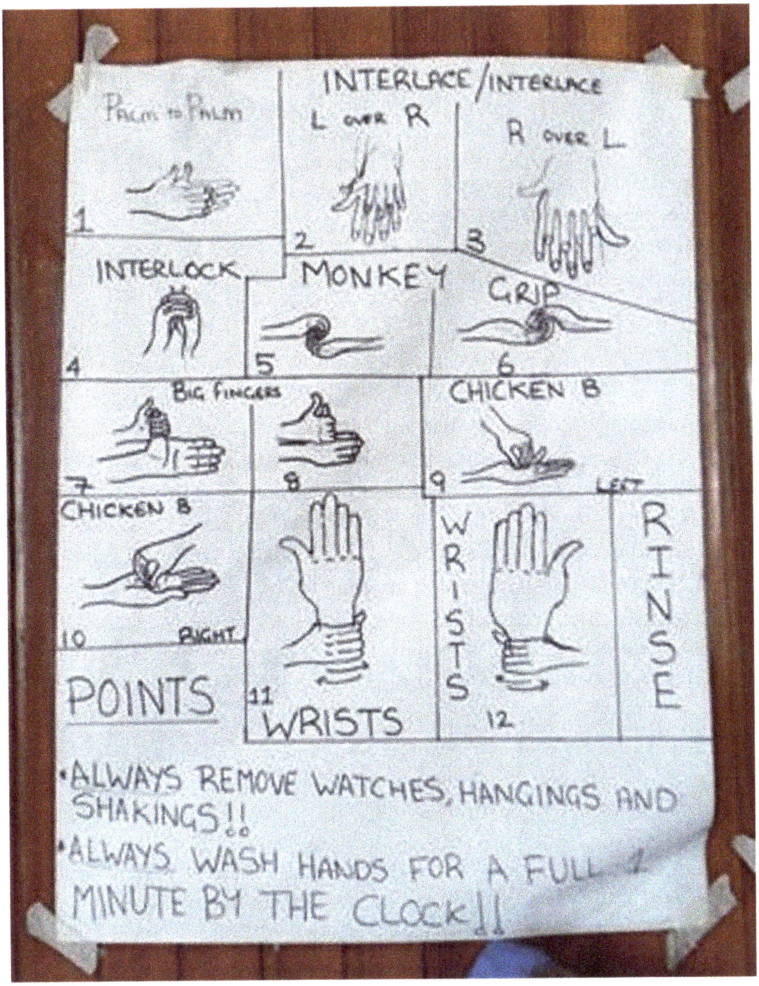

Fig. 3 A sign posted in a school in Sierra Leone 2015 giving detailed instructions for hand washing to prevent the spread of Ebola

8 Role of Public Health and International Aid Organizations in Outbreak Response and Preparedness

The objectives of the Ebola outbreak response are stopping transmission and clinical management of infected patients. [1] Preventing transmission requires case identification, laboratory capabilities, contact tracing (classically 10–15 persons per index case), capacity for isolation and monitoring, sensitization of populations to infection prevention measures, altered burial preparations, and resources for disinfection and waste management. In the most recent outbreaks, transmission prevention included additional human resources and equipment for vaccination

Table 1 Key activities for ebola outbreak response

Key activities
Case identification
Laboratory capabilities
Contact tracing
Isolation and monitoring
Community uptake of infection prevention measures
Altered burial preparation Combating stigma
Disinfection and waste management resources
Vaccination campaigned

campaigns that need to be conducted while fully staffing and funding all the above priorities.

Table 1 illustrates the key activities for the Ebola outbreak response.

Attention and resource allocation toward community engagement and education are mission-critical to get sick persons cared for in facilities with personal protective equipment, stop unsafe burials, and protect aid workers. There have been several attacks on aid workers in communities where mistrust and misunderstanding of medicine or governmental intervention are high.

There remains a need for multinational coalitions to fund research and implementation of Ebola therapeutics, vaccines, and upkeep of the workforce of local and international responders across all disciplines with training and equipment.

References

1. Feldmann, H., A. Sprecher, and T.W. Geisbert, *Ebola.* N Engl J Med, 2020. **382**(19): p. 1832–1842.
2. Broadhurst, M.J., T.J. Brooks, and N.R. Pollock, *Diagnosis of Ebola Virus Disease: Past, Present, and Future.* Clin Microbiol Rev, 2016. **29**(4): p. 773–93.
3. Rojas, M., et al., *Ebola virus disease: An emerging and re-emerging viral threat.* J Autoimmun, 2020. **106**: p. 102375.
4. Locke, M., et al., *Quantification of Type I Interferon Inhibition by Viral Proteins: Ebola Virus as a Case Study.* Viruses, 2021. **13**(12).
5. Lu, J., J.M. Gullett, and T.D. Kanneganti, *Filoviruses: Innate Immunity, Inflammatory Cell Death, and Cytokines.* Pathogens, 2022. **11**(12).
6. Centers for Disease Control and Prevention. *Ebola Disease.* 2023 March 23, 2023 [cited 2023 September 8]; Available from: https://www.cdc.gov/vhf/ebola/index.html.
7. Organization, W.H. *One year into the Ebola epidemic January 2015.* 2015 [cited 2023 September 8]; Available from: https://www.who.int/news-room/spotlight/one-year-into-the-ebola-epidemic.
8. Jansen van Vuren, P., et al., *Comparative Evaluation of the Diagnostic Performance of the Prototype Cepheid GeneXpert Ebola Assay.* J Clin Microbiol, 2016. **54**(2): p. 359–67.
9. Sealy TK, E.B., Taboy CH, et al.,, *Laboratory Response to Ebola — West Africa and United States.* MMWR, 2016. **65**: p. 44–49.

10. Mukadi-Bamuleka, D., et al., *Field performance of three Ebola rapid diagnostic tests used during the 2018–20 outbreak in the eastern Democratic Republic of the Congo: a retrospective, multicentre observational study.* Lancet Infect Dis, 2022. **22**(6): p. 891–900.
11. Mukadi-Bamuleka, D., et al., *Head-to-head comparison of diagnostic accuracy of four Ebola virus disease rapid diagnostic tests versus GeneXpert® in eastern Democratic Republic of the Congo outbreaks: a prospective observational study.* EBioMedicine, 2023. **91**: p. 104568.
12. Mukadi-Bamuleka, D., et al., *Postmortem Surveillance for Ebola Virus Using OraQuick Ebola Rapid Diagnostic Tests, Eastern Democratic Republic of the Congo, 2019–2020.* Emerg Infect Dis, 2022. **28**(2): p. 420–424.
13. Mulangu, S., et al., *A Randomized, Controlled Trial of Ebola Virus Disease Therapeutics.* N Engl J Med, 2019. **381**(24): p. 2293–2303.
14. Organization, W.H. *Neutralizing antibody cocktail for Ebola virus disease (EVD).* 2022 [cited 2023 September 8]; Available from: https://www.who.int/publications/i/item/WHO-EVD-therapeutics-REGN-EB3-Poster_A-2022.1.
15. Etard, J.F., et al., *Multidisciplinary assessment of post-Ebola sequelae in Guinea (Postebogui): an observational cohort study.* Lancet Infect Dis, 2017. **17**(5): p. 545–552.
16. Subissi, L., et al., *Ebola Virus Transmission Caused by Persistently Infected Survivors of the 2014–2016 Outbreak in West Africa.* J Infect Dis, 2018. **218**(suppl_5): p. S287-s291.
17. Amuzu, C., et al., *Post-Ebola sequelae among Ebola child survivors in Sierra Leone.* BMC Pediatr, 2021. **21**(1): p. 482.
18. Gunn, B.M., et al., *Survivors of Ebola Virus Disease Develop Polyfunctional Antibody Responses.* J Infect Dis, 2020. **221**(1): p. 156–161.
19. Canady, D., et al., *Ophthalmic and psychosocial sequelae in Ebola virus disease survivors: ongoing need for health systems strengthening across disciplines.* Expert Rev Anti Infect Ther, 2021. **19**(2): p. 117–119.
20. Henao-Restrepo, A.M., et al., *Efficacy and effectiveness of an rVSV-vectored vaccine in preventing Ebola virus disease: final results from the Guinea ring vaccination, open-label, cluster-randomised trial (Ebola Ça Suffit!).* Lancet, 2017. **389**(10068): p. 505–518.
21. Organization, W.H., *Preliminary results on the efficacy of rVSV-ZEBOV-GP Ebola vaccine using the ring vaccination strategy in the control of an Ebola outbreak in The Democratic Republic of the Congo: an example of integration of research into epidemic response.* Geneva: Organ, 2019.

A Clinical Approach to Novel Diagnostics and Therapeutics: The Challenge of Zika

Doniya Milani, Vidhya Sabapathy, and Beata Casanas

Abstract Zika virus is a mosquito-borne neurotropic virus that gained attention following the 2015 outbreak in Brazil. It often presents with nonspecific symptoms such as fever, headache, and myalgia. More specific symptoms include arthralgia in small joints, a maculopapular rash, and conjunctivitis. Zika-associated complications include Guillain-Barré syndrome and microcephaly. In this chapter, current and advanced diagnostic methods are discussed, including reverse transcriptase polymerase chain reaction (RT-PCR), reverse transcriptase loop-mediated isothermal amplification (RT-LAMP), microfluidic paper-based analytic devices (μPAD), nucleic acid sequence-based amplification (NASBA), recombinase polymerase amplification, and NASBA combined with a CRISPR/Cas9-based module. Additionally, vaccines and therapeutics are reviewed.

Keywords Zika virus · RT-PCR · RT-LAMP · NASBA · Zika diagnostics · Novel diagnostics and therapeutics · Novel detection of Zika infection

1 Introduction

Zika virus (ZIKV) is an arthropod-borne flavivirus that is primarily transmitted by *Aedes* mosquitos [1]. ZIKV, originally isolated in the Zika Forest of Uganda in 1947, is an 11-kb RNA virus that is related to other flaviviruses, such as dengue virus, yellow fever virus, and West Nile virus [2]. There are two chief lineages to classify ZIKV strains, African and Asian/American [4]. While vector-borne transmission from *Aedes* mosquitos is the principal source of human infection,

D. Milani (✉) · B. Casanas
University of South Florida, Morsani College of Medicine, Tampa, FL, USA
e-mail: dmilani@usf.edu; beata@usf.edu

V. Sabapathy
Infectious Disease Services of Georgia, P.C., Roswell, GA, USA

C. Somboonwit et al. (eds.), *Global Virology IV: Viral Disease Diagnosis and Treatment Delivery in the 21st Century*,
https://doi.org/10.1007/978-3-031-57369-9_4

nonvector-borne transmission has been reported as well. These routes include sexual transmission, saliva exchange, transplacental transmission, breastfeeding, and blood transfusions [3]. ZIKV has two noncoding regions in addition to one coding region that encodes three structural proteins (capsid, envelope, and membrane precursor) and seven nonstructural proteins (NS1, NS2A, NS2B, NS3, NS4A, NS4B, and NS5). The envelope structural protein is the protein specifically responsible for viral entry into the cell and has an important epitope targeted by neutralizing monoclonal antibodies. Additional antibody targets include the membrane precursor structural protein and the NS1 nonstructural protein [2].

Because of its self-limiting nature as an asymptomatic virus in approximately 80% of infected patients, relatively little attention was historically given to Zika after its initial discovery in the 1940s [2]. The virus was first discovered in a febrile sentinel rhesus monkey and from a pool of *Aedes africanus* mosquitos in the Ugandan Zika forest [1]. Five years later, it was detected in humans using neutralizing antibody testing. In the years following the discovery of ZIKV, the virus spread beyond Africa and Asia.

ZIKV commonly mimics symptoms of influenza infection, possibly contributing to under reporting [1]. Infection initially presents with fever, headache, myalgias, and rash. Some of the most common symptoms, such as arthralgia in small joints, maculopapular rash, and conjunctivitis, are often confused with dengue virus infection, further contributing to the estimated underreporting of Zika virus infection [1].

Zika virus infection has traditionally been seen as a relatively mild disease. However, global perception of Zika began to permanently shift during the 2015 outbreak in Brazil, as ZIKV distinguished itself as a unique mosquito-borne, sexually transmissible neurotropic virus associated with neurological complications and malformations. These are most notably Guillain-Barré syndrome and microencephaly [4]. ZIKV's fetal pathogenicity further distinguishes it from other flaviviruses [5]. Within one year of the Brazil outbreak, ZIKV developed into a global pandemic, spreading to 60 additional countries in the American, African, and Southeast Asian continents. ZIKV transmission was first confirmed in the United States and Brazil by the World Health Organization (WHO) in 2015 [3]. Two years later, Brazil was plagued with approximately 1.5 million confirmed ZIKV cases and 4000 unconfirmed cases of related microcephaly.

Beyond its rapid and explosive spread among humans, the ZIKV outbreak led to the birth of thousands of infants with congenital defects, including microcephaly, fetal demise, blindness, and hearing loss [4]. ZIKV can also trigger Guillain-Barré syndrome (GBS) in adults, an autoimmune disease that can lead to muscle weakness and paralysis. This association was elucidated during ZIKV global outbreaks, particularly in French Polynesia, where a 20- to 40-fold increase in GBS cases was reported during the 2015 outbreak [1].

Scientists considered gene sequence mutations, new routes of transmission, and genetic susceptibility of certain populations as possible explanations for ZIKV's transformed pathogenicity [1]. In 2016, the WHO declared ZIKV an international public health emergency and called for urgent research into advanced diagnostic methods, vaccines, and antiviral drug candidates and for the virus' molecular involvement in observed neurological complications. In this chapter, the role of bioinformatics and artificial intelligence in current/novel diagnostic methods and therapeutics will be discussed. It is important to note that the ZIVV pandemic was crucial in mobilizing methods of the global medical community, which assisted in rapid and successful efforts to combat the current SARS-2/COVID-19 pandemic.

2 Current Diagnostic Methods

The WHO currently recommends reverse-transcription polymerase chain reaction (RT-PCR) and serological testing for suspected cases of ZIKV [2]. While virus isolation through tissue or fluid culture is considered the gold standard for ZIKV diagnosis, this method is limited by its low sensitivity and inability to swiftly produce results. Similarly, serologic testing and virus isolation are time-consuming procedures [1]. Conversely, RT-PCR is a reliable testing method that has a high sensitivity, specificity, and a rapid response time. RT-PCR is preferred in the early, active stages of ZIKV infection and relies upon fluid, tissue, or serum samples. In recent years, new nucleic acid detection methods have been developed for ZIKV testing, including NASVA-CRISPR Cleavage RT-PCR, which enables strain differentiation at a single-base resolution within 4 hours, and reverse transcription loop-mediated isothermal amplification (RT-LAMP), which eliminates the need for RNA extraction and provides results within 40–90 minutes [1]. Next-generation sequencing of the complete viral genome is also being explored as a rapid and cost-effective diagnostic tool. It is important to note, however, that despite RT-PCR's high sensitivity, the increasing emergence of divergent ZIKV strains can potentially undermine this diagnostic method [6]. As with many other arboviral diseases, the ZIKV epidemic is further compounded by disproportionate vector risk near low-resource populations that are least equipped to fully detect and handle outbreaks.

Fast and accurate detection of ZIKV is vital for improved prognosis in patients, but clinical diagnosis can be challenging due to overlapping clinical features and the cross-reactivity of other flavivirus serum antibodies [3]. For example, several studies have found cross-reactive antibodies between dengue virus (DENV) and ZIKV [2]. This finding is important in the context of vaccine development because the presence of antibodies against DENV has been shown to worsen the consequences of ZIKV infection through a phenomenon known as antibody-dependent enhancement (ADE) [1]. Virus-specific antibodies characteristically play a protective role in the immune system's response to infection. Such antibodies can enhance viral entry and severity of infection in particular cases when they fail to neutralize a different viral serotype. This is the postulated explanation for the observed increase in ZIKV infection severity in patients with prior exposure to flaviviruses. For this reason, serological approaches can be complicated, especially in regions where ZIKV is co-circulating with other flaviviruses.

Despite this limitation, serological techniques are recommended following the acute phase of infection. Immunoassays such as ELISA and IFA can be used to detect IgM and IgG antibodies from serum and cerebrospinal fluid (CSF) samples [2]. The presence of IgM antibodies against the Zika virus, detected using ELISA, indicates previous infection within 2–5 months earlier [1]. These tests are highly sensitive and produce results quickly but are limited by potential cross-reactivity with other flavivirus serum antibodies, which can cause false positives. Thus, ZIKV infection is considered probable in patients with positive immunoassay results and clinical symptoms but cannot be confirmed without neutralizing antibody testing.

Since 2015, ZIKV detection has extended from using exclusively serum and CSF samples to include other bodily fluids as effective diagnostic specimens. Some studies have reported shedding of ZIKV RNA in saliva and urine for several weeks following the onset of symptoms [3]. In 2016, the Florida Department of Health Bureau of Public Health Laboratories (BPHL) performed RT-PCR on urine, saliva, and semen samples from 913 persons to determine the most sensitive specimen type in the context of ZIKV RNA detection in acute infections [7]. Their results indicated that urine specimens may be preferred over serum specimens in ZIKV detection.

Owing to the expense and uncertainty of serological testing, the syndromic overlap among arboviruses, and the higher prevalence of DENV, diagnosis is sometimes made on clinical grounds alone. Insufficient diagnostic capabilities are key limitations facing current Zika outbreak management strategies. Novel approaches to address this are discussed as follows (Fig. 1).

3 Advanced Techniques in Diagnostics

Modern technologies, including computational tools as well as experimental approaches, need to be utilized effectively to address the challenges surrounding Zika diagnostics. Computer-assisted and data-driven approaches have been used in producing specific point-of-care NAAT amplification systems, including reverse transcriptase loop-mediated isothermal amplification (RT-LAMP), microfluidic paper-based analytical devices (μPAD), nucleic acid sequence-based amplification (NASBA), and recombinase polymerase amplification (RPA) [8].

Loop-mediated isothermal amplification (LAMP) has been used as part of "smart molecular diagnostics," where it is capable of accurate point-of-care detection of variable nucleic targets. Research has shown simultaneous LAMP reactions can be used with multiple primers which can compute the presence of almost any Zika variant. Sequence-specificity of oligonucleotide strand exchange probes ensures signal specificity, while the redundancy of the molecular logic processors guards against false negatives [11].

Coupling of RT-LAMP with unincorporated amplification signal reporters (QUASR) technique via smartphone chromaticity algorithms has demonstrated highly sensitive and specific detection of Zika, chikungunya, and dengue viruses

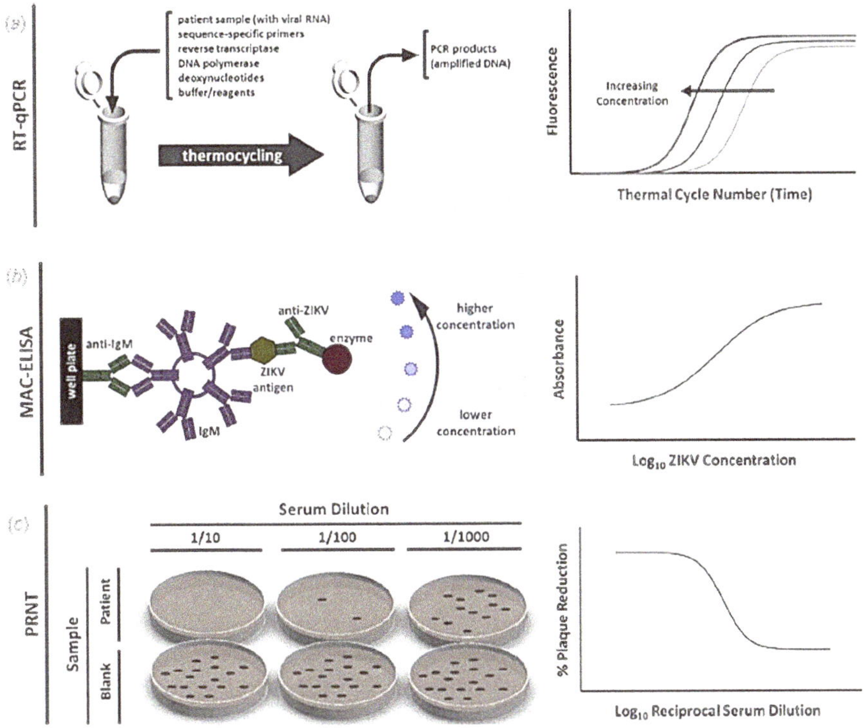

Fig. 1 CDC-recommended tests for ZIKV detection. (Reprinted with permission from Herrada [8])

[9]. Reactions are conducted in simple, inexpensive, and portable "LAMP boxes" supplemented with a consumer-class smartphone. In this platform, the RT-LAMP detection setup includes an isothermal heater with reaction tubes, LED excitation source, and Bluetooth microcontroller. A smartphone app wirelessly actuates the heater and excitation source to enable real-time monitoring of the temperature and illumination. The illuminated reagents are captured via the phone camera. An algorithm utilizes chromicity to analyze fluorescence signals, which improves the discrimination of positive/negative signals by fivefold when compared to RGB intensity sensors or the naked eye. Blood, saliva, and urine samples could be used in this system, which provides versatility in point-of-care molecular diagnostics.

μPAD platforms are being developed for point-of-care settings due to their low cost, flexibility, and availability around the world. Microporous papers are used in these devices, allowing for efficient absorption and sample flow without peripheral pumps [8]. This offers an attractive alternative to traditional lab-based assays, enables affordable and rapid disease testing, and is thus being investigated for ZIKV detection [10]. However, the paper-based device is only a prototype and requires FDA approval and testing of clinical samples.

NASBA combined with a CRISPR/Cas9-based module has shown promise in detecting ZIKV RNA, and its specificity allows for differentiation between closely related flaviviruses and even different ZIKV lineages [8]. The primers and toehold sensors are selectively designed to match specific ZIKV strains. These are infused and freeze-dried onto paper. When a patient's sample is added, the paper is activated, and a color change would indicate a positive reaction (Fig. 2).

Lastly, RPA can be used for nucleic acid amplification without the need for active temperature control in between runs. The main advantage of this device is the speed of extraction and purification. However, the lower RNA extraction performance may reduce sensitivity when detecting samples with lower viral load [9].

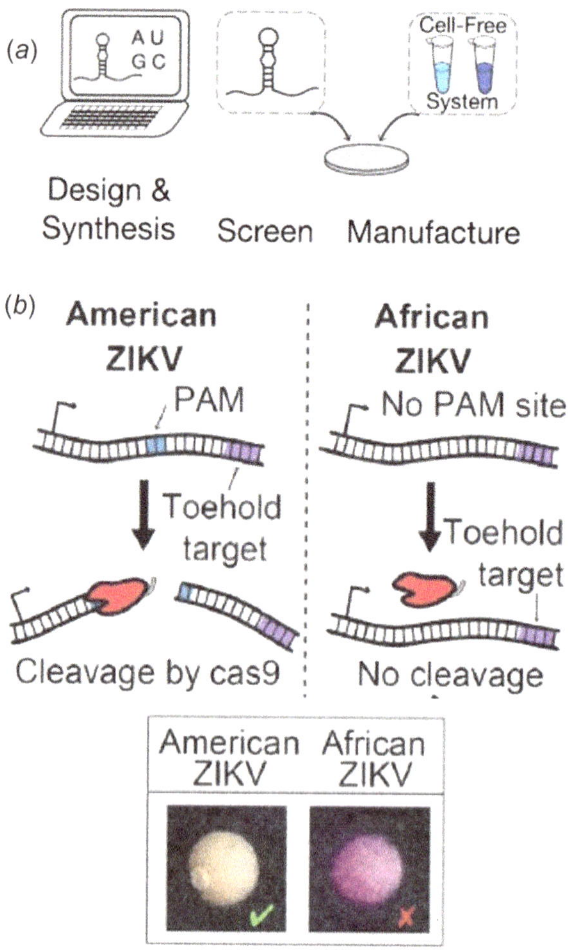

Fig. 2 CRISPR/CAS9 biomolecular sensors. (Reprinted with permission from Herrada [8])

4 Vaccine and Therapeutics

Monitoring virus genotype diversity is crucial to recognize the emergence and spread of outbreaks. This in turn will promote effective prevention and treatment strategies. Genomic surveillance of epidemics enables the identification of strains associated with greater epidemic potential and disease severity [12]. A new computational method for identification of ZIKV sequences with respect to species and sub-species has been evaluated. A tool was developed implementing a classification pipeline, which was assessed on a large dataset of envelope and whole genome sequences. Rapid, accurate, and high-throughput classification of DENV, CHIKV, and ZIKV species and genotypes were accomplished [12].

The association of ZIKV with congenital malformations and Guillain-Barré syndrome emphasizes the need for effective vaccines. Hurdles to be overcome include extending and preserving the longevity of live attenuated vaccines, sustaining the immunogenicity of inactive strains, and preventing relapse of vector-based vaccines that lead to unforeseen viral genetic and host immune responses [13].

There have been several studies that target the viral envelope protein; however, a risk of antibody-dependent enhancement remains. Recently, three different novel DNA vaccines encoding a secreted nonstructural protein (NS1) were described [14]. The ability of each to induce NS1-specific humoral and cell-mediated immune responses and protective efficacy against ZIKV were studied. Functional NS1-specific T cell responses critical for immune protection were discovered, prompting the need for further investigation using NS1 as a target for T cell-based ZIKV DNA vaccines [14].

Although no drug has yet been approved by the FDA for the treatment of ZIKV infection, computer-assisted approaches have been used to identify potential anti-ZIKV drugs. To reduce the cost of experimental assays, less expensive computer-assisted drug discovery (CADD) is used as a complementary approach to prioritize compounds for synthesis and/or biological evaluation [15].

In silico structural-bioinformatics-based methodologies have been developed for the design of ZIKV antibodies [14]. This also includes de novo design of humanized monoclonal antibodies targeting specific antigen epitopes. Although no anti-ZIKV drug has yet been discovered by pharmacophore modeling, many studies have indicated that this approach shows promise for the synthesis of multifunctional inhibitors, which will eliminate cross-resistance and toxicity and improve patient adherence [16].

5 Conclusions

Much progress has been made in the detection and treatment of Zika virus. More advanced diagnostic methods, coupling of RT-LAMP with unincorporated amplification signal reporters (QUASR) technique via smartphone chromaticity algorithms, are highly sensitive and specific for the detection of Zika virus. In the future, antiviral drugs and vaccines may be possibly used to treat and prevent Zika infection; however, no anti-ZIK drug has yet been discovered, indicating the need for more research and continued genomic surveillance.

Conflicts of Interest The authors have no financial disclosures or conflicts of interest.

References

1. Koppolu, V. and T.S. Raju, *Zika virus outbreak: a review of neurological complications, diagnosis, and treatment options*. Journal of Neurovirology, 2018. **24**(3): p. 255–272.
2. Karkhah, A., et al., *Zika virus: epidemiology, clinical aspects, diagnosis, and control of infection*. European Journal of Clinical Microbiology & Infectious Diseases, 2018. **37**(11): p. 2035.
3. Khurshid, Z., et al., *Human saliva can be a diagnostic tool for Zika virus detection*. Journal of Infection and Public Health, 2019. **12**(5): p. 601–604.
4. Diamond, M.S., J.E. Ledgerwood, and T.C. Pierson, *Zika Virus Vaccine Development: Progress in the Face of New Challenges*, in *Annual Review of Medicine, Vol 70*, M.E. Klotman, Editor. 2019, Annual Reviews: Palo Alto. p. 121–135.
5. Kauffman, E.B. and L.D. Kramer, *Zika Virus Mosquito Vectors: Competence, Biology, and Vector Control*. J Infect Dis, 2017. **216**(suppl_10): p. S976-S990.
6. Corman, V.M., et al., *Assay optimization for molecular detection of Zika virus*. Bulletin of the World Health Organization, 2016. **94**(12): p. 880–892.
7. Bingham, A.M., et al., *Comparison of Test Results for Zika Virus RNA in Urine, Serum, and Saliva Specimens from Persons with Travel-Associated Zika Virus Disease – Florida, 2016*. Mmwr-Morbidity and Mortality Weekly Report, 2016. **65**(18): p. 475–478.
8. Herrada CA, Kabir MA, Altamirano R, Asghar W. Advances in Diagnostic Methods for Zika Virus Infection. *J Med Device*. 2018;12(4):0408021–4080211. https://doi.org/10.1115/1.4041086
9. Priye, A., Bird, S., Light, Y. *et al.* A smartphone-based diagnostic platform for rapid detection of Zika, chikungunya, and dengue viruses. *Sci Rep* **7,** 44778 (2017). https://doi.org/10.1038/srep44778
10. Bedin, F., Boulet, L., Voilin, E., Theillet, G., Rubens, A., and Rozand, C., 2017, " Paper-Based Point-of-Care Testing for Cost-Effective Diagnosis of Acute Flavivirus Infections," J. Med. Virol., 89(9), pp. 1520–1527.https://doi.org/10.1002/jmv.24806
11. Bhadra S., Saldaña M.A., Han H.G., Hughes G.L., Ellington A.D. Simultaneous Detection of Different Zika Virus Lineages via Molecular Computation in a Point-of-Care Assay. Viruses. 2018;10:714. https://doi.org/10.3390/v10120714
12. Fonseca V, Libin PJK, Theys K, et al. A computational method for the identification of Dengue, Zika and Chikungunya virus species and genotypes. *PLoS Negl Trop Dis*. 2019;13(5):e0007231. Published 2019 May 8. https://doi.org/10.1371/journal.pntd.0007231
13. Basak SC, Majumdar S, Nandy A, et al. Computer-Assisted and Data Driven Approaches for Surveillance, Drug Discovery, and Vaccine Design for the Zika Virus. *Pharmaceuticals (Basel)*. 2019;12(4):157. Published 2019 Oct 16. https://doi.org/10.3390/ph12040157
14. Grubor-Bauk B, Wijesundara DK, Masavuli M, et al. NS1 DNA vaccination protects against Zika infection through T cell-mediated immunity in immunocompetent mice. *Sci Adv*. 2019;5(12):eaax2388. Published 2019 Dec 11. https://doi.org/10.1126/sciadv.aax2388

15. Mottin, M, Borba, J, Melo-Filho, C, et al. Computational drug discovery for the Zika virus. *Brazilian Journal of Pharmaceutical Sciences*, *54*(spe), e01002. Epub November 08, 2018. https://doi.org/10.1590/s2175-97902018000001002
16. Ncube N.B., Ramharack P., Soliman M.E.S. An “All-In-One” Pharmacophoric Architecture for the Discovery of Potential Broad-Spectrum Anti-Flavivirus Drugs. Appl. Biochem. Biotechnol. 2018;185:799–814. https://doi.org/10.1007/s12010-017-2690-2.

Oncogenic Human Papillomavirus

Ana P. Velez and Lynette J. Menezes

Abstract Human papillomavirus (HPV) is a sexually transmitted disease associated with multiple clinical presentations, including genital and nongenital warts (verrucae), anogenital malignancy, and some oropharyngeal cancers. Understanding the clinical manifestations, indications for screening, and treatment is important to prevent or reduce the consequences, including anogenital and oropharyngeal neoplasms. This chapter will review the most common clinical manifestations, screening, and treatment of HPV-related diseases, excluding detailed information about chemotherapy or radiation for more advanced HPV-related malignancies.

Keywords HPV · Common warts · Cervical cancer · Anal cancer · Oropharyngeal cancer · Penile cancer

1 Introduction

Human papillomaviruses (HPVs) are small DNA viruses that constitute the papillomavirus genus of the papillomaviridae family [1]. There are over 200 different types of known HPVs worldwide [2]. HPV infects humans via skin-to-skin contact. Grossly, they can be divided between genital and nongenital infections. Nongenital infections occur primarily in children 12–16 years old but can also occur less commonly in adults [3].

Recent estimates in the United States indicate that 77.3 million of 15–59-year-olds had a prevalent HPV infection in 2018 [4]. However, incidence was highest during adolescence, with 11.1 million of 15–24-years-olds acquiring a new

A. P. Velez (✉) · L. J. Menezes
University of South Florida, Morsani College of Medicine, Tampa, FL, USA
e-mail: Ana.Velez@moffitt.org; lmenezes@usf.edu

C. Somboonwit et al. (eds.), *Global Virology IV: Viral Disease Diagnosis and Treatment Delivery in the 21st Century*,
https://doi.org/10.1007/978-3-031-57369-9_5

infection [4]. Most HPV infections, including carcinogenic and noncarcinogenic, resolve within 12 months [5, 6]. The prevention, screening, and treatment vary depending on the age, location, and type of HPV and potential malignant association.

2 Clinical Manifestations

HPV has been associated with both malignant and nonmalignant disease (Table 1).

2.1 *Benign Disease*

2.1.1 Nongenital Warts

HPV can infect keratinocytes in the epidermis. The HPV infection leads to epidermal thickening and hyperkeratinization, resulting in a visible wart (common warts) [7]. Nongenital warts occur in 10% of children ages 12–16, but they can also be seen in adults [3, 8]. Warts can resolve spontaneously after 2 years in children but may persist for several years in adults [7, 8]. Decreased cell-mediated immunity can lead to extensive and refractory disease with widespread warts [9].

Infection is typically acquired by direct skin contact or the environment with wounds or minor lacerations facilitating the spread of the virus to deeper tissues [7]. The most common types of warts include plantar or verruca plantaris, common warts or verruca vulgaris, and flat warts. Commonly affected areas are hands and feet, but they can occur in any area of the body. Symptoms include painful papular lesions with associated pruritus and sometimes bleeding. Plantar warts are particularly disabling when located on pressure points [8].

Nongenital warts may resolve spontaneously. Treatment may be indicated for severe pain, discomfort, functional impairment, cosmesis, and persistent or resistant

Table 1 Clinical manifestations of HPV

Nonmalignant disease
Nongenital warts or common warts
Genital warts or condyloma acuminate
Malignant disease
Cervical cancer
Vulvar and vaginal cancers
Anal cancer
Penile cancer
Oropharyngeal cancer
Cutaneous disease with malignant potential; Bowen disease, Epidermodysplasia verruciformis

warts. Treatment modalities include resection or destruction of the lesion with salicylic acid, trichloroacetic acid, cryotherapy, cantharidin, or surgery. Other treatments include intralesional immunotherapy, imiquimod, topical bleomycin, or fluorouracil.

2.1.2 Genital Warts

Genital HPV is the leading sexually transmitted infection worldwide [10]. It is estimated that most sexually active individuals have been infected with a genital HPV type at some point in their lifetime [11]. However, in some countries, HPV infection prevalence is decreasing due to HPV vaccination [12].

Anogenital HPV can be further divided by risk types based on their carcinogenic potential. Low-risk HPV 6 and HPV 11 cause 90% of anogenital warts (condyloma acuminate) [13]. High-risk HPV 16 and 18 are commonly associated with cervical cancer, with type 16 seen in about 50% of cases and HPV 18 in 15% of cases. An additional eleven HPV types—HPV 31, 33, 35, 39, 45, 51, 52, 56, 58, 59, and 66 are characterized as high-risk based on the degree of carcinogenicity [11, 14]. Infections are acquired by contact with skin or genital mucosa. Anogenital HPV is almost always contracted by sexual contact. Subclinical and latent infection with HPV are common [10], but in patients who develop genital warts, the median incubation period is 2.9 months [15]. Immunosuppressed patients with poor cell-mediated immunity have more severe, extensive, and refractory disease, often with recurrent lesions, and a high risk for malignant transformation [16].

The appearance of the lesions can range from skin-colored, erythematous, or hyperpigmented to small papular, flat, or massive cauliflower-like. These lesions are typically located in the groin, penis, perineum, perianal skin, and mucosal surfaces. They are often asymptomatic but can cause pruritus, bleeding, and depending on the size, can obstruct the genital or anal canal. These lesions are typically transitory and resolve without treatment, but long-term remission rates are currently unknown [17].

Treatment includes imiquimod, sinecatechins, and imiquimod that the patient can apply. Other clinician-determined therapies include cryotherapy, trichloroacetic acid, and surgical removal with laser, electrosurgery, or excision [18].

2.2 *HPV-Associated Malignancies*

For more than 20 years, the association of HPV with different malignancies, particularly oropharyngeal, cervical, vulvar, vaginal, penile, and anal cancer, has led to greater attention on HPV-mediated oncogenesis. HPV infects epithelial cells in the skin, oropharyngeal, or anogenital mucosa. The viral replication in these well-differentiated epithelial cells may have less exposure to immunological interference allowing adaptation of new HPV variants to different types of differentiated epithelial cells. Despite the large number of HPV genotypes, approximately 40 subtypes have been detected in anogenital cancer biopsy specimens. Additionally, 70% of oropharyngeal cancers are caused by the same carcinogenic HPV [19], with HPV 16 accounting for 90–95% of these cancers [20]. These carcinogenic HPV types have cell immortalization and cell-transforming proprieties.

High-risk HPV genomes code for at least three proteins with growth-stimulating and transforming properties. The E5 protein is involved in the early expansion of an infected cell clone, and E6 and E7 jointly immortalize different human cell types. The oncogenic process and the interaction between high-risk HPV and the host are complex. It requires not only the expression of the E5, E6, and E7 proteins but also chemical and physical mutagens with the integration of the viral DNA, jointly creating the perfect storm to alter the host DNA [21, 22].

2.2.1 Cervical Cancer

Cervical cancer is the fourth leading cancer among women globally, representing an estimated 604,000 cases, and the fourth leading cause of mortality with 342,000 deaths in 2020 [23]. It remains the leading cause of cancer death among women in 36 countries located in Sub-Saharan Africa, Melanesia, South America, and Southeastern Asia [23]. Malawi has the highest incidence (40/100,000 women) and mortality (28.6/100,000 women) in the world. It is well-established that infection by carcinogenic HPV genotypes causes nearly all cases of cervical cancer [24]. Women with an HPV infection may develop low-grade epithelial lesions—cervical intraepithelial neoplasia (CIN) 1, but the majority will clear the infection with the lesions regressing spontaneously in 1–2 years [25]. However, in a small fraction of women, these infections persist, resulting in the development of high-grade epithelial lesions (CIN 2+) within 2–3 years [15]. The progression to cervical cancer can take decades, allowing for timely screening and treatment intervention to prevent further advancement. In contrast, women living with HIV experience a 2- to 25-fold risk for cervical cancer [26]. Because of HIV-related immunosuppression, the HPV infection in these women persists longer. It can rapidly progress to cervical dysplasia and invasive cancer at a younger age compared to women without HIV [27, 28].

2.2.2 Vulvar and Vaginal Cancers

Unlike cervical cancer, vulvar and vaginal cancers are not always associated with HPV. HPV causes 43% of vulvar carcinomas and 70% of vaginal cancers in women [29].

2.2.3 Anal Cancer

Anal cancer is uncommon overall, with HPV associated in 88–90% of anal cancers and high-grade squamous intraepithelial lesions (HSIL) [30, 31]. The risk of anal cancer and HSIL is higher in people living with HIV, particularly in women and men who have sex with men [31]. Similar to cervical cancer, anal cancer, is often preceded by a precancerous lesion (HSIL), Studies on preventive strategies to screen and prevent anal cancer are being established, emulating already established strategies to prevent cervical cancer [31].

2.2.4 Penile Cancer

Penile cancer is uncommon overall and is only associated with HPV 16 and 18 in 35–40% of all penile cancers and about 80% of all HPV-associated cancers [32]. High-risk HPV infection is more common in men with phimosis and without neonatal circumcision [33].

2.2.5 Oropharyngeal Cancer

Oropharyngeal cancers are further divided into HPV- related and non-HPV-related cancers. The incidence of HPV-related oropharyngeal carcinomas in the United States is estimated to be 70% of all cases [19]. HPV-associated oropharyngeal carcinomas are steadily increasing in the United States, surpassing non-HPV-associated carcinomas [19]. In the United States, there were 12,500 cases of HPV-associated oropharyngeal carcinoma in men, surpassing the 10,800 cases of cervical cancer in women [34].

HPV-associated oropharyngeal cancer has been strongly associated with oral-genital contact. Direct mouth-to-mouth contact as a form of HPV transmission cannot be ruled out at this time. Sexual behaviors and a prior history of HPV 16 infection have been associated with an increased risk for oropharyngeal cancer [20].

2.2.6 Cutaneous Diseases with Malignant Potential

Bowen disease is a squamous cell carcinoma in situ located in the genitals and outside the genitals. It is frequently sun-induced, but several types of high-risk HPV have also been associated [35].

Epidermodysplasia verruciformis is a rare genetic condition characterized by specific immunosuppression that facilitates certain HPV infections. The patients present earlier in life with several warts. Malignant transformation is often found in sun-exposed areas with warts infected with HPV 5 and HPV 8 [36].

3 Screening for HPV-Associated Malignancies

Regular cervical screening for early detection and treatment of cervical disease and timely HPV vaccination have been well-established as effective strategies to eliminate cervical cancer, an essential World Health Organization (WHO) goal. Cervical screening with cytology (the Papanicolaou test or Pap smear), and more recently, molecular HPV testing in adult women has dramatically reduced cervical cancer incidence and mortality in most high-income economies and some upper-middle-income economies [37, 38]. However, the lack of national screening guidelines and/or access to cervical screening continue to place a substantial burden on women in low- and middle-income countries (LMICs) [39]. Current guidelines in the United States by the American College of Gynecology and Obstetrics (ACOG) and the American Society for Colposcopy and Cervical Pathology (ASCCP) [40] recommend cervical screening of normal-risk women aged 21–29 years with Pap Cytology every 3 years. For those aged 30–65, an FDA-approved HPV test alone or co-testing with Pap cytology every 5 years or Pap cytology alone every 3 years is recommended for screening (Table 2). For women with a previous history of cervical disease, the frequency of screening varies based on the initial results and treatment efficacy. The American Society of Clinical Oncology (ASCO) recently published resource-stratified guidelines that cover the range of settings from basic, limited, enhanced, and maximal recourses respectively [41]. The suboptimal infrastructure and lack of trained cytopathologists have been an impediment to implementing Pap cytology in LMICs [39]. While molecular HPV testing is the screening test recommended by the WHO [42] and the American Cancer Society [43] as part of cervical screening guidelines (Table 2) for women worldwide and in the U.S., respectively, the high cost per test makes it prohibitive in LMICs. Addressing these gaps through the reduction of HPV test costs, developing low-cost screening technologies, and building human resource capacity and infrastructure to conduct screening with follow-up treatment are imperative to meet the WHO goal of eliminating cervical cancer by 2030.

Some public entities, such as the New York Department of Health and the AIDS Institute, proposed anal cytology for anal cancer screening for high-risk patients, including patients older than 35 living with HIV, men who have sex with men, cisgender women, transgender women, and transgender men [44].

Since no prospective randomized controlled trials have been done evaluating anal cancer screening, the AIDS Malignancy Consortium (AMC) proposed screening for anal cancer, including anal cytology, followed by high-resolution anoscopy (HRA) and HRSA-guided biopsy to find HSIL and rule out cancer. Areas with HSIL could be treated using procedures similar to HPV-associated cervical lesions [31].

To address the above concern, The ANCHOR study— a large-scale randomized study comparing different methods to prevent anal cancer in high-risk populations was developed. The idea was extrapolated from the similarities between HPV-associated cervical and anal cancers. The objective was to determine whether treating HSIL was as efficacious in reducing the incidence of anal cancer compared with

Table 2 Summary of cervical cancer screening guidelines for women by the American College of Gynecology and Obstetrics (ACOG) [40], American Society for Colposcopy and Cervical Pathology (ASCCP) [40], American Cancer Society [43], and World Health Organization (WHO) [42]

Population	ACOG, ASCCP	ACS	WHO
Aged <21 y Aged 21–25 years	No screening Cytology alone every 3 y	No screening No screening	No screening No screening
Aged 25–65 years	Cytology alone every 3 y until age 29 y Aged 30–65: Cytology alone every 3 y FDA-approved primary hrHPV testing alone every 5 y Cotesting (Cytology and HPV testing) every 5 y	Primary HPV test alone every 5 y (preferred) Use an FDA-approved HPV test for primary screening every 5 y Cotesting (Cytology and HPV testing) every 5 y (acceptable)* Cytology alone every 3 y (acceptable)* *Only if HPV testing is not available. In future guidelines, this will be eliminated after the US transitions to HPV testing	Recommends starting screening at 30 y Suggests starting screening at 25 y for women living with HIV Prioritize women 30–49 years* Screen with HPV DNA test in screen and treat approach starting at age 30 years with regular screening interval of 5–10 y Screen with HPV DNA test in screen, triage, and treat approach starting at age 30 y with regular screening interval of 5–10 y Screen with HPV DNA test in screen, triage, and treat approach starting at age 25 years for women living with HIV with regular screening interval of 3–5 y Suggest using genotyping, colposcopy, VIA or cytology to triage women with positive HPV DNA test If HPV DNA testing is not yet operational, then a screening interval of 3 years with VIA or cytology is acceptable for women from the general population and women living with HIV

(continued)

Table 2 (continued)

Population	ACOG, ASCCP	ACS	WHO
Aged >65 years	Discontinue screening if adequate negative screening (2 negative HPV tests or cotests, or 3 negative cytology tests) in past 10 y with the most recent occurring in the past 3–5 y	Discontinue screening if adequate negative screening (2 negative HPV tests or cotests or 3 negative cytology tests) in past 10 y with the most recent occurring in the past 3–5 y	Discontinue screening after age 50 years, if two consecutive negative results are consistent with regular screening intervals for women from the general population and women living with HIV
After hysterectomy	No screening for women without cervix and no history of CIN2+ or cervical cancer	No screening for women without cervix and no history of CIN2+ in past 25 y or cervical cancer	
HPV vaccinated	Same as age-specific recommendations for unvaccinated women	Same as age-specific recommendations for unvaccinated women	No change. Recommendations to be re-evaluated once more vaccinated women move into age ranges for cervical cancer screening

active monitoring. The ANCHOR study concluded that treating HSIL reduced the incidence of anal cancer and could be used as a standard of care for all populations at risk for anal cancer [31].

Currently, there are no effective screening tests for other HPV-associated cancers such as head and neck, and penile cancer; the detection of HPV as a screening method for cancer is only recommended as a part of routine cervical cancer screening and for oropharyngeal cancer that needs biopsies to assist with staging and prognosis. HPV testing to screen for vaginal, penile, anal, and oropharyngeal cancer has been only used for research purposes in the United States and other countries and is not US FDA-approved.

4 HPV Vaccination

Since 2006, when the first quadrivalent HPV vaccine, Gardasil® (HPV 6, 11, 16 and 18), and, later, the bivalent vaccine Cervarix® (HPV 16 and 18) were authorized for primary prevention of cervical cancer, numerous studies in high-income economies have established the safety and efficacy of these vaccines in decreasing high-risk HPV prevalence, the incidence of cervical neoplasia and genital warts [45–48]. HPV vaccines have the potential to reduce anal [49] and oropharyngeal cancers [50, 51]. A recent study among Danish women found a 70% reduction in risk of anal HSIL in women vaccinated at ages <17 years compared to unvaccinated women [52]. In the United States, only the nine-valent Gardasil®9 (HPV 6, 11, 16, 18, 31, 33, 45, 52, and 58) is available for use [53], albeit the quadrivalent Gardasil® and Cervarix® are available in other countries [54].

Table 3 HPV vaccine schedule and dosing by the US Advisory Committee on Immunization Practices (ACIP) [53, 55] and the World Health Organization (WHO) [54]

	US ACIP	WHO recommendations
Routine vaccination	2 doses at age 11–12 years for girls and boys; can be started at age 9 years	2 doses, one year apart for girls and boys aged 9–14 y; a single dose schedule can be used to increase uptake if 2 doses are not available or accessible
Catch-up vaccination	Ages 13–26 years if not adequately vaccinated; 3 doses for those 15 years and older	Vaccination of secondary target populations, e.g., females aged ≥15 years, boys, older males, or MSMs, is recommended only if this is feasible and affordable
Shared clinical decision-making	Some adults aged 27–45 years, if not adequately vaccinated	
People with HIV	3 doses up to age 26	At least 2 doses, 6 months apart, and three doses if available and accessible

These HPV vaccines have been recommended in the United States at two doses for girls and boys aged 11–12 years as part of routine vaccination and can be started at age 9 years [53, 55]. ACIP also recommends HPV vaccination for females ages 13 to 26 and males ages 13 to 21 not vaccinated previously [55]. Vaccination is also recommended up to 26 years for men who have sex with men, transgender persons, and for immunocompromised, including men and women living with HIV [53]. For those older than 15, a three-dose series is recommended [53, 55] (Table 3).

The WHO recommends routine vaccination at 2 doses for girls starting at 9 years and up to 26 years and at least a single dose to increase uptake if two doses are not available or accessible [54] (Table 3).

The HPV vaccine has demonstrated and is expected to achieve great success in preventing anogenital cancers and warts among both women and men and more recently with oropharyngeal cancers. The potential benefits of the HPV vaccine in preventing other HPV-associated cancers, including some head and neck, remains unclear at this time [56].

5 Conclusion

HPV is associated with nonmalignant infections but has been associated with several malignancies, including cervical, anal, penile, oropharyngeal, and some cutaneous lesions with malignant potential. Cervical cytology and more recently HPV testing have been critical for the dramatic decline in cervical cancer incidence and mortality. More recently, anal cytology targeting HSIL has shown reduced incidence of anal cancer through early detection. HPV vaccination has demonstrated the potential to dramatically reduce cervical, anal, and oropharyngeal cancer incidence, and mortality. An effective global HPV vaccine strategy could save millions of lives from HPV-associated cancers.

References

1. Brianti P, De Flammineis E, Mercuri SR. Review of HPV-related diseases and cancers. The new microbiologica. 2017;40(2):80–5.
2. de Villiers EM, Fauquet C, Broker TR, Bernard HU, zur Hausen H. Classification of papillomaviruses. Virology. 2004;324(1):17–27.
3. Beutner KR. Nongenital human papillomavirus infections. Clinics in laboratory medicine. 2000;20(2):423–30.
4. Lewis RM, Laprise JF, Gargano JW, Unger ER, Querec TD, Chesson HW, et al. Estimated Prevalence and Incidence of Disease-Associated Human Papillomavirus Types Among 15- to 59-Year-Olds in the United States. Sex Transm Dis. 2021;48(4):273–7.
5. Plummer M, Schiffman M, Castle PE, Maucort-Boulch D, Wheeler CM. A 2-year prospective study of human papillomavirus persistence among women with a cytological diagnosis of atypical squamous cells of undetermined significance or low-grade squamous intraepithelial lesion. The Journal of infectious diseases. 2007;195(11):1582–9.
6. Rodríguez AC, Schiffman M, Herrero R, Wacholder S, Hildesheim A, Castle PE, et al. Rapid clearance of human papillomavirus and implications for clinical focus on persistent infections. Journal of the National Cancer Institute. 2008;100(7):513–7.
7. Sterling JC, Gibbs S, Haque Hussain SS, Mohd Mustapa MF, Handfield-Jones SE. British Association of Dermatologists' guidelines for the management of cutaneous warts 2014. The British journal of dermatology. 2014;171(4):696–712.
8. Allen AL, Siegfried EC. What's new in human papillomavirus infection. Current opinion in pediatrics. 2000;12(4):365–9.
9. Barbosa P. Plantar verrucae and HIV infection. Clinics in podiatric medicine and surgery. 1998;15(2):317–27.
10. de Camargo CC, Tasca KI, Mendes MB, Miot HA, de Souza Ldo R. Prevalence of Anogenital Warts in Men with HIV/AIDS and Associated Factors. The open AIDS journal. 2014;8:25–30.
11. IARC Working Group on the Evaluation of Carcinogenic Risks to Humans. IARC monographs on the evaluation of carcinogenic risks to humans: Human Papillomaviruses. Biological Agents Volume 100 B A Review of Human Carcinogens. Lyon, France: International Agency for Research on Cancer (IARC); 2012.
12. King EM, Gilson R, Beddows S, Soldan K, Panwar K, Young C, et al. Human papillomavirus DNA in men who have sex with men: type-specific prevalence, risk factors and implications for vaccination strategies. British journal of cancer. 2015;112(9):1585–93.
13. Garland SM, Steben M, Sings HL, James M, Lu S, Railkar R, et al. Natural history of genital warts: analysis of the placebo arm of 2 randomized phase III trials of a quadrivalent human papillomavirus (types 6, 11, 16, and 18) vaccine. The Journal of infectious diseases. 2009;199(6):805–14.
14. de Sanjose S, Quint WG, Alemany L, Geraets DT, Klaustermeier JE, Lloveras B, et al. Human papillomavirus genotype attribution in invasive cervical cancer: a retrospective cross-sectional worldwide study. The Lancet Oncology. 2010;11(11):1048–56.
15. Winer RL, Kiviat NB, Hughes JP, Adam DE, Lee SK, Kuypers JM, et al. Development and duration of human papillomavirus lesions, after initial infection. The Journal of infectious diseases. 2005;191(5):731–8.
16. Gormley RH, Kovarik CL. Human papillomavirus-related genital disease in the immunocompromised host: Part I. Journal of the American Academy of Dermatology. 2012;66(6):867.e1–14; quiz 81–2.
17. Lynde C, Vender R, Bourcier M, Bhatia N. Clinical features of external genital warts. Journal of cutaneous medicine and surgery. 2013;17 Suppl 2:S55–60.
18. Workowski KA, Bachmann LH, Chan PA, Johnston CM, Muzny CA, Park I, et al. Sexually Transmitted Infections Treatment Guidelines, 2021. MMWR Recommendations and reports: Morbidity and mortality weekly report Recommendations and reports. 2021;70(4):1–187.

19. Chaturvedi AK, Engels EA, Pfeiffer RM, Hernandez BY, Xiao W, Kim E, et al. Human papillomavirus and rising oropharyngeal cancer incidence in the United States. J Clin Oncol. 2011;29(32):4294–301.
20. D'Souza G, Kreimer AR, Viscidi R, Pawlita M, Fakhry C, Koch WM, et al. Case-control study of human papillomavirus and oropharyngeal cancer. The New England journal of medicine. 2007;356(19):1944–56.
21. zur Hausen H. Papillomaviruses causing cancer: evasion from host-cell control in early events in carcinogenesis. Journal of the National Cancer Institute. 2000;92(9):690–8.
22. Winer RL, Hughes JP, Feng Q, Xi LF, Cherne S, O'Reilly S, et al. Early natural history of incident, type-specific human papillomavirus infections in newly sexually active young women. Cancer epidemiology, biomarkers & prevention: a publication of the American Association for Cancer Research, cosponsored by the American Society of Preventive Oncology. 2011;20(4):699–707.
23. Sung H, Ferlay J, Siegel RL, Laversanne M, Soerjomataram I, Jemal A, et al. Global Cancer Statistics 2020: GLOBOCAN Estimates of Incidence and Mortality Worldwide for 36 Cancers in 185 Countries. CA Cancer J Clin. 2021;71(3):209–49.
24. Walboomers JM, Jacobs MV, Manos MM, Bosch FX, Kummer JA, Shah KV, et al. Human papillomavirus is a necessary cause of invasive cervical cancer worldwide. J Pathol. 1999;189(1):12–9.
25. Moscicki AB, Shiboski S, Hills NK, Powell KJ, Jay N, Hanson EN, et al. Regression of low-grade squamous intra-epithelial lesions in young women. Lancet. 2004;364(9446):1678–83.
26. De Vuyst H, Lillo F, Broutet N, Smith JS. HIV, human papillomavirus, and cervical neoplasia and cancer in the era of highly active antiretroviral therapy. European Journal of Cancer Prevention. 2008;17(6):545–54.
27. Gichangi PB, Bwayo J, Estambale B, De Vuyst H, Ojwang S, Rogo K, et al. Impact of HIV infection on invasive cervical cancer in Kenyan women. AIDS. 2003;17(13):1963–8.
28. Hagensee ME, Cameron JE, Leigh JE, Clark RA. Human papillomavirus infection and disease in HIV-infected individuals. The American Journal of the Medical Sciences. 2004;328(1):57–63.
29. Forman D, de Martel C, Lacey CJ, Soerjomataram I, Lortet-Tieulent J, Bruni L, et al. Global burden of human papillomavirus and related diseases. Vaccine. 2012;30 Suppl 5:F12–23.
30. Joseph DA, Miller JW, Wu X, Chen VW, Morris CR, Goodman MT, et al. Understanding the burden of human papillomavirus-associated anal cancers in the US. Cancer. 2008;113(10 Suppl):2892–900.
31. Lee JY, Lensing SY, Berry-Lawhorn JM, Jay N, Darragh TM, Goldstone SE, et al. Design of the ANal Cancer/HSIL Outcomes Research study (ANCHOR study): A randomized study to prevent anal cancer among persons living with HIV. Contemporary clinical trials. 2022;113:106679.
32. Backes DM, Kurman RJ, Pimenta JM, Smith JS. Systematic review of human papillomavirus prevalence in invasive penile cancer. Cancer causes & control: CCC. 2009;20(4):449–57.
33. Afonso LA, Cordeiro TI, Carestiato FN, Ornellas AA, Alves G, Cavalcanti SM. High Risk Human Papillomavirus Infection of the Foreskin in Asymptomatic Men and Patients with Phimosis. The Journal of urology. 2016;195(6):1784–9.
34. Division of Cancer Prevention and Control Centers for Disease Control and Prevention. How Many Cancers Are Linked with HPV Each Year? 2023 [updated September 12, 2023. Available from: https://www.cdc.gov/cancer/hpv/statistics/cases.htm#print.
35. Mackenzie-Wood A, Kossard S, de Launey J, Wilkinson B, Owens ML. Imiquimod 5% cream in the treatment of Bowen's disease. Journal of the American Academy of Dermatology. 2001;44(3):462–70.
36. Karrer S, Szeimies RM, Abels C, Wlotzke U, Stolz W, Landthaler M. Epidermodysplasia verruciformis treated using topical 5-aminolaevulinic acid photodynamic therapy. The British journal of dermatology. 1999;140(5):935–8.

37. Dickinson JA, Stankiewicz A, Popadiuk C, Pogany L, Onysko J, Miller AB. Reduced cervical cancer incidence and mortality in Canada: national data from 1932 to 2006. BMC Public Health. 2012;12(1):992.
38. Landy R, Sasieni PD, Mathews C, Wiggins CL, Robertson M, McDonald YJ, et al. Impact of screening on cervical cancer incidence: A population-based case-control study in the United States. Int J Cancer. 2020;147(3):887–96.
39. Denny L. The prevention of cervical cancer in developing countries. BJOG. 2005;112(9):1204–12.
40. American College of Obstetricians and Gynecologists. Updated Cervical Cancer Screening Guidelines, Practice Advisory April 2021 2021 [updated Reaffirmed April 2023. Available from: https://www.acog.org/clinical/clinical-guidance/practice-advisory/articles/2021/04/updated-cervical-cancer-screening-guidelines#.
41. Shastri SS, Temin S, Almonte M, Basu P, Campos NG, Gravitt PE, et al. Secondary Prevention of Cervical Cancer: ASCO Resource-Stratified Guideline Update. JCO Glob Oncol. 2022;8:e2200217.
42. World Health Organization (WHO). WHO guideline for screening and treatment of cervical pre-cancer lesions for cervical cancer prevention, second edition. Geneva: World Health Organization; 2021.
43. Fontham ETH, Wolf AMD, Church TR, Etzioni R, Flowers CR, Herzig A, et al. Cervical cancer screening for individuals at average risk: 2020 guideline update from the American Cancer Society. CA Cancer J Clin. 2020;70(5):321–46.
44. Hirsch BE, McGowan JP, Fine SM, Vail R, Merrick ST, Radix A, et al. New York State Department of Health AIDS Institute Clinical Guidelines. Screening for Anal Dysplasia and Cancer in Adults With HIV. Baltimore (MD): Johns Hopkins University. Copyright © Johns Hopkins University HIV Clinical Guidelines Program 2000–2020. All Rights Reserved. The use, reproduction, and distribution of original documents and related graphics from this web site is encouraged provided that full credit of source accompanies all uses, in all forms. Please note that if we have adapted or reproduced copyrighted material from another source, with permission, we cannot extend permission to reproduce. Links to pages on this site are also encouraged and may be created without seeking permission.; 2022.
45. Ahrlund-Richter A, Cheng L, Hu YOO, Svensson M, Pennhag AAL, Ursu RG, et al. Changes in Cervical Human Papillomavirus (HPV) Prevalence at a Youth Clinic in Stockholm, Sweden, a Decade After the Introduction of the HPV Vaccine. Frontiers in cellular and infection microbiology. 2019;9:59.
46. Brotherton JML, Bloem PN. Population-based HPV vaccination programmes are safe and effective: 2017 update and the impetus for achieving better global coverage. Best Practice & Research: Clinical Obstetrics & Gynaecology. 2018;47:42–58.
47. Wright TC, Jr., Parvu V, Stoler MH, Kodsi S, Eckert K, Yanson K, et al. HPV infections and cytologic abnormalities in vaccinated women 21–34years of age: Results from the baseline phase of the Onclarity trial. Gynecologic oncology. 2019.
48. Drolet M, Bénard É, Pérez N, Brisson M. Population-level impact and herd effects following the introduction of human papillomavirus vaccination programmes: updated systematic review and meta-analysis. Lancet. 2019;394(10197):497–509.
49. Berenson AB, Guo F, Chang M. Association of Human Papillomavirus Vaccination With the Incidence of Squamous Cell Carcinomas of the Anus in the US. JAMA Oncol. 2022;8(4):1–3.
50. Macilwraith P, Malsem E, Dushyanthen S. The effectiveness of HPV vaccination on the incidence of oropharyngeal cancers in men: a review. Infectious Agents and Cancer. 2023;18(1):24.
51. Nielsen KJ, Jakobsen KK, Jensen JS, Grønhøj C, Von Buchwald C. The Effect of Prophylactic HPV Vaccines on Oral and Oropharyngeal HPV Infection-A Systematic Review. Viruses. 2021;13(7).
52. Baandrup L, Maltesen T, Dehlendorff C, Kjaer SK. HPV vaccination and anal high-grade precancerous lesions and cancer: a real-world effectiveness study. JNCI: Journal of the National Cancer Institute. 2023.

53. Meites E, Kempe A, Markowitz LE. Use of a 2-Dose Schedule for Human Papillomavirus Vaccination – Updated Recommendations of the Advisory Committee on Immunization Practices. MMWR Morbidity and mortality weekly report. 2016;65(49):1405–8.
54. World Health Organization (WHO). Human papillomavirus vaccines: WHO position paper, December 2022. WHO, editor. Geneva: World Health Organization; 2022. 645–72 p.
55. Meites E, Szilagyi PG, Chesson HW, Unger ER, Romero JR, Markowitz LE. Human Papillomavirus Vaccination for Adults: Updated Recommendations of the Advisory Committee on Immunization Practices. MMWR Morbidity and mortality weekly report. 2019;68(32):698–702.
56. Mariani L, Preti M, Cristoforoni P, Stigliano CM, Perino A. Overview of the benefits and potential issues of the nonavalent HPV vaccine. Int J Gynaecol Obstet. 2017;136(3):258–65.

Hepatitis C Virus Infection in People Who Inject Drugs

Jacqueline E. Sherbuk

Abstract The United States is experiencing rising rates of hepatitis C infections. New infections are most commonly transmitted through injection drug use. Treating hepatitis C in people who inject drugs benefits (1) the individual by improving health and preventing complications and (2) community health by preventing ongoing transmission to others. Hepatitis C treatment is proven to be effective in people with recent or active injection drug use, yet only a minority of people who inject drugs infected with hepatitis C have been treated. This chapter provides an overview of the current state of hepatitis C testing, linkage to care, and treatment of people who inject drugs through a harm reduction and patient-centered comprehensive approach to care.

Keywords Hepatitis C virus · People who inject drugs · Injection drug use · Substance use · Syndemic · Antiviral treatment · Harm reduction · Treatment as prevention · HCV cascade of care · HCV testing · HCV treatment

1 Key Concepts

1. Injection drug use is the most common mode of hepatitis C transmission.
2. People who inject drugs have a high prevalence of hepatitis C infections and low rates of hepatitis C treatment.
3. Hepatitis C treatment is highly effective in people who use drugs.
4. Novel interventions have been developed to improve testing, linkage to care, and treatment for people who inject drugs.

J. E. Sherbuk (✉)
University of South Florida, Morsani College of Medicine, Tampa, FL, USA
e-mail: sherbuk@usf.edu

C. Somboonwit et al. (eds.), *Global Virology IV: Viral Disease Diagnosis and Treatment Delivery in the 21st Century*,
https://doi.org/10.1007/978-3-031-57369-9_6

5. Hepatitis C care for people who inject drugs is best provided with a focus on comprehensive healthcare and utilizing harm reduction strategies.

2 Overview of Hepatitis C in People Who Inject Drugs

2.1 Overview of Hepatitis C in the United States

An estimated 2.4 million people in the United States are living with hepatitis C virus (HCV) [1]. Hepatitis C cases nationwide have been rising since 2013, with an estimated 66,700 new cases in 2020 [2]. The surge of new cases has been driven primarily by injection drug use. Injection drug use accounts for nearly two-thirds of new HCV infections in the United States [2].

Highly effective and well-tolerated treatment for hepatitis C is available, with >= 95% of people with hepatitis C cured with current treatment regimens [3]. Over time, untreated hepatitis C infection can lead to cirrhosis, liver cancer, and death [3]. HCV treatment improves health, prevents transmission, and is cost-effective [4]. Unfortunately, only a third of people diagnosed with hepatitis C received treatment within 1 year of diagnosis, with even lower treatment rates among young people and people who use drugs [4]. Without treatment, a person with hepatitis C is at risk for complications of chronic hepatitis C infection and can transmit hepatitis C to others.

The viral hepatitis national strategic plan defines a national goal of HCV elimination by 2030, with a 90% reduction in new infections and a 65% decrease in mortality through prevention, diagnosis, and treatment [5]. People who inject drugs (PWID) are identified as a priority population for focused testing and treatment efforts to move towards HCV elimination [5].

2.2 Hepatitis C Transmission Among People Who Inject Drugs

Hepatitis C is transmitted through exposure to the blood of a person infected with hepatitis C. Exposures that can lead to transmission include shared injection equipment for drug use, vertical transmission from mother to baby at birth, and sexual transmission, particularly in men who have sex with men [6]. Less common modes of transmission include tattoos or body piercings with non-sterile materials, blood transfusions or organ transplants, and shared personal items such as razors or nail clippers [6].

People who inject drugs are at high risk for transmission through exposure to blood infected with Hepatitis C from shared injection equipment, including syringes, cookers, spoons, needles, or cotton, or from shared snorting straws [7]. The hepatitis C virus remains stable outside the human body for days, increasing the risk of transmission when supplies are shared over time [8]. PWID can also acquire HCV through other modes of transmission, such as sexual transmission in men who have sex with men [9].

The overall prevalence of HCV among PWID in the United States is estimated to be 55% [10]. Prevalence is age dependent, with one-third of young PWID infected with HCV and as many as 70–90% of older PWID infected with HCV [4]. Young people tend to inject drugs as part of a social group and are more likely to share injection equipment with others [9], which has contributed to young people having the highest rates of new HCV infections [4].

2.3 *Syndemic Approach to Hepatitis C*

Hepatitis C, human immunodeficiency virus (HIV), substance use disorder, and sexually transmitted infections form a syndemic of co-occurring public health concerns [5] (Fig. 1). Syndemics occur when multiple disease states cluster together and negatively interact with each other. For example, both HIV and HCV can be transmitted by the same behaviors of injection drug use and sexual transmission. Co-infection with HIV and HCV results in more severe liver disease and increased mortality compared to infection with HCV alone [11, 12].

Addressing syndemics requires a holistic approach: meticulous attention to the concerns of each individual and an all-encompassing strategy to tackle the root structural factors. For syndemics such as HCV, HIV, sexually transmitted infections, and substance use, these factors encompass stigma, mental health disorders, poverty, and underemployment [13, 14].

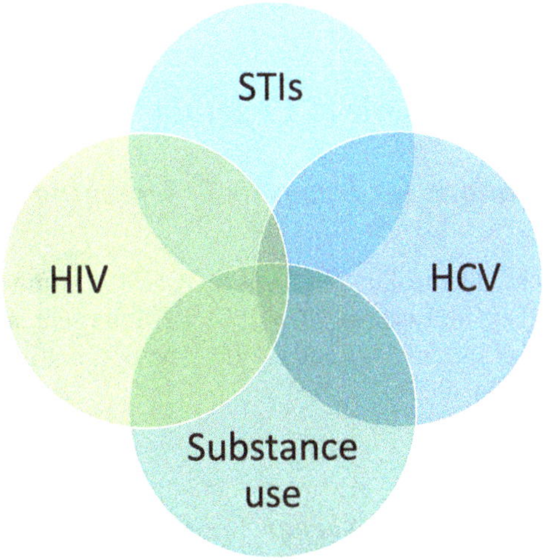

Fig. 1 Syndemic of hepatitis C, HIV, substance use, and sexually transmitted infections. A syndemic approach to hepatitis C, HIV, substance use, and sexually transmitted infections develops an understanding of how these elements interact to impact health and the underlying factors contributing to the development of all of these conditions

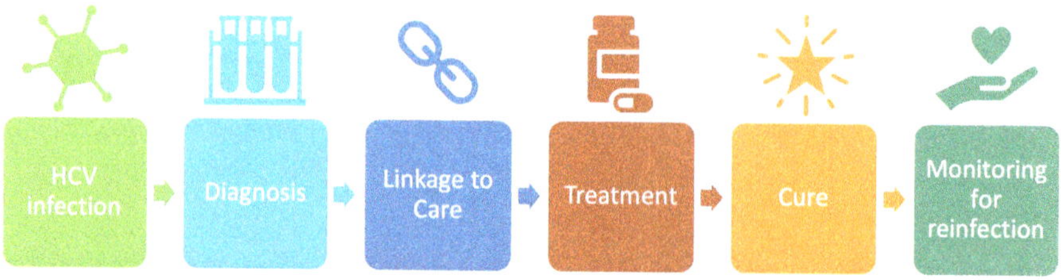

Fig. 2 Steps in the hepatitis C cascade of care. The hepatitis C cascade of care defines the steps needed to move from hepatitis C infection through diagnosis, linkage to care, treatment, and post-treatment monitoring. Each of the steps included may be further broken down into additional, more specific steps. For example, diagnosis of hepatitis C typically involves a screening antibody test followed by a confirmatory hepatitis C RNA test. The cascade of care can be utilized to identify gaps in care and develop interventions

2.4 Hepatitis C Care Cascade

The hepatitis C care cascade identifies the key steps in HCV care (Fig. 2). The cascade starts with all people infected with HCV and progresses from diagnoses to linkage to care to treatment to cure and to ongoing monitoring for reinfection [15–19]. Each step in the cascade may be made up of multiple components. For example, diagnosis of hepatitis C often occurs through a screening antibody test, followed by viral confirmatory testing. The HCV care cascade can be used to identify gaps in current care models, lead to interventions to improve care and assess the impact of public health interventions by monitoring changes in the cascade over time [16, 19].

3 Screening and Diagnosis of Hepatitis C in PWID

3.1 Recommendations for Hepatitis C Screening

The Centers for Disease Control and Prevention (CDC) recommends that all adults be screened once in their lifetime for HCV [20]. Periodic screening is indicated for those with ongoing risk factors, such as PWID. The optimal interval for screening has yet to be clearly defined [3].

3.2 Identifying Patients to Screen for Hepatitis C

Nearly half of people with HCV in the United States are unaware of their diagnosis [21], so improved screening and testing are needed. PWID have a high rate of HCV and are a priority population for testing initiatives [5]. PWID may not access routine healthcare, so screening must occur in places where PWID congregate. Testing services provided in community-based settings or co-located with substance use treatment or syringe services programs have demonstrated increased screening rates [22, 23]. In settings such as those with a high HCV prevalence, opt-out screening is preferred [3]. In opt-out screening, the standard of care is to screen for HCV, and people must actively request not to be screened.

3.3 Interpretation of Hepatitis C Screening Tests

The diagnosis of hepatitis C requires a blood test. Rapid tests are available using blood obtained from a fingerstick or venipuncture. Most commonly, a serologic test for antibodies to HCV is performed as a screening test (Fig. 3). A positive result demonstrates prior exposure to HCV. This must be followed by a confirmatory RNA viral load test to diagnose chronic HCV infection. If the serology test is positive, but no virus is detectable, the patient may have been previously treated, experienced

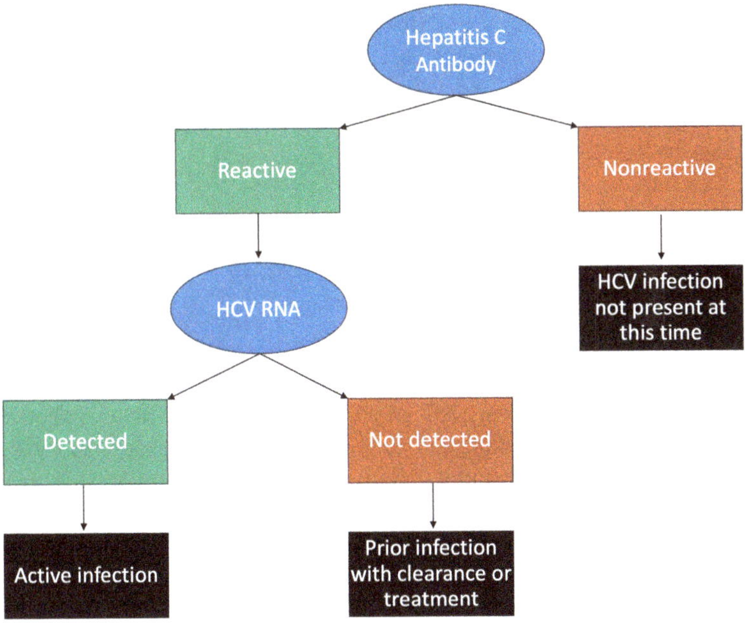

Fig. 3 Hepatitis C testing algorithm. A hepatitis C antibody test is typically used for screening. If the antibody test is nonreactive, the person does not have hepatitis C infection and no further testing is required. A reactive antibody test signifies prior exposure to hepatitis C. A reactive antibody test must be followed by a confirmatory hepatitis C RNA level to evaluate for active infection. If the HCV RNA is detectable at any level, the person has active hepatitis C infection and should be connected to hepatitis C care. If the hepatitis C RNA is not detectable, the person either cleared the virus spontaneously or was successfully treated and no further evaluation is indicated

spontaneous clearance, or have a false positive antibody test. In the case of no detectable viral load, no treatment is indicated.

While viral testing is needed to confirm a diagnosis of active HCV infection, approximately 1 in 9 people with positive antibody tests do not have confirmatory viral testing [18]. The CDC now recommends simplifying testing to require only a single blood draw using reflex laboratory testing [20]. With reflex testing, the laboratory will automatically test HCV RNA level if the HCV antibody is positive, allowing a diagnosis of HCV infection to be confirmed with only a single blood draw [3, 20].

3.4 Counseling Regarding Hepatitis C at the Time of Diagnosis

Testing for HCV requires a plan for counseling in the case of a new diagnosis of HCV infection. Counseling should address HCV care and treatment, how to mitigate the risk of ongoing HCV transmission, and an acknowledgment of an individual's emotional response to a new diagnosis. People can experience a range of emotions after a diagnosis of hepatitis C. For some, the diagnosis may lead to a reliving of past traumas [24]. Counseling should include a clear pathway for linkage to care, as significant drop-offs are seen in the HCV care cascade between diagnosis and linkage to care [25, 26].

3.5 Models for Improving Uptake of Testing and Linkage to Care

Screening and linkage to care are key opportunities for improvement in the HCV cascade. It is estimated that only half of people in the United States infected with HCV are aware of their infection, and diagnosis rates are even lower in PWID [19, 25]. Among people who use drugs, linkage to care rates have improved over the years, but remain quite low [27]. Recognizing that current models of care are not leading to successful HCV care in PWID, interventions have been developed to improve uptake of testing and linkage to care. Models to improve testing access include expanding testing initiatives at sites where PWID may already exist, such as syringe services programs, substance use treatment centers, street medicine programs, and emergency departments [27]. Dried blood spot testing and point-of-care testing may also be helpful in broadening opportunities for screening [28, 29].

Patient education and patient navigation have demonstrated success in improving testing and linkage to care by helping people with HCV navigate the healthcare system and by providing education on HCV infection, treatment, and prevention

[29–31]. Patient navigators may assist with various needs, including addressing insurance and financial barriers, facilitating appointments and referrals, and linking to existing resources to meet patient needs [32]. In a clinic using a nurse navigator, 76% of the 824 patients referred for treatment were successfully linked to care [33]. Peer navigators, who are often people with lived experience of injection drug use, demonstrate effective linkage to HCV care in conjunction with point-of-care testing and linkage to nursing support [34].

4 Treatment of Hepatitis C in PWID

4.1 *Indications for Treatment of Hepatitis C in PWID*

Treatment is recommended for all people with HCV infection unless the person has a limited life expectancy due to an unrelated condition [3]. Treatment is recommended for both chronic and acute infections. Current or prior substance use is not an absolute contraindication to treatment, and abstinence requirements for treatment are not medically indicated [3, 35].

Direct-acting antivirals (DAAs) are used for the treatment of HCV. The specific DAA treatment regimen used for an individual is determined based on treatment history and the presence of cirrhosis. Pangenotypic DAA regimens allow for the treatment of all HCV genotypes with the same regimens. Treatment has been simplified over the years. In the past, interferon-based regimens were used with high toxicity and only moderate effectiveness, and careful monitoring was required. DAAs, introduced in 2014, are safe and effective, and simplified protocols for minimal visits and monitoring of therapy have been developed. For most treatment-naïve patients without cirrhosis, HCV treatment duration is typically 8 to 12 weeks of oral medications with follow-up 12 weeks after completion of treatment to assess or sustain virologic response, or cure, of HCV infection. Updated treatment guidelines are available at HCVguidelines.org [3].

4.2 *Efficacy of Hepatitis C Treatment Among PWID*

Hepatitis C treatment is highly effective, with more than 95% of people treated achieving a cure [18, 36, 37]. Among PWID, studies have demonstrated high adherence to HCV treatment [38–40]. HCV treatment is proven to be highly effective in current and former PWID [39, 41–43].

4.3 Disparities in Hepatitis C Care and Barriers to Care Among PWID

Among PWID, hepatitis C treatment rates are low [25, 44]. Disparities in hepatitis C treatment are seen with lower treatment rates among young people, people who use drugs, and those without insurance [18, 33, 42, 45–47].

Understanding the barriers to successful HCV care is critical to determine interventions to improve care. Common barriers to care may be structural, provider-focused, or patient-focused [48]. Structural barriers include access to medical care, access to specialty care, cost, and transportation [14, 26, 49]. Some states continue to require abstinence as a prerequisite for HCV treatment, even though this is not medically based and not recommended in guidelines [50]. Provider-focused barriers include limited provider knowledge or experience treating HCV and patient-provider relationships [49]. Patient-focused barriers include limited knowledge regarding HCV, HCV as an asymptomatic disease for many, competing priorities, stigma, and concern for side effects of treatment [51, 52].

4.4 Expanding Access to HCV Care for PWID

Historically, HCV treatment was provided through a specialist physician, most commonly a gastroenterologist or infectious diseases specialist. However, as treatment is simplified with the introduction of highly safe and efficacious DAAs, the provision of HCV treatment has begun to expand to generalist providers and alternative sites of treatment [53]. Efforts to decentralize care away from specialists, co-locate services, and integrate harm reduction into HCV treatment have the potential to expand HCV treatment access for PWID [54].

Co-locating HCV treatment with sites treating SUD or providing harm reduction services has been shown to be acceptable to people with HCV [44, 55] and effective in providing treatment [52, 56]. Syringe service programs have successfully treated HCV with a randomized controlled trial showing improved HCV treatment outcomes when people are treated at a syringe service program instead of receiving a referral to a specialized provider [57]. Alternative sites for treatment include primary care providers, including HIV providers, telehealth, substance treatment sites, or prisons [26, 58–60].

Project ECHO is a long-standing alternative model to improve treatment access by increasing medical providers' capacity to provide HCV care and treatment [61]. In the ECHO model, local providers are trained to treat HCV with ongoing support from expert consultants. Similar models have demonstrated efficacy in using telehealth and provider training to expand HCV treatment access for PWID [52].

4.5 *Comprehensive Care for PWID Using a Syndemic Framework*

Optimal treatment for HCV in PWID requires an understanding of HCV as part of a syndemic associated with intersecting health concerns. Complications associated with injection drug use include transmission of blood-borne pathogens such as HIV, HCV, and hepatitis B; development of severe bacterial and fungal infections; acquisition of sexually transmitted infections; and risk of overdose. As HCV does not occur in isolation, HCV care should be part of comprehensive patient-centered healthcare. Comprehensive care includes primary care, treatment for substance use disorders, mental health disorders, reproductive/sexual health care, and harm reduction services. Preventive care should include vaccines for preventable infections such as Hepatitis A and Hepatitis B as well as access to pre-exposure prophylaxis for HIV. Table 1 outlines the components of comprehensive care for PWID [35].

Table 1 Comprehensive healthcare for people who inject drugs

Components of comprehensive care	Examples
Vaccination	Hepatitis A vaccine
	Hepatitis B vaccine
Sexual and reproductive health	Condoms
	STI testing and treatment
	Family planning and contraception
HIV prevention	Pre-exposure prophylaxis
	Post-exposure prophylaxis
	Condoms
Monitoring for bloodborne pathogens	HIV and viral hepatitis testing
	Linkage to treatment for HIV
	Linkage to viral hepatitis treatment
Substance use disorder treatment	Medications for opioid use disorder
	Counseling
Harm reduction	Syringe service programs
Overdose prevention	Drug checking
	Fentanyl tests
	Naloxone kits and training
	Overdose prevention sites
Case management	Connection to community resources for housing, food, or other psychosocial needs

5 Ongoing Care for PWID

5.1 *Hepatitis C Reinfection Risk and Monitoring*

Treatment of HCV does not result in immunity, and therefore, reinfection is possible. The risk of reinfection depends on behaviors associated with HCV transmission. The risk of reinfection is extremely low for people who do not engage in activities associated with HCV transmission [62]. In PWID, the reported risk of reinfection has varied in studies, ranging from 1.4/100-person years to 9.8/100 person-years [40, 63–68]. Higher rates of reinfection occur in young people, those with recent injection drug use, and those who inject stimulants [40, 63, 64, 68]. Lower rates of reinfection have been seen in people receiving medications for opioid use disorder, highlighting the importance of a syndemic approach to care [67]. Current guidelines recommend at least annual HCV RNA screening in PWID following successful HCV treatment [3].

5.2 *Harm Reduction Strategies*

Harm reduction promotes health by preventing harm associated with injection drug use. Harm reduction aims to meet PWID where they are and involve PWID in identifying and working toward patient-centered goals. Pragmatic, individualized plans are developed in conjunction with PWID and carried out in a nonputative manner, recognizing that not every person who injects drugs aims for abstinence [35, 69].

Components of harm reduction may include medications for opioid use disorder, syringe service programs, education on the prevention of infections, and overdose prevention strategies, including naloxone [3]. Table 2 provides further details on potential components of harm reduction programs [69]. In modeling studies, aiming for HCV elimination requires the incorporation of harm reduction strategies in addition to increasing HCV testing and treatment [70, 71].

5.3 *Treatment as Prevention for Hepatitis C Transmission*

Treating HCV infection benefits the health of the individual and has public health benefits. If a person is cured of HCV, they can no longer transmit HCV infection to others. As injection drug use is the most common mode of transmission for HCV at this time, treating PWID for HCV can prevent new infections [2]. This strategy is referred to as "treatment as prevention." Modeling studies have demonstrated the impact of HCV treatment on decreasing the prevalence of HCV infections. HCV elimination is most likely to occur with treatment in conjunction with harm

Table 2 Hepatitis C prevention, screening, and treatment utilizing a harm reduction approach

Hepatitis C care	Examples of interventions
Prevention	Education about hepatitis C transmission, such as avoiding shared injecting equipment, including syringes, needles, straws, spoons, pipes, cotton swabs
	Access to syringe service programs and injection equipment
Screening	Accessible testing in settings acceptable to people who inject drugs
	Hepatitis C testing co-located with other services sought by people who inject drugs, such as harm reduction organizations, syringe service programs, community health centers, substance use treatment programs, emergency rooms
	Simplified testing utilizing a single blood sample for screening and confirmatory testing
	Point-of-care testing
Treatment	Co-locating treatment with other services for people who inject drugs, such as syringe service programs, substance use treatment programs, or HIV clinics
	Addressing financial barriers to treatment such as lack of insurance, prior authorizations
	Addressing other medical needs along with hepatitis C care, including vaccinations, reproductive health, care for substance use disorder
	Offering repeat testing at regular intervals following treatment for those at risk of reinfection

reductions, such as medications for opioid use disorder and syringe service programs, to prevent new infections [70–72].

6 Future Directions

HCV is a treatable, curable infection. Prevalence of HCV is high in PWID, and injection drug use is the primary mode of HCV transmission. National goals for eliminating HCV as a public health problem cannot be achieved without changes in care structures to better reach PWID with HCV. The CDC has identified recommendations to improve HCV care [18]. These include reducing barriers by removing eligibility requirements and prior authorizations from medications; providing treatment co-located with other services people are receiving; providing safe and effective treatment in as few medical visits as possible; and expanding access to treatment providers by empowering primary care providers to be able to treat HCV. Incorporation of harm reduction strategies can promote health among PWID, including prevention of HCV reinfection. To improve health outcomes and to reach goals for hepatitis C elimination, PWID must be a priority population for prevention, testing, and treatment interventions.

References

1. Hofmeister MG, Rosenthal EM, Barker LK, Rosenberg ES, Barranco MA, Hall EW, et al. Estimating Prevalence of Hepatitis C Virus Infection in the United States, 2013–2016. Hepatology [Internet]. 2019 Mar 1 [cited 2022 Sep 12];69(3):1020–31. Available from: https://pubmed.ncbi.nlm.nih.gov/30398671/
2. Centers for Disease Control and Prevention. Viral Hepatitis Surveillance Report—United States, 2020 [Internet]. 2022 [cited 2023 Sep 5]. Available from: https://www.cdc.gov/hepatitis/statistics/2020surveillance/index.htm
3. AASLD-IDSA. http://www.hcvguidelines.org. Recommendations for testing, managing, and treating hepatitis C.
4. Thompson WW, Symum H, Sandul A, Gupta N, Patel P, Nelson N, et al. Vital Signs: Hepatitis C Treatment Among Insured Adults — United States, 2019–2020. MMWR Morb Mortal Wkly Rep [Internet]. 2022 Aug 12 [cited 2022 Sep 13];71(32):1011–7. Available from: https://www.cdc.gov/mmwr/volumes/71/wr/mm7132e1.htm
5. U.S. Department of Health and Human Services. Viral Hepatitis National Strategic Plan for the United States: A Roadmap to Elimination (2021–2025) [Internet]. Washington D.C.; 2020 [cited 2023 Aug 22]. Available from: www.hhs.gov/hepatitis.
6. Centers for Disease Control and Prevention. Hepatitis C [Internet]. [cited 2023 Sep 5]. Available from: https://www.cdc.gov/hepatitis/hcv/cfaq.htm#transmission
7. Fernandez N, Towers C V., Wolfe L, Hennessy MD, Weitz B, Porter S. Sharing of Snorting Straws and Hepatitis C Virus Infection in Pregnant Women. Obstetrics and Gynecology [Internet]. 2016 Aug 1 [cited 2022 Sep 13];128(2):234–7. Available from: https://journals.lww.com/greenjournal/Fulltext/2016/08000/Sharing_of_Snorting_Straws_and_Hepatitis_C_Virus.3.aspx
8. Doerrbecker J, Friesland M, Ciesek S, Erichsen TJ, Mateu-Gelabert P, Steinmann J, et al. Inactivation and Survival of Hepatitis C Virus on Inanimate Surfaces. Available from: https://academic.oup.com/jid/article/204/12/1830/1020812
9. World Health Organization. HIV and young people who inject drugs: Technical brief [Internet]. 2015 [cited 2023 Sep 5]. Available from: https://www.unaids.org/sites/default/files/media_asset/2015_young_people_drugs_en.pdf
10. Degenhardt L, Peacock A, Colledge S, Leung J, Grebely J, Vickerman P, et al. Global prevalence of injecting drug use and sociodemographic characteristics and prevalence of HIV, HBV, and HCV in people who inject drugs: a multistage systematic review. Lancet Glob Health. 2017 Dec 1;5(12):e1192–207.
11. Chen TY, Ding EL, Seage-Iii G, Kim AY. Meta-analysis: increased mortality associated with hepatitis C in HIV-infected persons is unrelated to HIV disease progression. Clin Infect Dis. 2009;49(10).
12. Brookmeyer KA, Haderxhanaj LT, Hogben M, Leichliter J. Sexual risk behaviors and STDs among persons who inject drugs: A national study. Prev Med (Baltim) [Internet]. 2019 Sep 1 [cited 2023 Aug 29];126:105779. Available from: /pmc/articles/PMC6816039/
13. Perlman DC, Jordan AE. The Syndemic of Opioid Misuse, Overdose, HCV, and HIV: Structural-Level Causes and Interventions. Curr HIV/AIDS Rep [Internet]. 2018 Apr 1 [cited 2022 Sep 13];15(2):96–112. Available from: https://link.springer.com/article/10.1007/s11904-018-0390-3
14. World Health Organization. Consolidated guidelines on HIV, viral hepatitis and STI prevention, diagnosis, treatment and care for key populations: policy brief [Internet]. Geneva; 2022 [cited 2023 Sep 5]. Available from: https://www.who.int/publications/i/item/9789240053274
15. Lazarus J V., Villota-Rivas M, Fernández I, Gea F, Ryan P, López SA, et al. A cascade of care analysis on the elimination of hepatitis C from public hospitals in Madrid. Communications Medicine 2022 2:1 [Internet]. 2022 Feb 24 [cited 2023 Sep 5];2(1):1–8. Available from: https://www.nature.com/articles/s43856-022-00077-9

16. Maier MM, Ross DB, Chartier M, Belperio PS, Backus LI. Cascade of Care for Hepatitis C Virus Infection Within the US Veterans Health Administration. Am J Public Health [Internet]. 2016 Feb 1 [cited 2023 Sep 5];106(2):353. Available from: /pmc/articles/PMC4815568/
17. Thomas DL. State of the Hepatitis C Virus Care Cascade. Clin Liver Dis (Hoboken) [Internet]. 2020 Jul 1 [cited 2023 Sep 5];16(1):8–11. Available from: https://onlinelibrary.wiley.com/doi/full/10.1002/cld.915
18. Wester C, Osinubi A, Kaufman HW, Symum H, Meyer WA, Huang X, et al. Hepatitis C Virus Clearance Cascade — United States, 2013–2022. MMWR Morb Mortal Wkly Rep [Internet]. 2023 Jun 30 [cited 2023 Aug 26];72(26):716–20. Available from: https://www.cdc.gov/mmwr/volumes/72/wr/mm7226a3.htm
19. Yehia BR, Schranz AJ, Umscheid CA, Lo Re V. The Treatment Cascade for Chronic Hepatitis C Virus Infection in the United States: A Systematic Review and Meta-Analysis. PLoS One [Internet]. 2014 Jul 2 [cited 2023 Sep 5];9(7). Available from: /pmc/articles/PMC4079454/
20. Schillie S, Wester C, Osborne M, Wesolowski L, Ryerson AB. CDC Recommendations for Hepatitis C Screening Among Adults — United States, 2020. MMWR Recommendations and Reports [Internet]. 2023 Apr 1 [cited 2023 Sep 7];69(2):1–17. Available from: https://www.cdc.gov/mmwr/volumes/69/rr/rr6902a1.htm
21. Kim H seok, Yang JD, El-Serag HB, Kanwal F. Awareness of chronic viral hepatitis in the United States: An update from the National Health and Nutrition Examination Survey. J Viral Hepat [Internet]. 2019 May 1 [cited 2023 Aug 29];26(5):596–602. Available from: https://pubmed.ncbi.nlm.nih.gov/30629790/
22. Coyle C, Viner K, Hughes E, Kwakwa H, Zibbell JE, Vellozzi C, et al. Identification and linkage to care of HCV-infected persons in five health centers — Philadelphia, Pennsylvania, 2012–2014. Morbidity and Mortality Weekly Report. 2015 May 8;64(17):459–63.
23. Hsieh MH, Tsai JJ, Hsieh MY, Huang CF, Yeh ML, Yang JF, et al. Hepatitis C virus infection among injection drug users with and without human immunodeficiency virus co-infection. PLoS One [Internet]. 2014 Apr 10 [cited 2023 Sep 5];9(4). Available from: https://pubmed-ncbi-nlm-nih-gov.ezproxy.hsc.usf.edu/24722534/
24. Burraway J, Helbert B, Schexnayder J, Knick T, Dillingham R, Scherz C. Reliving it All Over Again: Uncanny Temporalities of Injection Drug Use and Hepatitis C Diagnosis in Southwest Virginia, USA. https://doi.org/101080/0145974020222110090 [Internet]. 2022 [cited 2023 Sep 5];42(1):21–34. Available from: https://www.informahealthcare.com/doi/abs/10.1080/01459740.2022.2110090
25. Jordan AE, Perlman DC, Reed J, Smith DJ, Hagan H. Patterns and Gaps Identified in a Systematic Review of the Hepatitis C Virus Care Continuum in Studies among People Who Use Drugs. Front Public Health [Internet]. 2017 Dec 18 [cited 2023 Aug 24];5. Available from: https://pubmed.ncbi.nlm.nih.gov/29326922/
26. Sherbuk JE, McManus KA, Rogawski McQuade ET, Knick T, Henry Z, Dillingham R. Hepatitis C Within a Single Health System: Progression Along the Cascade to Cure Is Higher for Those With Substance Misuse When Linked to a Clinic With Embedded Support Services. Open Forum Infect Dis [Internet]. 2018 Sep 1 [cited 2023 Sep 5];5(9). Available from: /pmc/articles/PMC6147287/
27. Jordan AE, Cleland CM, Schackman BR, Wyka K, Perlman DC, Nash D. Hepatitis C Virus (HCV) Care Continuum Outcomes and HCV Community Viral Loads Among Patients in an Opioid Treatment Program. J Infect Dis [Internet]. 2020 Oct 10 [cited 2023 Sep 7];222(Suppl 5):S335. Available from: /pmc/articles/PMC7566623/
28. Calvaruso V, Bronte F, Ferraro D, Reina G, Conte E, Rini F, et al. Point-of-care HCV RNA testing in the setting of DAA therapy: HCV-FiS (HEpatitis C Virus Fingerstick Study). Liver Int [Internet]. 2019 Dec 1 [cited 2023 Sep 7];39(12):2240–3. Available from: https://pubmed.ncbi.nlm.nih.gov/31502755/
29. Cunningham EB, Wheeler A, Hajarizadeh B, French CE, Roche R, Marshall AD, et al. Interventions to enhance testing and linkage to treatment for hepatitis C infection for people

who inject drugs: A systematic review and meta-analysis. International Journal of Drug Policy. 2023 Jan 1;111:103917.
30. Coyle C, Moorman AC, Bartholomew T, Klein G, Kwakwa H, Mehta SH, et al. The Hepatitis C Virus Care Continuum: Linkage to Hepatitis C Virus Care and Treatment Among Patients at an Urban Health Network, Philadelphia, PA. Hepatology. 2019;
31. Henderson C, Madden A, Kelsall J. "Beyond the willing & the waiting" – The role of peer-based approaches in hepatitis C diagnosis & treatment. Int J Drug Policy [Internet]. 2017 Dec 1 [cited 2023 Aug 24];50:111–5. Available from: https://pubmed.ncbi.nlm.nih.gov/28927831/
32. Ward KM, Falade-Nwulia O, Moon J, Sutcliffe CG, Brinkley S, Haselhuhn T, et al. A Randomized Controlled Trial of Cash Incentives or Peer Support to Increase HCV Treatment for Persons With HIV Who Use Drugs: The CHAMPS Study. Open Forum Infect Dis [Internet]. 2019 Apr 1 [cited 2023 Sep 7];6(4). Available from: https://pubmed.ncbi.nlm.nih.gov/31049365/
33. Sherbuk JE, McManus KA, Kemp Knick T, Canan CE, Flickinger TE, Dillingham R. Disparities in Hepatitis C Linkage to Care in the Direct Acting Antiviral Era: Findings from a Referral Clinic with an Embedded Nurse Navigator Model. Front Public Health. 2019;7:362.
34. Conway A, Valerio H, Alavi M, Silk D, Treloar C, Hajarizadeh B, et al. A Testing Campaign Intervention Consisting of Peer-Facilitated Engagement, Point-of-Care HCV RNA Testing, and Linkage to Nursing Support to Enhance Hepatitis C Treatment Uptake among People Who Inject Drugs: The ETHOS Engage Study. Viruses [Internet]. 2022 Jul 1 [cited 2023 Aug 24];14(7). Available from: https://pubmed.ncbi.nlm.nih.gov/35891535/
35. United Nations Office on Drugs and Crime, International Network of People Who Use Drugs, Joint United Nations Programme on HIV/AIDS, United Nations Development Programme, United Nations Population Fund, World Health Organization, et al. Implementing comprehensive HIV and HCV programmes with people who inject drugs: practical guidance for collaborative interventions [Internet]. Vienna; 2017 [cited 2023 Sep 5]. Available from: https://www.undp.org/publications/implementing-comprehensive-hiv-and-hcv-programmes-people-who-inject-drugs
36. Zeuzem S, Foster GR, Wang S, Asatryan A, Gane E, Feld JJ, et al. Glecaprevir-Pibrentasvir for 8 or 12 Weeks in HCV Genotype 1 or 3 Infection. N Engl J Med [Internet]. 2018 Jan 25 [cited 2023 Aug 24];378(4):354–69. Available from: https://pubmed.ncbi.nlm.nih.gov/29365309/
37. Falade-Nwulia O, Suarez-Cuervo C, Nelson DR, Fried MW, Segal JB, Sulkowski MS. Oral Direct-Acting Agent Therapy for Hepatitis C Virus Infection: A Systematic Review. Ann Intern Med [Internet]. 2017 May 2 [cited 2023 Aug 24];166(9):637–48. Available from: https://pubmed.ncbi.nlm.nih.gov/28319996/
38. Grebely J, Dalgard O, Conway B, Cunningham EB, Bruggmann P, Hajarizadeh B, et al. Sofosbuvir and velpatasvir for hepatitis C virus infection in people with recent injection drug use (SIMPLIFY): an open-label, single-arm, phase 4, multicentre trial. Lancet Gastroenterol Hepatol [Internet]. 2018 Mar 1 [cited 2023 Sep 5];3(3):153–61. Available from: https://pubmed.ncbi.nlm.nih.gov/29310928/
39. Dore GJ, Altice F, Litwin AH, Dalgard O, Gane EJ, Shibolet O, et al. Elbasvir-Grazoprevir to Treat Hepatitis C Virus Infection in Persons Receiving Opioid Agonist Therapy: A Randomized Trial. Ann Intern Med [Internet]. 2016 Nov 1 [cited 2023 Sep 5];165(9):625–34. Available from: https://pubmed.ncbi.nlm.nih.gov/27537841/
40. Cunningham EB, Hajarizadeh B, Amin J, Hellard M, Bruneau J, Feld JJ, et al. Reinfection Following Successful Direct-acting Antiviral Therapy for Hepatitis C Virus Infection Among People Who Inject Drugs. Clin Infect Dis [Internet]. 2021 Apr 15 [cited 2023 Aug 27];72(8):1392–400. Available from: https://pubmed.ncbi.nlm.nih.gov/32166305/
41. Graf C, Mücke MM, Dultz G, Peiffer KH, Kubesch A, Ingiliz P, et al. Clinical Infectious Diseases Efficacy of Direct-acting Antivirals for Chronic Hepatitis C Virus Infection in People Who Inject Drugs or Receive Opioid Substitution Therapy: A Systematic Review and Meta-analysis. Clinical Infectious Diseases ®. 2020;70(11):2355–65.

42. Frankova S, Jandova Z, Jinochova G, Kreidlova M, Merta D, Sperl J. Therapy of chronic hepatitis C in people who inject drugs: focus on adherence. Harm Reduct J [Internet]. 2021 Dec 1 [cited 2023 Aug 26];18(1):1–9. Available from: https://harmreductionjournal.biomedcentral.com/articles/10.1186/s12954-021-00519-y
43. Coffin P, Santos G, Behar E, et al. Randomized feasibility trial of directly observed therapy versus unobserved hepatitis C treatment with ledipasvir-sofosbuvir among people who inject drugs. PLoS One. 2019;14(6).
44. Falade-Nwulia O, Gicquelais RE, Astemborski J, McCormick SD, Kirk G, Sulkowski M, et al. Hepatitis C Treatment Uptake among People who Inject Drugs in the Oral Direct-Acting Antiviral Era. Liver Int [Internet]. 2020 Oct 1 [cited 2023 Sep 5];40(10):2407. Available from: /pmc/articles/PMC7706292/
45. Falade-Nwulia O, McAdams-Mahmoud A, Irvin R, Niculescu A, Page KR, Mix M, et al. Primary Care Providers Knowledge, Attitude and Practices Related to Hepatitis C Screening and Treatment in the Oral Direct Acting Antiviral Agents Era. J Community Med Health Educ [Internet]. 2016 Oct [cited 2019 Sep 26];6(5). Available from: http://www.ncbi.nlm.nih.gov/pubmed/28083156
46. Cachay ER, Hill L, Torriani F, Ballard C, Grelotti D, Aquino A, et al. Predictors of Missed Hepatitis C Intake Appointments and Failure to Establish Hepatitis C Care Among Patients Living With HIV. Open Forum Infect Dis [Internet]. 2018 Jul 1 [cited 2019 Nov 21];5(7). Available from: https://academic.oup.com/ofid/article/doi/10.1093/ofid/ofy173/5062980
47. Socías ME, Ti L, Wood E, Nosova E, Hull M, Hayashi K, et al. Disparities in uptake of direct-acting antiviral therapy for hepatitis C among people who inject drugs in a Canadian setting. Liver International [Internet]. 2019 Aug [cited 2020 Feb 20];39(8):1400–7. Available from: http://www.ncbi.nlm.nih.gov/pubmed/30653809
48. Grebely J, Petoumenos K, Matthews G V., Haber P, Marks P, Lloyd AR, et al. Factors associated with uptake of treatment for recent hepatitis C virus infection in a predominantly injecting drug user cohort: the ATAHC Study. Drug Alcohol Depend [Internet]. 2010 Mar 3 [cited 2023 Aug 27];107(2–3):244. Available from: /pmc/articles/PMC2853739/
49. Sherbuk JE, Tabackman A, McManus KA, Kemp Knick T, Schexnayder J, Flickinger TE, et al. A qualitative study of perceived barriers to hepatitis C care among people who did not attend appointments in the non-urban US South. Harm Reduct J [Internet]. 2020 Sep 18 [cited 2023 Sep 5];17(1):1–11. Available from: https://harmreductionjournal.biomedcentral.com/articles/10.1186/s12954-020-00409-9
50. Center for Health Law and Policy Innovation & National Viral Hepatitis Roundtable. Hepatitis C: State of Medicaid Access [Internet]. [cited 2023 Sep 5]. Available from: https://stateofhepc.org/
51. Amoako A, Ortiz-Paredes D, Engler K, Lebouché B, Klein MB. Patient and provider perceived barriers and facilitators to direct acting antiviral hepatitis C treatment among priority populations in high income countries: A knowledge synthesis. International Journal of Drug Policy. 2021 Oct 1;96:103247.
52. Sherbuk JE, Knick TK, Canan C, Ross P, Helbert B, Cantrell ES, et al. Development of an Interdisciplinary Telehealth Model of Provider Training and Comprehensive Care for Hepatitis C and Opioid Use Disorder in a High-Burden Region. J Infect Dis [Internet]. 2020 Oct 1 [cited 2023 Aug 27];222(Suppl 5):S354–64. Available from: https://pubmed.ncbi.nlm.nih.gov/32877562/
53. Biondi MJ, Feld JJ. Hepatitis C models of care: approaches to elimination. Canadian Liver Journal [Internet]. 2020 Jun [cited 2023 Sep 7];3(2):165. Available from: /pmc/articles/PMC9202783/
54. Kattakuzhy S, Gross C, Emmanuel B, Teferi G, Jenkins V, Silk R, et al. Expansion of Treatment for Hepatitis C Virus Infection by Task Shifting to Community-Based Nonspecialist Providers: A Nonrandomized Clinical Trial. Ann Intern Med [Internet]. 2017 Sep 5 [cited 2023 Sep 7];167(5):311–8. Available from: https://pubmed.ncbi.nlm.nih.gov/28785771/

55. Socías ME, Karamouzian M, Parent S, Barletta J, Bird K, Ti L. Integrated models of care for people who inject drugs and live with hepatitis C virus: A systematic review. Int J Drug Policy [Internet]. 2019 Oct 1 [cited 2023 Sep 7];72:146–59. Available from: https://pubmed.ncbi.nlm.nih.gov/31147142/
56. Talal AH, Andrews P, McLeod A, Chen Y, Sylvester C, Markatou M, et al. Integrated, Co-located, Telemedicine-based Treatment Approaches for Hepatitis C Virus Management in Opioid Use Disorder Patients on Methadone. Clin Infect Dis [Internet]. 2019 Jul 2 [cited 2023 Sep 7];69(2):323–31. Available from: https://pubmed.ncbi.nlm.nih.gov/30329042/
57. Eckhardt B, Mateu-Gelabert P, Aponte-Melendez Y, Fong C, Kapadia S, Smith M, et al. Accessible Hepatitis C Care for People Who Inject Drugs: A Randomized Clinical Trial. JAMA Intern Med [Internet]. 2022 May 1 [cited 2022 Sep 11];182(5):494–502. Available from: https://jamanetwork.com/journals/jamainternalmedicine/fullarticle/2790054
58. Jones QB, Giemza K, Anglin L, Cottingham S. Treating Hepatitis C in a Substance Use Recovery Program: A Limited-Resource Group Model. NEJM Catal Innov Care Deliv [Internet]. 2022 Apr 20 [cited 2023 Sep 5];3(5). Available from: https://catalyst.nejm.org/doi/abs/10.1056/CAT.21.0453
59. Kronfli N, Linthwaite B, Kouyoumdjian F, Klein MB, Lebouché B, Sebastiani G, et al. Interventions to increase testing, linkage to care and treatment of hepatitis C virus (HCV) infection among people in prisons: A systematic review. International Journal of Drug Policy. 2018 Jul 1;57:95–103.
60. Morey S, Hamoodi A, Jones D, Young T, Thompson C, Dhuny J, et al. Increased diagnosis and treatment of hepatitis C in prison by universal offer of testing and use of telemedicine. J Viral Hepat [Internet]. 2019 Jan 1 [cited 2023 Sep 7];26(1):101–8. Available from: https://onlinelibrary.wiley.com/doi/full/10.1111/jvh.13017
61. Arora S, Thornton K, Murata G, Deming P, Kalishman S, Dion D, et al. Outcomes of treatment for hepatitis C virus infection by primary care providers. N Engl J Med [Internet]. 2011 Jun 9 [cited 2023 Aug 27];364(23):2199–207. Available from: https://pubmed.ncbi.nlm.nih.gov/21631316/
62. Huang P, Wang Y, Yue M, Ge Z, Xia X, Jeyarajan AJ, et al. The risk of hepatitis C virus recurrence in hepatitis C virus-infected patients treated with direct-acting antivirals after achieving a sustained virological response: A comprehensive analysis. Liver Int [Internet]. 2021 Oct 1 [cited 2023 Aug 27];41(10):2341–57. Available from: https://pubmed.ncbi.nlm.nih.gov/34051040/
63. Johannesson JM, Fridriksdottir RH, Löve TJ, Runarsdottir V, Hansdóttir I, Löve A, et al. High Rate of Hepatitis C Virus Reinfection Among Recently Injecting Drug Users: Results From the TraP Hep C Program—A Prospective Nationwide, Population-Based Study. Clinical Infectious Diseases [Internet]. 2022 Nov 14 [cited 2023 Aug 27];75(10):1732–9. Available from: https://doi.org/10.1093/cid/ciac272
64. Rossi C, Butt ZA, Wong S, Buxton JA, Islam N, Yu A, et al. Hepatitis C virus reinfection after successful treatment with direct-acting antiviral therapy in a large population-based cohort. J Hepatol [Internet]. 2018 Nov 1 [cited 2023 Aug 27];69(5):1007–14. Available from: https://pubmed.ncbi.nlm.nih.gov/30142429/
65. Latham NH, Doyle JS, Palmer AY, Vanhommerig JW, Agius P, Goutzamanis S, et al. Staying hepatitis C negative: A systematic review and meta-analysis of cure and reinfection in people who inject drugs. Liver Int [Internet]. 2019 Dec 1 [cited 2023 Aug 27];39(12):2244–60. Available from: https://pubmed.ncbi.nlm.nih.gov/31125496/
66. Valencia J, Alvaro-Meca A, Troya J, Cuevas G, Gutiérrez J, Morro A, et al. High rates of early HCV reinfection after DAA treatment in people with recent drug use attended at mobile harm reduction units. Int J Drug Policy [Internet]. 2019 Oct 1 [cited 2023 Aug 27];72:181–8. Available from: https://pubmed.ncbi.nlm.nih.gov/31253391/
67. Hajarizadeh B, Cunningham EB, Valerio H, Martinello M, Law M, Janjua NZ, et al. Hepatitis C reinfection after successful antiviral treatment among people who inject drugs: A

meta-analysis. J Hepatol [Internet]. 2020 Apr 1 [cited 2023 Aug 27];72(4):643–57. Available from: https://pubmed.ncbi.nlm.nih.gov/31785345/

68. Lindqvist K, Thorin Z, Kåberg M. Real-world hepatitis C treatment outcomes and reinfections among people who inject drugs at a needle and syringe program in Stockholm, Sweden. Harm Reduct J [Internet]. 2023 Dec 1 [cited 2023 Aug 27];20(1):1–11. Available from: https://harmreductionjournal.biomedcentral.com/articles/10.1186/s12954-023-00801-1
69. Substance Abuse and Mental Health Services Administration. Harm Reduction Framework (Draft) [Internet]. 2023 [cited 2023 Aug 29]. Available from: https://www.samhsa.gov/sites/default/files/harm-reduction-framework.pdf
70. Fraser H, Zibbell J, Hoerger T, Hariri S, Vellozzi C, Martin NK, et al. Scaling-up HCV prevention and treatment interventions in rural United States-model projections for tackling an increasing epidemic. Addiction (Abingdon, England) [Internet]. 2018 [cited 2023 Sep 5];113(1):173–82. Available from: https://pubmed.ncbi.nlm.nih.gov/28734093/
71. Blake A, Smith JE. Modeling Hepatitis C Elimination Among People Who Inject Drugs in New Hampshire. JAMA Netw Open [Internet]. 2021 [cited 2023 Sep 5];4(8). Available from: https://pubmed.ncbi.nlm.nih.gov/34342652/
72. Heffernan A, Cooke GS, Nayagam S, Thursz M, Hallett TB. Scaling up prevention and treatment towards the elimination of hepatitis C: a global mathematical model. Lancet [Internet]. 2019 Mar 30 [cited 2023 Sep 5];393(10178):1319–29. Available from: https://pubmed.ncbi.nlm.nih.gov/30704789/

Hepatitis C: Updates in Epidemiology and for Treatment with Direct-Acting Antivirals

Kaley Tash, Victor Javier Rivera-Santiago, and Jamie P. Morano

Abstract Hepatitis C (HCV) is the most common blood-borne viral infection in the United States and continues to be a major cause of hepatic cirrhosis and hepatocellular carcinoma. Since 2013, direct-acting antivirals have revolutionized HCV care. Older therapies were poorly tolerated and had limited efficacy; current therapies are well tolerated and cure the vast majority of patients who initiate them and remain adherent. Widespread HCV screening, implementation of behavioral prevention strategies and risk mitigation, and treatment of infected individuals with direct-acting antiviral therapy continue worldwide to reduce related comorbidities.

Keywords Hepatitis C virus · Direct-acting antiviral therapy · Hepatocellular carcinoma · Sustained virological response

---

K. Tash (✉)
Division of Infectious Disease and International Medicine, Department of Internal Medicine, University of South Florida, Morsani College of Medicine, Tampa, FL, USA

Department of Veterans Affairs, Bay Pines VA Healthcare System, Bay Pines, FL, USA
e-mail: kaley.tash@va.gov

V. J. Rivera-Santiago
Department of Veterans Affairs, Bay Pines VA Healthcare System, Bay Pines, FL, USA

J. P. Morano
Division of Infectious Disease and International Medicine, Department of Internal Medicine, University of South Florida, Morsani College of Medicine, Tampa, FL, USA

Department of Veterans Affairs, Tampa VA Medical Center, Tampa, FL, USA
e-mail: Jamie.Morano@va.gov

C. Somboonwit et al. (eds.), *Global Virology IV: Viral Disease Diagnosis and Treatment Delivery in the 21st Century*,
https://doi.org/10.1007/978-3-031-57369-9_7

1 Introduction

Hepatitis C (HCV) is the most common blood-borne virus infection in the United States. Even though approximately 30% (ranging from 15% to 45% of individuals clear HCV without treatment or additional sequelae, nearly 70% progress to chronic disease, with approximately 15–30% developing hepatic cirrhosis within 20 years, and up to 5% transforming to hepatocellular carcinoma. [1]

Globally, the World Health Organization (WHO) estimates that as of 2022, nearly 58 million persons are estimated to have hepatitis C viral infection worldwide, of whom 3.2 million are children and adolescents. Internationally, there is an incidence of 1.5 million new infections each year [1].

In the United States, there are approximately 3.5 million (estimated range of 2.4 to 4.7 million) people infected with HCV [2, 3]. Acute cases have increased fourfold from 2010 to 2018, likely due to rises in injection drug use and increased screening campaigns uncovering new cases [4].

Direct-acting antiviral agents (DAAs) capitalized upon the scientific advancements made with the development of targeted HIV antiretroviral therapies and were first introduced in 2011 (widespread in 2013) and resulted in a rapid momentum within the United States to eliminate HCV, especially with all oral therapies becoming available in 2014. Namely, the Veterans Health Administration (VHA) launched one of the largest and most comprehensive national campaigns to screen and treat Veterans, known to be a sector of the US population at the highest prevalence of HCV, resulting in an impressive over 100,000 cases eliminated with a 90% and above cure rate [5], with screening and treatment efforts ongoing.

WHO has set the goal for the elimination of HCV infection by 2030 [6]. Current treatment with HCV DAAs is effective and curative, and there is ongoing work in US health systems to continue to implement broad-based HCV screening and referral to treatment [7].

2 Hepatitis C Screening Guidelines

The United States Preventative Task Force (USPSTF) published updated HCV screening guidelines in the Journal of the American Medical Association (JAMA) in 2020 and expanded the recommended age of screening for HCV virus infection in adults aged 18 to 79 years [8]. Additional testing was indicated in high-risk populations or those in whom treatment is most urgent, such as pregnant women, who receive HCV screening at each pregnancy. Because over 80% of HCV transmission occurs among people who inject drugs (PWID), the Center for Disease Control (CDC) recommends that as part of monitoring opioid use disorder, one should include screening for HCV and referral of positive patients for treatment, as HCV cure prevents transmission through the population [9–12].

3 Extrahepatic Manifestations of Hepatitis C

HCV-infected individuals often present a wide range of extrahepatic symptoms. Nonspecific neurologic manifestations are common and include cognitive decline, fatigue, and depression [13]. These present regardless of the liver stage of the disease, and the mechanism of pathogenesis is under investigation. Some potential mechanisms include virus neuroinvasion, immune-mediated damage, alteration of neurotransmitters, and cryoglobulinemia [14]. Improvement of these manifestations is observed after treatment with DAA as measured by the Memorial Symptoms Assessment Scale (MSAS) and Patient-Reported Outcomes Measurement Information System (PROMIS) [15]. HCV infection is also associated with cardiovascular disease, insulin resistance and Type II diabetes, cryoglobulinemic vasculitis, and extrahepatic malignancies like B-cell Non-Hodgkin's Lymphoma. Patients with HCV have a higher risk of chronic kidney disease [16]. Studies have shown that HCV-infected individuals have improvement in GFR after treatment with DAA but not necessarily an improvement in renal function [17]. Moreover, early treatment for HCV should be considered in all individuals to slow progression of kidney disease [18].

4 Direct-Acting Antiviral (DAA) Therapy Development

DAA therapy for Hepatitis C first emerged in 2011, albeit with complex treatment start and stop algorithms for general population use ranging from 8 to 24 weeks with high cure rate (>90%) depending on the algorithm followed. DAA's have definitively replaced the previous regimen of subcutaneous pegylated interferon-alpha (Peg-INF-alpha), which broadly stimulates the immune system, and ribavirin, a guanosine analog that interferes broadly with viral replication. Unlike the DAAs, which target a specific replication protein of the HCV viral genome, the older interferon and ribavirin-based regimens targeted very broad mechanisms and for this reason were not only more poorly tolerated with significant side effects (nausea, flu-like symptoms, headache, weakness, neuropathy) but also were substantially less effective (only 50% compared to 99% with DAA's) and required a longer duration of nearly 52 weeks, as compared now to current DAAs of 8 weeks for noncirrhotic HCV for all genotypes.

DAA medications revolutionized HCV treatment. These agents were designed based on prior knowledge of HIV antiretroviral mechanisms and thus were designed to act more strategically at specific viral replication pointswithin the hepatocyte, thus resulting in greater anti-viral efficacy and fewer clinical side effects; this remarkably improved both adherence and tolerability.

Telaprevir and Boceprevir were the first HCV protease inhibitor DAAs that were FDA approved in 2011 for general use; these initially were combined with ribavirin and interferon in complicated start and stop algorithms and were promptly discontinued due to side effects and difficulty in practical clinical use. As additional targeted DAA's were developed, sofosbuvir/ledipasvir for HCV genotypes (GT) 1 and 3 were synergistic in clearing Hepatitis C and became widely implemented: Ledipasvir inhibits nonstructural protein 5A (NS5A), while sofosbuvir is an inhibitor of HCV NS5B RNA-dependent RNA polymerase, which acts as an RNA chain terminator [19]. Other prominent DAAs were daclatasvir (GT3), and ombitasvir/paritaprevir/ritonavir (GT 1). In 2016, DAAs evolved to be pan-genotypic, therefore, sofosbuvir/velpatasvir quickly became the standard of care. In 2017, sofosbuvir/velpatasvir/voxilaprevir was FDA approved in 2017, and pangenotypic (glecaprevir/pibrentasvir) was FDA approved in August 2017 for a short duration of 8 weeks without cirrhosis [19].

The goal of DAA therapy is to eliminate HCV infection, prevent transmission, prevent the progression of liver disease, and prevent associated with morbidity and mortality associated with HCV. The benefits of therapy also include improved extrahepatic manifestations of HCV, such as fatigue and insulin resistance [20, 21].

5 Additional Hepatitis C Prevention Tools

The most well-known prevention tools for the deterrence of HCV acquisition include harm-reduction services for people who inject drugs, including evidence-based syringe services programs and medication-assisted treatment (MAT) for opioid replacement therapy (ORT). These two interventions have been shown to reduce the risk of transmission by meta-analysis [11, 12]. Multiple models have been used to compare how effective these interventions are in various communities, including rural and urban communities, criminal justice-involved individuals, and men who have sex with men [22].

In addition, increased mandatory opioid education and guidelines for medical providers have led to a decreased opioid prescription for non-cancer-related pain, and providers are now incorporating alternative therapies for chronic, non-cancer-related pain (acupuncture, meditation, yoga, and tai chi) [23]. Evidence has shown that chronically prescribed opioids can lead to higher rates of intravenous drug use initiation and 4.7-fold higher risk of HCV acquisition [10].

6 Hepatitis C Direct-Acting Antiviral (HCV DAA) Selection

The selection of HCV DAA hinges upon the algorithm of treatment naïve vs. treatment experienced and the presence or absence of hepatic cirrhosis. Patients who are treatment naïve (never received treatment for HCV) and noncirrhotic may be eligible for simplified regimens of 8 weeks. Unlike HIV therapy in which one pill daily or an injectable medication is preferred for treatment to increase adherence, in HCV DAA therapy, the current regimens have already been maximally simplified and thus are selected based on existing drug-drug interactions, presence of past resistance, and presence or absence of hepatic cirrhosis (Fig. 1).

The most frequent drug interactions associated with DAA affect drug metabolism by CYP450 induction or inhibition, leading to reduced or increased drug metabolism. It is recommended to reconcile all prescribed and over-the-counter medications and supplements prior to initiation of therapy with DAA. These medications should then be individually evaluated for potential drug interactions. The most common interactions with DAA are statins, acid-suppressing medications like proton pump inhibitors and H2 Blockers, and anti-epileptics. Clinical resources that may be of help are drug interaction online tools from the University of Liverpool [24] and UpToDate.com [25].

For individuals who have partially completed an HCV DAA regimen (i.e., 4 weeks of a total 8- or 12-week regimen) or those who have unfortunately become reinfected due to relapsed intravenous substance use after a documented sustained virologic response (SVR), one is advised to consult with a Hepatitis C treatment expert and to seek advice through the current guidelines. Current and updated Hepatitis C treatment guidelines can be found in the joint American Association for the Study of Liver Diseases-Infectious Diseases Society of America Recommendations for Testing, Managing and Treating Hepatitis C Virus Infection [26], updated October 2022, as well as helpful graphics and tables from the

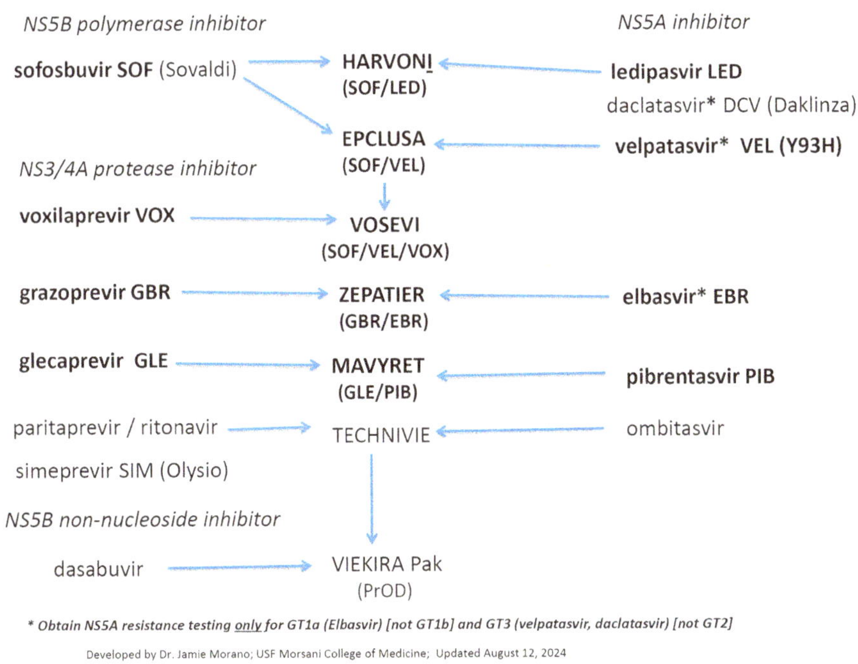

Fig. 1 Hepatitis C direct acting antiviral agents with categories of action

Department of Veteran Affairs HIV, Hepatitis, and Related Conditions Program, within the Office of Specialty Care Services, last updated 2021 [27] (Table 1).

7 HCV Treatment and Posttreatment Monitoring

The goal of HCV treatment is to achieve sustained virologic response (SVR), or hepatitis C "cure". With modern DAAs, SVR rates are generally in the 95% or higher range. Patients who received DAA therapy have an excellent prognosis for virological cure and most tolerate treatment well. After initiation of DAA therapy,

Table 1 Simplified pangenotypic treatment regimen spectrum for hepatitis C treatment

<table>
<tr><th><u>Treatment naïve</u></th><th colspan="4"><u>Treatment experienced</u></th></tr>
<tr><td></td><td>PEG-IFN</td><td>SOF
(NS3/4A naïve)</td><td>Any NS3/4A protease inhibitor (glecaprevir, simepravir, grazoprevir, voxilaprevir) (SOF naïve)</td><td>Any NS5A inhibitor (velpatasvir, pibrentasvir, daclastasvir) (NS3/4A PI c)</td></tr>
<tr><td colspan="5"><u>Non cirrhotic</u></td></tr>
<tr><td colspan="3">GLE/PIB × 8 weeks</td><td>GLE/PIB × 12 weeks</td><td>GLE/PIB × 16 weeks</td></tr>
<tr><td colspan="2">SOF/VEL × 12 weeks</td><td>SOF/VEL ×12 weeks (GT1b only)</td><td>SOF/VEL × 12 weeks</td><td>SOF/VEL/VOX × 12 weeks</td></tr>
<tr><td colspan="2"></td><td>SOF/VEL/VOX x 12 weeks (GT1a, GT3)</td><td colspan="2"></td></tr>
<tr><td colspan="5"><u>Compensated cirrhotic:</u>
CTP A</td></tr>
<tr><td colspan="4">GLE/PIB × 12 weeks</td><td></td></tr>
<tr><td colspan="2">SOF/VEL × 12 weeks
+ RBV if Y93H</td><td>SOF/VEL/VOX × 12 weeks (GT1a, GT3)</td><td>SOF/VEL x 12 weeks</td><td>SOF/VEL/VOX × 12 weeks</td></tr>
<tr><td colspan="5"><u>Decompensated cirrhotic:</u>
CTP B or C
(DO NOT USE INTERFERON; DO NOT USE HCV PROTEASE INHIBITORS)</td></tr>
<tr><td colspan="4">SOF/VEL + RBV × 12 weeks</td><td>SOF/VEL + RBV × 24 weeks (not FDA approved)</td></tr>
<tr><td colspan="4">SOF/VEL (without RBV) × 24 weeks</td><td></td></tr>
</table>

Table Compiled by Jamie P Morano, MD, MPH, 2023 Please note that the above examples are not intended to address all clinical situations.

Adapted from: Bhattacharya, D., et al., *Hepatitis C Guidance 2023 Update: AASLD-IDSA Recommendations for Testing, Managing, and Treating Hepatitis C Virus Infection.* Clin Infect Dis, 2023.

Table Key: *PEG-INF* pegylated interferon, *SOF* sofosbuvir, *VEL* velpatasvir, *VOX* voxilaprevir, *GLE* glecaprevir, *PIB* pibrentasvir, *RBV* ribavirin patients benefit from treatment monitoring, including follow-up HCV RNA quantitative assay, complete blood count, and the complete metabolic panel, which are indicated after 1 month, at the end of treatment, and at 3 months post treatment. A virological cure (sustained virological response, week 12) is declared if the viral load remains negative at the 3-month (12-week or 90-day) posttreatment mark [28].

After SVR 12 has been achieved, patients with risk factors for reinfection with HCV, such as intravenous drug use or high-risk sexual encounters, should be offered at least yearly rescreening with HCV RNA (HCV viral load). Note that the HCV antibody will remain positive throughout life and does not necessarily indicate active reinfection or viral replication, unless the HCV viral load also returns positive.

For non-cirrhotic individuals, who achieve a DAA SVR 12, there are no specific liver-related follow-up screenings necessary at the time of this review beyond the routine risk assessment and if needed harm reduction interventions/referrals. However, cirrhotic individuals obtaining a DAA SVR12 will need yearly screening for hepatocellular carcinoma (HCC) and esophageal varices utilizing a twice-yearly liver ultrasound, ideally with alpha fetoprotein laboratory measurement (optional but recommended). Esophageal varices monitoring is also recommended every 3 years in this setting [28].

8 HCC Association with HCV: DAA Role in Prevention of Hepatocellular Carcinoma

Hepatic cirrhosis due to years of cumulative inflammation from hepatitis C (and also hepatitis B) can lead to hepatocellular carcinoma (HCC), which is the most frequent malignant tumor of the liver. Uncontrolled hepatitis C or B along with HIV infection can also accelerate formation of HCC. Curative treatment for HCV with DAA still remains an effective strategy to reduce the risk of development of HCC. However, unfortunately, once an individual develops HCV-related cirrhosis, the elevated risk for progression to HCC will remain and thus necessitate frequent screening.

The mechanism of hepato-carcinogenesis is a complex process characterized by progressive molecular and cellular dedifferentiation of hepatocytes [29]. The molecular changes can be genetic changes (mutations in the actual DNA sequence of hepatocytes which leads to changes in protein structure and function) or epigenetic changes (modifications of DNA chromatin without altering the genetic sequence which leads to increased or decreased expression levels in specific proteins). Curative-intent therapy for HCC includes liver resection and ablation. In treatment-naïve patients who are candidates for HCC, curative therapy should proceed with cancer treatment prior to initiation of DAA [30]. The timing of DAA initiation once the tumor has been treated is under investigation [29].

Cirrhosis from any cause is a risk factor for HCC. Chronic inflammation causes cellular injury, which is often associated with genetic or epigenetic mutations. HCV plays a role in persistent inflammation, and treatment of the virus seems to be associated with risk reduction. There is ongoing debate regarding whether patients treated with DAA therapy have different long-term risks of HCC compared to patients who achieved SVR using older immune-based therapies [30]. Usually, once HCC is present, it is difficult to clear or "cure" HCV using DAAs.

9 Impact of DAA Treatment on Diabetes

While HCV infection is associated with increased insulin resistance, the impact of treatment with DAA varies among studied populations. Some studies suggested no significant changes in HgA1c [31]. Other studies suggest that patients with diabetes or prediabetes treated with DAA and who achieved SVR12 weeks had improvements in glycemic control at the end of 12 weeks and up to 3 years after the initial treatment. According to this study, these individuals had fewer requirements in insulin doses and oral hypoglycemic agents [21]. Another showed overall improvement in glycemic control in patients with diabetes after DAA-induced SVR in 12 weeks. These patients were less likely to require insulin. As patients with HCV may have an element of comorbid metabolic liver disease, improved glycemic metabolism may be one mechanism by which HCV clearance improves liver disease [32].

Treatment of hepatitis C can occasionally lead to sufficient decreases in insulin requirement that patients on insulin or secretagogues may develop hypoglycemia, thus alerting patients to possible glycemic changes and encouraging them to alert their provider if hypoglycemia occurs is recommended.

10 Hepatitis C Pretreatment Considerations

Diagnosis of HCV might be the result of screening or clinical evaluation in symptomatic individuals. HCV screening involves the use of a serologic assay that measures human antibodies generated in response to HCV. This process usually takes between 1 and 6 months, but in some people, this seroconversion might take up to 9 months. A nonreactive HCV antibody result generally indicates that the individual has no infection but could also represent that window period or severe immune deficiency [33].

All exposures to HCV should lead to at least transient seroconversion. Around 30% of infected persons spontaneously clear the virus. The remaining 70% will develop chronic infection. To differentiate both groups, it is recommended to test for HCV RNA. In those with detection of virus in serum, further quantitation is possible and helpful to monitor progression and response to therapy. The CDC recommends that laboratories use procedures where every positive HCV antibody sample be reflexively tested for HCV RNA using a sample of blood. For recently exposed individuals or immunocompromised patients, it is recommended to screen with HCV RNA.

Once the diagnosis of chronic hepatitis C infection is made, it is important to inform the patient of the diagnosis and offer counseling and treatment options. HCV genotype testing classifies the virus as GT 1 to 6, with GT 1 being the most common in the United States.

DAA therapy is the current recommended therapy for all genotypes of HCV. Current DAA drug classes include: NS5A inhibitors, NS5B polymerase inhibitors, and NS3/4A protease inhibitors. HCV can develop resistance to one or more of these classes and these mutations are referred to as resistance associated variants (RAVs). In order to test for these mutations, there are HCV resistance panels [28].

As part of pretreatment evaluation, it is recommended to obtain a complete blood count, the metabolic panel with liver enzymes, coagulation panel, and sero-biomarkers for fibrosis. It is very important to test for hepatitis B co-infection with HBsAg, HBsAb, and Anti-HBc (known as hepatitis B core antibody) as well as to screen for HIV and hepatitis A infection. Often screening for latent tuberculosis is also wise in high-risk populations exposed to shelters or incarcerated settings. Imaging the liver is usually done with ultrasound imaging to see if cirrhosis is present.

11 Hepatitis C Treatment Strategies: Consideration of Cirrhosis and Treatment History with the Role of APRI, FIB-4, and Hepatic Imaging

Liver fibrosis is the accumulation of extracellular matrix proteins like collagen, which results from chronic inflammation. Fibrosis usually is a progressive process that leads to cirrhosis and carries an increase in morbidity and mortality in comparison to healthy individuals [27]. The gold standard for the diagnosis of liver fibrosis is a biopsy, but that depends on the sample being reflective of the overall process. In one study using core biopsies for the staging of liver fibrosis, the size of the sample was a factor for accurate staging and even those larger than 25 mm were able to accurately diagnose 75% of cases [34].

In the last decade, noninvasive alternatives have been developed to diagnose and stage fibrosis [35, 36]. There are four categories of fibrosis F1, F2, F3, F4. All these tests have been validated in relation to liver biopsy. Noninvasive tests perform well at the extremes of no fibrosis (F0) or severe cirrhosis (F4). More accurate results are obtained when are used in correlation with laboratory findings, markers, and clinical pictures consistent with liver disease (ex: splenomegaly).

Aspartate Aminotransferase-to Platelet Ratio Index (APRI) is a practical biomarker panel for the estimation of liver fibrosis which has been validated for chronic Hepatitis C in multiple studies [37]. False positives are often seen in acute hepatitis.

FIB-4 (Fibrosis 4) was developed as a surrogate marker of cirrhosis by using the liver biopsy as the standard for which the score was developed. A score of <1.45 is considered to have a negative 90% predictive value for advanced cirrhosis as compared to >3.25 which has a positive predictive value of 65% for advanced cirrhosis (defined as an Ishak fibrosis score of 4–6 with early bridging fibrosis to cirrhosis) [28].

12 DAA Treatment Failure Management

Current treatment for chronic hepatitis C is highly effective but is not infallible. Depending upon the population being treated and the regimen, 2–10% of patients will experience failure to achieve SVR. These patients should be managed with input from a specialist, and some considerations lie outside the scope of this chapter.

Resistance is the most alarming mechanism for treatment failure. Viral resistance generally occurs when drugs exhibit selective pressure against wild-type variants but fail to sufficiently "shut down" viral replication. This situation can occur due to some combination of missed doses, reduced absorption, drug interactions causing reduced antiviral exposure, baseline resistance to one or more components of a multidrug regimen, a high baseline viral load that takes a long time to suppress, et cetera. There is concern that resistant virus could spread among at-risk populations without appropriate interventions, including appropriate retreatment. Retreatment should be guided by genotyping and resistance testing [38], and agents with a high intrinsic barrier to resistance (e.g., sofosbuvir) are generally preferred.

Another common situation is reinfection. Risk factors for infection or reinfection with hepatitis C are intravenous drug use, unsafe sex, and exposure to medical procedures including multiple syringe use, dental treatment, hemodialysis, and surgery.

While expert input is key, most patients experiencing treatment failure have a good prognosis for SVR following retreatment of an appropriately selected regimen, with between 90% and 99% achieving SVR in recent studies involving sofosbuvir-based regimens with or without ribavirin. At referral to specialty care, patients should understand that they likely have an excellent prognosis for SVR upon retreatment; this shared understanding may motivate patients to follow up appropriately. In most situations, even a patient with cirrhosis *and* prior treatment failure has a ~9 in 10 chance of HCV cure when retreated [39].

13 Acute Hepatitis Management Considerations

Acute Hepatitis C is defined as the 6-month time following exposure to the Hepatitis C virus. In the first weeks after initial infection, the viral load increases exponentially [40]. More than two-thirds of patients with acute hepatitis C are asymptomatic, which makes diagnosis very challenging [2]. Symptoms usually develop after

2 weeks and can last up to 6 months [6]. In symptomatic patients, jaundice is the most common presentation but abdominal pain, generalized malaise, low-grade fever, and confusion can also be seen [41].

During acute infection, there is a high risk of HCV transmission. It is important to discuss with newly diagnosed patients different strategies to reduce viral transmission to others; those include not sharing needles and not engaging in high-risk sexual activities. The risk of sexual transmission is higher among men who have sex with men particularly those with HIV [42]. The barriers to early treatment during acute HCV infection include the lack of diagnosis and asymptomatic disease, poor awareness and education on hepatitis C, lack of medical coverage, high cost of treatment, and medical failure to screen patients as well as concerns about the risk of HCV reinfection. Some providers would like to wait 6 months to determine whether the patient has developed a chronic infection.

Diagnosis of Hepatitis C depends on the timing of the infection. Antibodies for Hepatitis C can be negative for the first 6 to 12 weeks. According to the CDC, acute hepatitis C can be diagnosed when a patient meets clinical (symptoms or elevated liver enzymes) and laboratory criteria (positive HCV RNA). According to WHO, around 30% of infected persons spontaneously (without treatment) clear the virus within 6 months. Most infections will progress to chronic hepatitis C and carry the risk of fibrosis of the liver and further complications.

Treatment of acute hepatitis C should be considered in all patients. Early treatment has been shown to prevent complications associated with chronic hepatitis C infection. It has been shown to reduce transmission among high-risk group populations, and multiple studies have shown that is also cost-effective [43]. Since diagnosis is essential for treatment for acute hepatitis C infection, current recommendations are for routine testing for high-risk groups, including men who have sex with men, HIV-positive patients, IV drug users, and patients on hemodialysis among other groups.

Early treatment for Acute Hepatitis C with DAA is highly effective; in some studies, sustained virologic response (SVR) has been achieved in 100% of patients [44]. Additionally, HIV patients who developed acute hepatitis C and are treated with DAA have similar response rates to treatment [45]. Treatment regimens recommended for acute HCV infections are the same as those for chronic infection. Please refer to the treatment of hepatitis C infection with the previous DAA section for further details.

While the prognosis in treatment-experienced patients remains excellent, optimal timing of treatment in patients at high risk for nonadherence and reinfection should be individualized. In some patients with asymptomatic HCV, the authors have occasionally offered a temporary delay in treatment while optimization of mental health or medical comorbidities is underway, where said optimization was clinically expected to improve treatment tolerability and success. Expert consultation may be advisable in situations where the prognosis for treatment completion and freedom from reinfection is felt to be poor.

14 HBV/HCV Co-Iinfection Treatment Considerations

According to the Hepatitis B Foundation, co-infection of hepatitis B and hepatitis C varies between 9% and 30% [46]. Co-infected patients tend to have more severe liver disease and are at an increased risk for progression to liver cancer. Treatment of HBV/HCV co-infected patients is challenging but possible.

Both HBV and HCV replicate primarily in the liver, and even though the pathogenesis of the coinfection is poorly understood, some studies suggest that both viruses can replicate in the same hepatocyte without evidence of interference [47]. Both viruses share similar risk factors for infection, including IV drug use and high-risk sexual activities. Usually, one virus is dominant over the other and replicates more actively, suppressing the replication of the nondominant virus [3]. Patients with coinfection can have spontaneous clearance of one or both viruses, develop acute fulminant hepatitis, or progress to chronic hepatitis [47].

Current recommendations are to test all patients with hepatitis C for hepatitis B virus and vice versa immediately after diagnosis and prior to treatment. This should include evaluation for the presence of HBsAg, anti-HBs, and anti-HBc. Patients with positive HBsAg should be tested for HBV DNA viral load prior to initiation with DAA therapy. Those who were found to have active hepatitis B infection should be treated before initiation of hepatitis C treatment with DAA. Those with undetectable or very low HBV DNA levels should be monitored regularly. If reactivation of HBV occurs, treatment for HBV should follow. There is no definitive data about continuing HBV therapy after the HCV cure [48].

There is very limited data regarding treatment with DAAs in hepatitis B and hepatitis C coinfection. Treatment for hepatitis C with DAAs is highly efficacious even in those affected with hepatitis B, but this population is at risk for reactivation of HBV while on treatment [49]. At least one recent study showed that HBV/HCV coinfection persons have higher overall SVR rates with near or DAA regimens than those with mono-infection [50].

15 Hepatitis C/HIV Coinfection Treatment Considerations

According to the CDC, approximately 21% of people with HIV in the United States also have HCV. Infection with HIV and Hepatitis C share the same risk factors, including sharing needles and high-risk sexual activity. People with HIV/HCV coinfection have higher HCV viral loads, higher rates of progression to cirrhosis, lower rates of HCV clearance after acute infection, and higher rates of hepatocellular carcinoma than those with HCV mono-infection [51]. It is unclear if HCV causes HIV to advance faster, but studies have shown an increase in inflammatory markers and persistent levels of HIV viremia even in those on ART [52]. The current recommendation is for all HIV patients to be tested for HCV with an HCV antibody test and if positive to be followed by HVC RNA Viral Load.

Treatment with DAAs in the HIV/HCV coinfection is possible but providers need to be aware of drug-to-drug interactions between DAAs and antiretroviral therapy. DAA therapies have resulted in similarly high rate of SVR for HCV infection regardless of HIV status [53]. Patients with HIV with undetectable viral load who are started on DAAs usually have persistently suppressed levels of viremia while on treatment, but virological failure occurs and close monitoring of HIV viral load on selected patients (older age, previous HIV virological failures or blips and those with residual viremia at DAA's initiation) might be required during hepatitis C treatment [54].

Data suggest that the DAA therapy may improve outcomes associated with liver transplant in those with HIV/HCV coinfection. A recent study concluded that HIV/HCV co-infected patients have comparable outcomes to HCV mono-infection after the introduction of DAA therapy [53].

16 HCV DAA in Pregnancy

In the last decade, hepatitis C infections have increased especially among injection drug users between the ages of 20 and 40 years of age, many of whom are women of childbearing age [55, 56]. It is recommended for all pregnant women to be screened for hepatitis C [57], and there has been a fivefold increase in the diagnosis of pregnant women from 1998 to 2011 [28]. If screening is positive, then discussions about treatment timing with direct-acting antivirals should follow.

Consideration must be given to the natural history of HCV among women, including the fact that, at least in observational studies, women appear to have a higher rate of spontaneous clearance of HCV infection when compared to men. The mechanism by which women seem to clear the virus more often than men is unclear but might be related to sex-related steroids binding to expressed specific receptors and immune cells [28]. The progression of hepatitis C infection in women also seems to be directly associated with estrogen, and some studies have shown accelerated progression after menopause.

Based on current studies, there is no significant effect on hepatitis C infection and the progression of the disease during pregnancy. Hence, the current recommendation is for no treatment with DAAs during pregnancy or post-delivery breastfeeding period. Nonetheless, there are many other risk factors to consider during pregnancies in women with hepatitis C infection, including poor perinatal care and concurrent substance use. Some studies have shown an association between HCV infection and miscarriages and some reduction in such complications after successful treatments with DAAs [56]. There is also evidence that HCV-infected pregnant women might have preterm deliveries and delivered infants with low birth weights [58, 59].

Multiple studies have also evaluated the perinatal transmission risk of hepatitis C viremic women. There is no evidence that individuals with anti-HCV antibodies who have cleared the virus can transmit it. Among pregnancies complicated by maternal HCV viremia, an approximately 5–10% risk of vertical transmission has been described, with maternal HIV coinfection associated with higher HCV transmission risk [60]. Although most transmissions occurred around the time of delivery, there is no clear benefit of C-section in the reduction of HCV infection [56]. Although HCV RNA has rarely been detected in breast milk [59], breastfeeding has not been associated with an increased risk of infant infection [60]. It is unknown whether breastfeeding women with HCV viremia and cracked or bleeding nipples are at increased risk of spreading HCV to their infants, but the Centers for Disease Control (CDC) recommends that mothers in this situation temporarily "express and discard" milk until the nipples heal ([56]).

17 Transplant Medicine: The Hepatitis C Positive Donor and the Role of Direct-Acting Antivirals

According to the United Network for Organ Sharing (UNOS), the United States in 2022 performed 41,112 total organ transplants across all centers, of which 9002 were liver transplants and 24,797 were renal transplants [61]. It is estimated that 5–6% of deceased organ donors in the United States are infected with HCV [32]. For decades, individuals with HCV were considered donor candidates for other HCV-infected patients but not considered donor candidates to HCV naïve patients due to the difficulty in treating the virus and high associated morbidities in the setting of immunosuppressive medications with non-DAA medications and the risk of transmission from a donor with detectable HCV RNA is close to 100% [62]. For example, between 2005 and 2014, approximately 2698 kidneys from HCV-infected intended donors were deemed not eligible for transplant [63]. However, the development of DAA in 2011 created an opportunity for the transplantation of HCV-infected organs to HCV-naive recipients.

Since approximately 2015, transplant centers have begun using HCV-infected organs for all recipients as a strategy to reduce the waitlist mortality associated with increased organ demand [64]. All organ donors are tested for HCV antibody and HCV nucleic acid test (NAT). These organs are stratified between anti-HCV positive organs and HCV-viremic organs. Particular attention is given to evaluating the window period for each group since the risk of transmission depends on the viral load of the donor at the time of transplant, but even those with undetectable levels of RNA can have a transmission rate as high as 20% [32].

After detailed discussions and obtaining consent, transplant centers have developed protocols to treat the organ with DAA before transplantation [63] or the recipient prior to or immediately after transplantation. Depending on risk factors, some recipients are closely monitored after the transplant and treated only if they develop viremia [65].

Recipients who develop detectable HCV RNA are treated with DAA for 12 weeks with a cure rate of 99% a year after transplant. The most common regimens currently used after transplantation are Sofosbuvir/Velpatasvir and Glecaprevir/Pibretensavir, both of which are effective for all six genotypes and have excellent sustained virologic response [62].

DAAs do not tend to interact with standard immunotherapy used after transplantation. Immunosuppressants such as mycophenolic acid, calcineurin inhibitors (tacrolimus), and corticosteroids are usually safe. The minority of recipients who have persistent HCV viremia after standard treatment with DAA should be evaluated for HCV resistance testing and treated accordingly [62].

18 Conclusions

Patients living with HCV who undergo screening and therapy prior to the onset of advanced liver disease have an excellent prognosis. DAA therapy is well tolerated and cures HCV in the vast majority of treated patients, including those with comorbid HIV infection and even prior treatment failure. As noted above, HCV science is evolving. The Infectious Disease Society of America (IDSA) and the American Association for the Study of Liver Diseases (AASLD) publish updated guidance under the joint HCV Guidance Project at https://www.hcvguidelines.org [28, 66]. The authors recommend that clinicians treating HCV refer to this resource regularly. American guidelines for HCV screening, as well as other screening recommendations, are published by the United States Preventative Task Force, at https://uspreventiveservicestaskforce.org.

Conflicts of Interest The authors state no conflicts of interest.

References

1. Organization, W.H. *Hepatitis C: Key Facts*. 2023 July 18 2023 September 10, 2023]; Available from: https://www.who.int/news-room/fact-sheets/detail/hepatitis-c.
2. Blackard, J.T., et al., *Acute hepatitis C virus infection: a chronic problem*. Hepatology, 2008. **47**(1): p. 321–31.
3. Edlin, B.R., et al., *Toward a more accurate estimate of the prevalence of hepatitis C in the United States*. Hepatology, 2015. **62**(5): p. 1353–63.
4. Services, U.S.D.o.H.a.H. *Viral Hepatitis in the United States: Data and Trends*. [cited September 10, 2023; Available from: https://www.hhs.gov/hepatitis/learn-about-viral-hepatitis/data-and-trends/index.html.
5. Belperio, P.S., et al., *Curing Hepatitis C Virus Infection: Best Practices From the U.S. Department of Veterans Affairs*. Ann Intern Med, 2017. **167**(7): p. 499–504.
6. Seeff, L.B., *Natural history of hepatitis C*. Hepatology, 1997. **26**(3 Suppl 1): p. 21s-28s.
7. Fraser, H., et al., *Scaling Up Hepatitis C Prevention and Treatment Interventions for Achieving Elimination in the United States: A Rural and Urban Comparison*. Am J Epidemiol, 2019. **188**(8): p. 1539–1551.

8. Owens, D.K., et al., *Screening for Hepatitis C Virus Infection in Adolescents and Adults: US Preventive Services Task Force Recommendation Statement.* Jama, 2020. **323**(10): p. 970–975.
9. Schillie, S., et al., *CDC Recommendations for Hepatitis C Screening Among Adults - United States, 2020.* MMWR Recomm Rep, 2020. **69**(2): p. 1–17.
10. Wilton, J., et al., *Association Between Prescription Opioid Therapy for Noncancer Pain and Hepatitis C Virus Seroconversion.* JAMA Netw Open, 2022. **5**(1): p. e2143050.
11. Platt, L., et al., *Needle and syringe programmes and opioid substitution therapy for preventing HCV transmission among people who inject drugs: findings from a Cochrane Review and meta-analysis.* Addiction, 2018. **113**(3): p. 545–563.
12. Turner, K.M., et al., *The impact of needle and syringe provision and opiate substitution therapy on the incidence of hepatitis C virus in injecting drug users: pooling of UK evidence.* Addiction, 2011. **106**(11): p. 1978–88.
13. Cacoub, P. and D. Saadoun, *Extrahepatic Manifestations of Chronic HCV Infection. Reply.* N Engl J Med, 2021. **385**(1): p. 95–96.
14. Faccioli, J., et al., *Neurological and psychiatric effects of hepatitis C virus infection.* World J Gastroenterol, 2021. **27**(29): p. 4846–4861.
15. Kim, H.P., et al., *Symptoms and functioning improve after chronic hepatitis C cure as assessed by the Memorial Symptom Assessment Scale and PROMIS measures.* J Viral Hepat, 2021. **28**(10): p. 1491–1495.
16. Fabrizi, F., et al., *Hepatitis C Virus Infection Increases the Risk of Developing Chronic Kidney Disease: A Systematic Review and Meta-Analysis.* Dig Dis Sci, 2015. **60**(12): p. 3801–13.
17. Aby, E.S., et al., *Impact of sustained virologic response on chronic kidney disease progression in hepatitis C.* World J Hepatol, 2017. **9**(36): p. 1352–1360.
18. Cohen-Bucay, A., J.M. Francis, and C.E. Gordon, *Progress in hepatitis C virus management in chronic kidney disease.* Curr Opin Nephrol Hypertens, 2021. **30**(5): p. 493–500.
19. Gritsenko, D. and G. Hughes, *Ledipasvir/Sofosbuvir (harvoni): improving options for hepatitis C virus infection.* P t, 2015. **40**(4): p. 256–76.
20. Stine, J.G., et al., *Effect of Treatment with Direct Acting Antiviral on Glycemic Control in Patients with Diabetes Mellitus and Chronic Hepatitis C.* Ann Hepatol, 2017. **16**(2): p. 215–220.
21. El Sayed, O.A.E.-f., et al., *Effect of hepatitis C eradication with DAA on type II diabetes mellitus control as regard insulin resistance & lipid profile.* QJM: An International Journal of Medicine, 2021. **114**(Supplement_1).
22. Braun, D.L., et al., *A Treatment-as-Prevention Trial to Eliminate Hepatitis C Among Men Who Have Sex With Men Living With Human Immunodeficiency Virus (HIV) in the Swiss HIV Cohort Study.* Clin Infect Dis, 2021. **73**(7): p. e2194–e2202.
23. Urits, I., et al., *A Comprehensive Review of Alternative Therapies for the Management of Chronic Pain Patients: Acupuncture, Tai Chi, Osteopathic Manipulative Medicine, and Chiropractic Care.* Adv Ther, 2021. **38**(1): p. 76–89.
24. Liverpool, U.o. *HEP Drug Interactions*. Available from: https://www.hep-druginteractions.org/.
25. Kluwer, W. Up To Date; Available from: http://www.uptodate.com.
26. Ghany, M.G. and T.R. Morgan, *Hepatitis C Guidance 2019 Update: American Association for the Study of Liver Diseases-Infectious Diseases Society of America Recommendations for Testing, Managing, and Treating Hepatitis C Virus Infection.* Hepatology, 2020. **71**(2): p. 686–721.
27. Affairs, D.o.V., *Chronic Hepatitis C (HCV) Infection: Treatment Considerations*, H. HIV, and Related Conditions Program, within the Office of Specialty Care Services, Editor. March 18, 2021.
28. Bhattacharya, D., et al., *Hepatitis C Guidance 2023 Update: AASLD-IDSA Recommendations for Testing, Managing, and Treating Hepatitis C Virus Infection.* Clin Infect Dis, 2023.
29. Sirlin, C., et al., *Chapter 6: Hepatocarcinogenesis*. 2018: American College of Radiology.
30. Gao, X., et al., *Timing of DAA Initiation After Curative Treatment and Its Relationship with the Recurrence of HCV-Related HCC.* J Hepatocell Carcinoma, 2020. **7**: p. 347–360.

31. Andres, J., et al., *Treatment of Hepatitis C Virus and Long-Term Effect on Glycemic Control.* J Manag Care Spec Pharm, 2020. **26**(6): p. 775–781.
32. (UNOS), U.N.f.O.S. September 10, 2023]; Available from: https://optn.transplant.hrsa.gov/data/view-data-reports/national-data/.
33. *Laboratory Tests and Hepatitis C - Viral Hepatitis and Liver Disease* Department of Veterans Affairs.
34. Poynard, T., et al., *Prospective analysis of discordant results between biochemical markers and biopsy in patients with chronic hepatitis C.* Clin Chem, 2004. **50**(8): p. 1344–55.
35. Castera, L., *Noninvasive methods to assess liver disease in patients with hepatitis B or C.* Gastroenterology, 2012. **142**(6): p. 1293–1302.e4.
36. Li, C., R. Li, and W. Zhang, *Progress in non-invasive detection of liver fibrosis.* Cancer Biol Med, 2018. **15**(2): p. 124–136.
37. Zhou, Y., et al., *Hypoglycemia associated with direct-acting anti-hepatitis C virus drugs: An epidemiologic surveillance study of the FDA adverse event reporting system (FAERS).* Clin Endocrinol (Oxf), 2022. **96**(5): p. 690–697.
38. Sterling, R.K., et al., *Development of a simple noninvasive index to predict significant fibrosis in patients with HIV/HCV coinfection.* Hepatology, 2006. **43**(6): p. 1317–25.
39. Ridruejo E, P.M., Flichman DM, Di Lello FA., *Hepatitis C virus treatment failure: Clinical utility for testing resistance-associated substitutions.* World J Hepatol, 2021. **13**(9): p. 1069–1078.
40. Parigi, T.L., M.C.P. Torres, and A. Aghemo, *Upcoming direct acting antivirals for hepatitis C patients with a prior treatment failure.* Clin Mol Hepatol, 2019. **25**(4): p. 360–365.
41. Marcellin, P., *Hepatitis C: the clinical spectrum of the disease.* J Hepatol, 1999. **31 Suppl 1**: p. 9–16.
42. Gerlach, J.T., et al., *Acute hepatitis C: high rate of both spontaneous and treatment-induced viral clearance.* Gastroenterology, 2003. **125**(1): p. 80–8.
43. Werner, R.N., et al., *Incidence of sexually transmitted infections in men who have sex with men and who are at substantial risk of HIV infection - A meta-analysis of data from trials and observational studies of HIV pre-exposure prophylaxis.* PLoS One, 2018. **13**(12): p. e0208107.
44. Bethea, E.D., et al., *Should we treat acute hepatitis C? A decision and cost-effectiveness analysis.* Hepatology, 2018. **67**(3): p. 837–846.
45. Rockstroh, J. and C. Boesecke, *Treatment of acute hepatitis C in HIV coinfection: Is this a chance for achieving microelimination?* United European Gastroenterol J, 2019. **7**(4): p. 465–466.
46. Chromy, D., et al., *High efficacy of interferon-free therapy for acute hepatitis C in HIV-positive patients.* United European Gastroenterol J, 2019. **7**(4): p. 507–516.
47. Jamma, S., G. Hussain, and D.T. Lau, *Current Concepts of HBV/HCV Coinfection: Coexistence, but Not Necessarily in Harmony.* Curr Hepat Rep, 2010. **9**(4): p. 260–269.
48. Mavilia, M.G. and G.Y. Wu, *HBV-HCV Coinfection: Viral Interactions, Management, and Viral Reactivation.* J Clin Transl Hepatol, 2018. **6**(3): p. 296–305.
49. Abdelaal, R., B. Yanny, and M. El Kabany, *HBV/HCV Coinfection in the Era of HCV-DAAs.* Clin Liver Dis, 2019. **23**(3): p. 463–472.
50. Papadopoulos, N., et al., *Hepatitis B and C coinfection in a real-life setting: viral interactions and treatment issues.* Ann Gastroenterol, 2018. **31**(3): p. 365–370.
51. Butt, A.A., et al., *Hepatitis C virologic response in hepatitis B and C coinfected persons treated with directly acting antiviral agents: Results from ERCHIVES.* Int J Infect Dis, 2020. **92**: p. 184–188.
52. Thein, H.H., et al., *Natural history of hepatitis C virus infection in HIV-infected individuals and the impact of HIV in the era of highly active antiretroviral therapy: a meta-analysis.* Aids, 2008. **22**(15): p. 1979–91.
53. Morsica, G., et al., *Risk of HIV viral rebound in HIV infected patients on direct acting antivirals (DAAs) treatment for HCV.* PLoS One, 2022. **17**(2): p. e0262917.

54. Sikavi, C., et al., *Hepatitis C and human immunodeficiency virus coinfection in the era of direct-acting antiviral agents: No longer a difficult-to-treat population.* Hepatology, 2018. **67**(3): p. 847–857.
55. Shimada, S., et al., *Improvements in liver transplant outcomes in patients with HCV/HIV coinfection after the introduction of direct-acting antiviral therapies.* Transpl Infect Dis, 2022. **24**(2): p. e13808.
56. Kushner, T. and N.A. Terrault, *Hepatitis C in Pregnancy: A Unique Opportunity to Improve the Hepatitis C Cascade of Care.* Hepatol Commun, 2019. **3**(1): p. 20–28.
57. Ly, K.N., et al., *Hepatitis C Virus Infection Among Reproductive-Aged Women and Children in the United States, 2006 to 2014.* Ann Intern Med, 2017. **166**(11): p. 775–782.
58. Karampatou, A., et al., *Premature ovarian senescence and a high miscarriage rate impair fertility in women with HCV.* J Hepatol, 2017.
59. Cottrell, E.B., et al., *Reducing risk for mother-to-infant transmission of hepatitis C virus: a systematic review for the U.S. Preventive Services Task Force.* Ann Intern Med, 2013. **158**(2): p. 109–13.
60. Polywka, S., et al., *Low risk of vertical transmission of hepatitis C virus by breast milk.* Clin Infect Dis, 1999. **29**(5): p. 1327–9.
61. Prevention, U.S.C.f.D.C.a. *Hepatitis B or C Infections.* September 10, 2023]; Available from: https://www.cdc.gov/breastfeeding/breastfeeding-special-circumstances/maternal-or-infant-illnesses/hepatitis.html.
62. Diseases, A.A.f.t.S.o.L. *Treatment of HCV-Uninfected Transplant Receipients Receiving Organs from HCV-Viremic Donors.* Available from: https://www.hcvguidelines.org/unique-populations/organs-from-hcv-viremic-donors.
63. Levitsky, J., et al., *The American Society of Transplantation Consensus Conference on the Use of Hepatitis C Viremic Donors in Solid Organ Transplantation.* Am J Transplant, 2017. **17**(11): p. 2790–2802.
64. Reese, P.P., et al., *Transplanting Hepatitis C-Positive Kidneys.* N Engl J Med, 2015. **373**(4): p. 303–5.
65. Cypel, M., et al., *Prevention of viral transmission during lung transplantation with hepatitis C-viraemic donors: an open-label, single-centre, pilot trial.* Lancet Respir Med, 2020. **8**(2): p. 192–201.
66. Patnaik, R. and E. Tsai, *Hepatitis C Virus Treatment and Solid Organ Transplantation.* Gastroenterol Hepatol (N Y), 2022. **18**(2): p. 85–94.

NeuroAIDS, Comorbid Mood Disorders and Neuropsychiatric Effects of Antiretroviral Therapy

Hector E. Rodriguez, Yeisell Duconge, Gianfranco Molfetto, Syeb Ahmad, Nazish Bais, and Gerardo Ferrer

Abstract Mental illnesses may be directly associated with the neuropathology of the human immunodeficiency virus (HIV). Comorbid psychiatric conditions in persons living with HIV(PLWH) are a major health concern. HIV has been shown to directly affect the central and peripheral nervous system, leading to neurocognitive disorders collectively termed NeuroAIDS. In addition to the virus' direct impacts on the brain, the medications used in the treatment of HIV have also been shown to have a direct link to neuropsychiatric dysfunction. The combination of these two intrusions has led to substantial morbidity and mortality within the HIV population. This chapter reviews the literature surrounding NeuroAIDS, along with the literature related to the medications used for HIV treatment, specifically in relation to mood dysfunction.

Keywords NeuroAIDS · Psychiatric HIV comorbidities · Neuropsychiatric disorders · Mood dysfunction · PTSD · Bipolar manic disorders · Depression · Anxiety · Major depression disorders · Combined antiretroviral therapy (cART) · NRTI · Dotutegravir · Raltegravir · Efavirenz

1 Introduction

NeuroAIDS is defined as a group of neurological disorders caused primarily by HIV's direct damage to the central and peripheral nervous systems, leading to a high prevalence of neuropsychiatric and neurocognitive disorders (including depression 7%, anxiety 6.5%, and mania 2.7%) [1]. Among such conditions, mood disorders are among the most common mental health conditions in the general population,

H. E. Rodriguez (✉) · Y. Duconge · N. Bais · G. Ferrer
Larkin Community Hospital, South Miami, FL, USA

G. Molfetto · S. Ahmad
Nova Southeastern University College of Osteopathic Medicine, Davie, FL, USA

C. Somboonwit et al. (eds.), *Global Virology IV: Viral Disease Diagnosis and Treatment Delivery in the 21st Century*,
https://doi.org/10.1007/978-3-031-57369-9_8

suggesting a strong association between HIV and neuropathology. The World Health Organization (WHO) estimates that 36.7 million people were infected with HIV worldwide in 2015 [1]. Due to the advanced treatment methods available, the life expectancy of persons living with HIV (PLWH) has increased. Epidemiologic studies examining subsets of HIV/AIDS patients indicate that the rates of psychiatric symptoms, including substance abuse, depression, PTSD, sleep disturbance, and psychosis, are between 1.5 and 8 times higher in the HIV-infected population compared with the general or noninfected population [1]; the full effects of these comorbid psychiatric conditions are unknown. In this narrative review, we aim to provide new insight into how they affect patients with HIV and NeuroAIDS.

Furthermore, global combined antiretroviral therapy (cART) coverage is increasing at a rate of 4–5% per year with widespread access, leading to increased life expectancy. Widespread access to cART may elevate the lifetime prevalence of chronic mental illness [1]. Psychiatric conditions comorbid with HIV infection can potentially worsen clinical outcomes for treating psychiatric conditions and HIV infection. The use of cART in this population has been associated with various neuropsychiatric symptoms covered here. Of note, cART medications have also been associated with other psychiatric and neuropsychiatric symptoms not covered in this article; our focus was mood dysfunction-related therapy. A recent investigation of hospital admissions showed that 9% of hospitalizations across five continents were attributable to psychiatric disorders [1]. Psychiatric comorbid conditions can contribute to poor clinical outcomes and higher mortality rates that are due, in part, to reduced medication adherence. Several studies have found that treating mental health symptoms can improve cART adherence, highlighting the importance of considering psychiatric comorbidities within the diagnosis and treatment framework of NeuroAIDS [2, 3]. The current review aimed to elaborate on the psychiatric mood disorders present in persons with HIV to provide new insight as to why this population is affected more severely than the general population.

2 Methods

Several procedures were followed to ensure a high-quality literature review, including the following five steps: (1) identifying the research question, (2) search methods for identifying relevant studies, (3) study selection, (4) data analysis, and (5) reporting the results.

2.1 Identifying the Research Question

What psychiatric mood disorder comorbidity is more prevalent in NeuroAIDS?

2.2 *Search Methods for Identifying Relevant Studies*

This narrative review was conducted using comprehensive peer-reviewed journals, which included cross-sectional epidemiologic studies, case reports, case-control studies, and meta-analyses. The databases searched for relevant articles were the NIMH, PubMed, and Google Scholar. To increase precision, we used Boolean terms (AND/OR) to separate the keywords and medical subject headings (MeSH). The keywords used to search resources for data collection were: "NeuroAIDS," "psychiatry," "neuropsychiatric disorders," "combined antiretroviral therapy (cART)," "depression," "AIDS virus," "human immunodeficiency virus," and "HIV-1." A selection of relevant articles from 1998 to 2019 was used for the study.

2.3 *Study Selection*

Following an initial search on Google Scholar, 159 papers were identified, and the titles and abstracts for those papers were reviewed. In the initial review, 68 articles were included, which required further exclusion of other psychiatric comorbidities. After reviewing all the articles, data from 27 of them were utilized to create this chapter.

2.3.1 Inclusion Criteria

The inclusion criteria for this chapter were papers that described the association between psychiatric mood disorders (major depressive disorder, bipolar manic disorder, anxiety disorder, posttraumatic stress disorder) and NeuroAIDS. The articles with study patient population ≥18 years of age, HIV diagnosis and CD4 count<200 (late onset), or >or =200 (early onset). The studies included were cross-sectional epidemiologic studies, case reports, case-control studies, and meta-analyses.

2.3.2 Exclusion Criteria

The exclusion criteria for this chapter were papers using patients aged 17 and below with psychotic disorders and psychiatric symptoms due to another medical condition.

2.4 *Data Extraction*

The authors reviewed full-text articles for this narrative review, which analyzed the main contributions of each paper. For each study, we analyzed the target population and the comorbid mood disorders, the country where the studies were held, the methodology used that led to the relevant outcome presented here in this narrative review, and the type of study. A summary of the data extracted is provided in Table 1.

2.5 Results

A total of 159 articles published between 1998 and 2019 were screened for titles and abstracts. From those articles, 8 were excluded as they were duplicated, 10 were excluded after the abstracts were screened, and 11 were omitted due to the lack of access to the full text of the articles. Finally, a total of 27 articles were included in the review. Other characteristics of the studies included in this review are: Of the included studies, 5 were conducted in the United States, 2 in Western Cape, South Africa, 1 in Uganda, and 1 in Nigeria. The total number of participants in all included studies was 3655. Of the included articles, 9 were cross-sectional epidemiologic studies, 7 were case reports, 6 were case-control studies, 3 were clinical trials, and 2 were meta-analysis.

As reported in Table 1, there were comparable findings that suggested that 3 of the studies had significant reported depression by the sample size. Of the studies analyzed [6, 11, 13], 3 reported relevant outcomes in bipolar mania [6, 19, 20], as opposed to bipolar depression, in the population studied, and 2 had anxiety as a comorbid neuropsychiatric condition [6, 18]. Of the studies reported, 3 had PTSD as a predominant comorbid, with the addition of stigma and pain as contributing factors leading to PTSD or worsening the PTSD symptoms [22, 24, 26].

3 Discussion

3.1 Overview of NeuroAIDS and the Associated Psychiatric Disorders

This narrative review includes mood disorders associated with NeuroAIDS (Major depressive disorder, anxiety disorders, bipolar manic disorders, posttraumatic stress disorder) and highlights major HIV medication-related neuropsychiatric disorders, with exclusion of viral association with psychotic disorders, psychiatric disorders caused by another medical condition, and neuroAIDS in children. Overview of NeuroAIDS and associated psychiatric disorders in PLWH has been shown to directly infect the central and peripheral nervous systems, leading to a wide spectrum of neuropsychiatric and neurocognitive disorders as previously mentioned; this is termed NeuroAIDS, which includes HIV-associated dementia, minor neurocognitive disorder, mania, psychosis, anxiety, depression, seizures, myelopathy, and neuropathy with accompanying chronic neuropathic pain and physical disabilities [4]. Due to the complex nature of the HIV virus, the combination of its effects on

Table 1 Characteristics and relevant outcomes of the included studies in this review

Source	Location	Sample size	Methodology	Type of study	Relevant outcome
Kilbourne [6]	United States	881	Patient report in semi-structured interviews	Cross-sectional cohort study	46% of the participants admitted that they experienced significant *depressive* symptoms, 32% of patients reported *anxiety*, 4% reported *mania*.
Schroecksnad [13]	United States	152	Monitored using viral load, CD4+ counts, urinary/ plasma neopterin, and tryptophan concentrations. Using self-reported questionnaires (BDI and MQoL-HIV)	Cross-sectional cohort study	Lower self-reported *depression* (41 patients presented with mild, 22 with moderate, and 14 with severe depression) in patients with lower neuropterin concentrations and higher CD4+ cell counts.
Fuchs [11]	United States	23	Obtained CSF fluid	Correlational	Degradation of *tryptophan*, a precursor of serotonin, is enhanced in PLWH;
Robertson [18]	Canada	2863	Hospital Anxiety and Depression Scale (HADS)	Cross-sectional epidemiology	33.3% screened positive for *anxiety*.
Lyketsos [20]	United States	20	Grouped according to whether their first manic episode occurred when CD4 count was <200 (late onset) or > or = 200 (early onset)	Case-control	*Manic* symptom profiles associated with early-onset and late-onset mania; late-onset patients had more manic symptoms.
Nakimuli-Mpungu [19]	Uganda	141	HIV-negative and HIV-positive patients admitted to psychiatric wards with acute mania; compared the patients' psychiatric, physical, and immunological (CD4 cell counts)	comparative cross-sectional	PLWH were found to be more irritable, *aggressive*, and hyperverbal in their acute manic episode.

(continued)

Table 1 (continued)

Source	Location	Sample size	Methodology	Type of study	Relevant outcome
Martin [22]	Western Cape, South Africa	85	PTSD module of the Composite International Diagnostic Interview (CIDI)	Cross-sectional	Receiving an HIV-positive diagnosis and/or being HIV-positive may be considered a stressor that frequently results in HIV-related *PTSD*.
Smith [24]	United States	145	Stressful Life Events Checklist (SLE), Posttraumatic Stress Disorder Checklist-Civilian (PCL-C), the Mental Health Inventory (MHI), and the Brief Pain Inventory (BPI)	Clinical trial	Pain may be a contributing factor to the prognosis of HIV-1 in individuals with *PTSD*, who reported significantly higher pain intensity and greater interference with daily activities due to pain.
Breet [26]	Western Cape, South Africa	210	Self-report questionnaires that assessed HIV-related stigma, social support, posttraumatic stress disorder (PTSD) and depression	Cross-sectional	Perceived HIV-related stigma may decrease patient's perceived level of social support, which in turn may increase *PTSD* symptoms.
Onyebuchi-Iwudibia [27]	Nigeria	105	Subscales of the questionnaire for personalized stigma, disclosure, negative self-image, and public attitudes	Cross-sectional	Depressive symptoms with a positive correlation regarding the presence of *stigma*.

the brain, coupled with individual social factors, makes it complicated to recognize the psychiatric manifestations of HIV disease. As a result, the psychiatric symptoms often go unrecognized and are left untreated. During the primary phase of HIV infection, the virus enters the nervous system early, infecting glial cells and subsequently leading to a variety of symptoms, including headaches, fevers, cognitive changes, dementia, and other motor dysfunction [5]. A multisite cross-sectional cohort study, comprised of 881 adult patients with HIV with a mean age of 49 years old, was performed by Kilbourne et al. The study examined veterans diagnosed with HIV for the prevalence of associated psychiatric comorbidities. Results showed that 46% of the participants admitted that they experienced significant depressive symptoms, 21% reported at-risk drinking, 32% of patients reported anxiety, 4% reported mania, 4% reported schizophrenia, and 11% experienced cognitive impairment with associated dementia. The study was unable to determine if these comorbidities were a result of antiretroviral drug toxicity or lifestyle-associated conditions. However, they suggested the importance of effective overall management with respect to PLWH [6].

3.1.1 Major Depressive Disorder (MDD)

Of the psychiatric conditions listed, mood disorders are among the most common mental health conditions affecting the general world population [7]. The manifestation of mood disorders includes generalized anxiety disorder, acute anxiety disorder, posttraumatic stress disorder (PTSD), phobias, panic disorder, major depressive disorder (MDD), bipolar illness, and other mood dysfunctions. We will concentrate on MDD, anxiety, mania, and PTSD regarding their association with NeuroAIDS. MDD, as defined in the Diagnostic and Statistical Manual of Mental Disorders 5 (DSM-5), includes the following symptoms: depressed mood, loss of interest or pleasure in all or most daily activities, significant weight loss or gain, sleeping disturbances, fatigue or loss of energy, feelings of worthlessness or excessive or inappropriate guilt, diminished ability to think or concentrate, or recurrent thoughts of death or suicidal ideation [8]. PLWH are more likely to be diagnosed with MDD than HIV-negative individuals [10], which can result as a risk factor for the infection. Of those living with HIV, 20–30% have been shown to have MDD [9]. The combination of HIV with MDD may lead to poor overall outcomes, as studies have shown that those treated for their depression are more likely to be medically compliant with their antiretroviral therapy and have sustained viral suppression along with improved quality of life. Depression and psychological stress have been shown to have negative biological effects, such as impairment of cell-mediated immunity [11]. The brain suffers from serotonin (5-HT) neuron and neurotransmitter loss with normal aging, which manifests as depression in the elderly population [12]. In a study by Schroecksnadel et al., the disease progression of 152 PLWH was monitored using viral load, CD4+ counts, urinary/plasma neopterin, and tryptophan concentrations. Using self-reported questionnaires, the study concluded lower self-reported depression (41 patients presented with mild, 22 with moderate, and 14 with severe depression) in patients with lower neuropterin concentrations and higher CD4+ cell counts [13]. In addition to the aforementioned association, disturbances in tryptophan metabolism may also be linked with HIV-related depression due to the chronic nature of the infection [14]. The degradation of tryptophan, a precursor of serotonin, is enhanced in PLWH. Decreased levels of tryptophan were noted in the cerebrospinal fluid of 22 patients when compared to the cerebrospinal fluid of individuals without HIV. It may be surmised that decreased tryptophan levels are associated with chronic immune stimulation, such as that seen in patients with HIV-1 infection [11]. The proposed mechanism is the immune degradation of tryptophan via the kynurenine pathway resulting in a diminished production of serotonin in the brain [15]. The exact pathogenesis, however, remains elusive. The direct effect on the brain causing mood disorders and cognitive impairment is deleterious to the quality of life of PLWH and NeuroAIDS. It is recommended that PLWH be screened for mood disorders and accordingly treated with antidepressants [16]. Additionally, PLWH who have a purported stigma are at a higher risk of encountering depression; this highlights the importance of incorporating methods to decrease the propagation of HIV stigmas.

3.1.2 Anxiety Disorders

Multiple studies have demonstrated elevated anxiety rates, compared to the general population, among both PLWH and those with risk factors for HIV infection. The study by Sewell et al. also revealed that these patients expressed symptoms of fatigue, physical dysfunction, persistent preoccupation with death, and mortality fears. Anxiety has also been reported to be prevalent at the early stages of HIV presentation; this may be due to the discovery of their positive infective status. Anxiety has also been reported at other times throughout the progression of the illness to AIDS [17]. In the CRANIum (sCReen for Anxiety, Depression, and Neurocognitive Impairment in PLWH) study, 2863 patients with HIV from Europe and Canada participated in a multinational, multicenter, cross-sectional epidemiologic study regarding mental health [18]. The patient population comprised 1766 males (61.7%) and 1096 females (38.3%). These PLWH were ≥18 years of age and were able to attend routine medical follow-up visits and were able to complete the designated screening tools. The Hospital Anxiety and Depression Scale (HADS), a self-administered questionnaire in the hospital setting, revealed that out of the 2884 participants, 33.3% screened positive for anxiety [18].

3.1.3 Bipolar Manic Disorders

Mania is classically characterized by an elevated, expansive, or irritable mood alongside an inflated self-esteem or grandiosity with an accompanying decreased need for sleep, flight of ideas, and hyperverbal or pressured speech; these classical findings are seen in bipolar disorder and are subclassified as primary mania [8]. Primary mania has been reported to be considered different from the secondary mania seen in PLWH individuals, which is often referred to as "AIDS mania." This subtype of mania is markedly different from the mania associated with bipolar disorder with regard to symptomatology and severity. Often linked with the pathophysiology of NeuroAIDS, it is primarily characterized by the presence of irritability rather than the classical euphoria seen in primary mania that manifests in bipolar disorder [19]. Lyketsos et al. reported in their study of 20 HIV patients, who were grouped according to whether their first manic episode occurred when CD4 count was <200 (late onset) or > or = 200 (early onset), that the late onset HIV diagnosis individuals are less likely to present with a past history of mood disorders than the early onset [20]. Notably, HIV mania and the secondary mania seen in HIV/AIDS are used interchangeably in the research. In a comparative cross-sectional study, Nakimuli-Mpungu et al. compared 141 patients, of which 64 (45.4%) without HIV and 77 (54.6%) with HIV, admitted to the psychiatric ward of Mulago and Butabika Hospitals due to acute mania. The HIV-positive patients were older, more cognitively impaired, less educated, and had a higher female-to-male ratio. They reported that the patients with HIV were found to be more irritable, aggressive, and hyperverbal; these patients also displayed higher rates of paranoia and visual and auditory hallucinations. Previous studies have reported contradicting results, but it is important to note that those studies contained a control group of individuals without HIV with primary mania [19]. Increased frequency of mania has been reported during the initial onset of AIDS, with a total of 4–8% of AIDS patients experiencing these symptoms. Furthermore, other studies suggest that the HIV-related mania seen may be directly correlated with the CNS effects of HIV. This was reported in a 17-month chart review study examining HIV-positive patients, which demonstrated that 8% of patients with HIV with a previous manic episode had significantly higher inducer lymphocyte (CD3+/CD4+ cells) counts compared to patients with HIV without prior episodes of mania [21]. Additionally, AIDS dementia was significantly more common in patients with HIV and mania as reported by a case-control study of 19 participants with HIV-associated mania versus 57 HIV-seropositive controls performed by Dubé et al. [21]. This study posits that there is a strong association between HIV neuropathology and related manic symptoms [21].

3.1.4 Posttraumatic Stress Disorder (PTSD)

Posttraumatic stress disorder (PTSD) is a common sequel to traumatic events that may include psychologically impacting events. Such severe stressors are conceptualized by the individual as traumatic in nature as per the fifth edition of the Diagnostic and Statistical Manual of Mental Disorders (DSM-V) [8]. The exposure to actual/threatened death or sexual violence is directly listed in the DSM-V as examples of traumatic events including residual effects such as the exposure to cues that bring about memories of the traumatic event. The traumatic nature of being diagnosed with a disease, which to some individuals may be perceived as life-threatening, can consequently result in the development of PTSD. A cross-sectional study of individuals performed by Martin et al. suggested that receiving the diagnosis of HIV is considered a stressor that can lead to HIV-related PTSD [22]. As per the study, the lifetime rate of PTSD in HIV-positive patients was demonstrated as 54.1% (95% CI: 43.6–64.3%) and HIV-related PTSD as 40% (95% CI: 30.2–50.6% [22]. In a meta-analysis performed by Machtinger et al., it was estimated that HIV-positive women had a 30% rate of PTSD when compared to non-HIV infected women [23]. The studies of comorbid HIV/AIDS and PTSD have focused on a diverse demographic with various cultural and socioeconomic subgroups. Pain can also be a detrimental symptom and stressor accompanying any chronic condition. Smith et al. set out to determine if there was any association of pain as a common and pervasive symptom affecting PLWH and those with PTSD. In the study, 145 patients completed the self-report Stressful Life Events Checklist (SLE), Posttraumatic Stress Disorder Checklist-Civilian (PCL-C), the Mental Health Inventory (MHI), and the Brief Pain Inventory (BPI). Those with PTSD admitted to experiencing an increased pain intensity and pain disturbance with respect to activities of daily living [24]. Pain may be a contributing factor to the prognosis of the disease and it can be a burden impeding the clinical improvement of patients with HIV. HIV-related PTSD has also been associated with the development of MDD, as reported by Kelly et al., in homosexual plus bisexual men 6 months after the initial diagnosis of HIV infection. Kelly et al. used a modified DIS-III-R to suggest that a PTSD response has clinical validity, allowing research to further investigate on the reported population and its significance on the diagnosis and treatment of PTSD [25].

PTSD may not be isolated to specific traumatic events; it may also be experienced in patients with intense stigmatizing events or situations. Stigmatization, isolation, and a lack of social support are all risk factors for the development of major depressive disorder in PLWH. In a cross-sectional study by Breet et al., 210 HIV+ participants completed a self-reported questionnaire to assess HIV-related stigma [26]. The results displayed that social support had an integral role in the relationship between PTSD and the stigma related to HIV. The study concluded that although social support can be a protective factor in the reduction of PTSD, a perpetuation of the stigma by social support systems may still increase PTSD symptoms [26]. A separate study by Onyebuchi-Iwudibia et al., encompassing 105 participants from Eastern Nigeria, also used a questionnaire and the Berger HIV stigma scale and reported that 33.3% of patients were thought to have depressive symptoms with a positive correlation regarding the presence of stigma [27]. Both depression and stigma have been shown to correspond with an increased negative prognosis within this patient population [27]. The overall result may be a mood disorder exacerbated by PTSD and pain.

3.2 HIV Medication Related Neuropsychiatric Disorders

Combined antiretroviral therapy (cART) recommendations are in a constant state of flux due to the nature of the virus itself; cART may be comprised of nucleoside or nucleotide reverse transcriptase inhibitors (NRTI), nonnucleoside reverse transcriptase inhibitors (NNRTI) protease inhibitors, and integrase inhibitors. Both types of RTIs, inhibit viral reverse transcriptase and therefore inhibit DNA viral synthesis [28]. More specifically, they function by terminating the elongation of viral cDNA, this is achieved by binding to viral RNA-dependent DNA polymerase acting as false substrates [29]. NRTIs specifically lack an essential 3′-hydroxyl group motif whereas the NNRTIs cause allosteric inhibition [30]. Consequently, NRTIs/NNRTIs have been also found to inhibit DNA polymerase γ, which results in decreased mitochondrial DNA [31]. Polymerase γ is essential for mitochondrial repair and replication. This effectively results in the loss of membrane potential of mitochondria and capacity for oxidative phosphorylation, all of which induce stress [32, 33]. Studies performed on the mitochondrial side effects have not gone unnoticed by governing bodies; the World Health Organization (WHO) decided to remove stavudine from the list of first-line agents used in HIV therapy [34]. On top of their mitochondrial effects, cART drugs have been shown to cause significant central nervous system (CNS) neuronal damage [35]. The peripheral neurotoxicity, resulting in neuropathy is already a well-documented side effect of cART [36]. Akay et al. postulated that similar mechanisms may result in CNS damage. Their research focused on cART-inducing oxidative stress and neuronal damage by way of cellular stress. The study, which was conducted on rats and macaques in vivo, found that there was an accumulation of reactive oxygen species correlating to oxidative stress culminating in neuronal cell death [35]. The animal group treated with cART was found to have decreased synaptophysin expression in the hippocampus as well as decreased calmodulin kinase II, a marker of synaptodendritic integrity, in the frontal cortex of this same group [36]. Interestingly enough, the damage may be prevented with the addition of fumaric acid esters. Akay et al. discerned that prolonged cART regimen comprised of known neurotoxic/mitochondrial toxic agents over a prolonged period of time caused mounting damage. Campbell et al., by way of a meta-analysis, found evidence of a 4–6% volume reduction of the hippocampus in patients suffering major depressive disorder. The frontal cortex may also play a role in regulating mood and therefore the changes brought on by cART may cause mood dysfunctions by affecting both the frontal cortex and hippocampus [37]. The possible mood dysfunctions that are seen with cART are not just limited to the reasons. Fumaz et al. performed a cross-sectional study regarding the correlation between psychological stress and circulating IL-6 levels in HIV patients on cART. They reported a relationship between increased plasma IL-6 levels and psychological stress as well as depression and anxiety [38]. In another study, Shah et al. went further to propose that if negative neurocognitive performance can be associated with a high plasma IL-6 level then it is feasible that the inflammation caused by cART and comorbid conditions affecting the CNS may present with mood and behavior dysfunctions [39]. As proposed in the studies above, mood dysfunction and cognitive effects with cART therapy may be due to the summation of multiple causes and cannot be isolated as a singular cause.

3.2.1 Efavirenz

Efavirenz is a nonnucleoside reverse transcriptase inhibitor (NNRTI) that acts as a non-competitive inhibitor of reverse transcriptase [35]. Efavirenz is still prescribed medication for HIV therapy; it is not the preferred treatment regimen; that title is given to integrase inhibitors such as dolutegravir [40, 41]. Its known side effects of

PNS and CNS toxicity have been well studied. Although there is no definitive reason for the drug's CNS toxicity, several theories have been postulated, ranging from mitochondrial toxicity to the drug's ability to cross the blood-brain barrier (BBB) due to its lipophilic nature and small molecular size [40]. Efavirenz has also been used in pediatric antiretroviral therapy for HIV. Studies show that this drug may present with severe psychiatric sequelae brought about by the drug's use. De Wijer et al. performed an observational, cross-sectional study with respect to cART and its effects on school performance of children who have been treated for a minimum of 6 months [42]. Results indicated that the Efavirenz group had an overall lower performance mark than those in the non-efavirenz cART group, the overall difference in the percentage of students performing in each group was 30% and 22%, respectively [42]. It was also discovered that the efavirenz group had a higher incidence of behavioral problems when compared to their non-efavirenz counterparts, although the study determined that these were not significant. It was concluded that efavirenz-based treatment was found to be associated with a mild increase in neuropsychiatric symptoms, specifically mood dysfunction, particularly at higher doses when compared with non-efavirenz cART. Examples of these symptoms include increased internalizing problems as well as increases in depression, anxiety, and feelings of uneasiness. Although the study was performed in children with low risk of comorbid diseases causing the neurocognitive changes, it should be noted that different factors may affect cognitive performance in various patient age populations. For example, Meeker et at. suggest that cognitive decline may not just be isolated to the primary HIV disease and medication use but that it is secondary to systemic pathologic change. Changes involving the cardiovascular system, lipodystrophy, and diabetes in the PLWH have all been associated with decreased cognitive function.

Of note, efavirenz may also present with a psychosis manifestation secondary to increased plasma levels of the drug. The CYP system, specifically *CYP2B6* and *CYP3A4/5*, is responsible for the hepatic degradation of efavirenz [44]. As expected, any genetic condition or drug interaction that affects the CYP system should therefore decrease or increase plasma clearance of the drug. Theoretically, this can result in a greater incidence and severity of the known neuropsychiatric symptoms. Mugusi et al. performed a study examining this principle regarding *CYP* system changes and subsequent neuropsychiatric symptomatology [43]. His group performed a prospective cohort study on patients taking efavirenz-based cART with tuberculosis taking rifampin, a known *CYP450 3A* inducer. It was found that psychiatric symptoms were more common in patients on efavirenz-cART alone (incidence rate of 66.7%) versus those on cART plus rifampin [43]. Moving over to the opposite end of the spectrum and turning focus to *CYP* system inhibition, Hasse et al. describes a case report involving a homozygous *CYP2B6* G516T mutation [44]. As a result of excessively high plasma levels of efavirenz due to the mutation, the female patient presented with new-onset psychosis and no past medical history of mental illness before this mutation. These symptoms immediately resolved once the dosage was adjusted to account for her genetic mutation, despite her medications being dosed appropriately per guidelines [44].

3.2.2 Dolutegravir and Raltegravir

Dolutegravir and raltegravir are integrase inhibitors that function by limiting viral replication and latency by way of preventing HIV integration into host cell DNA [45, 46]. As mentioned in the previous section, integrase inhibitor-based cART is the first line in the treatment of HIV per federal guidelines for infected children and adults [41]. In a phase 3, double-blind study with adults commencing a cART regimen for the first time; dolutegravir was shown to be more efficacious than efavirenz. In addition, Walmsley et al., discovered fewer neuropsychiatric side effects with dolutegravir when compared to efavirenz [46]. Separate studies are in agreeance with the aforementioned finding, showing an improvement in patients with efavirenz-associated depression when changed to dolutegravir therapy; Hoffman et al. demonstrated a marked improvement pertaining to levels of depression, anxiety, and dizziness in their patients [47]. Despite this, it should still be noted that recent studies show a tendency for the dolutegravir to cause mood dysfunctions. Hoffman et al. performed a retrospective analysis of patients taking dolutegravir over 9 years and found that the most common side effect leading to discontinuation of the drug was sleep dysfunction; the neurocognitive decline was also reported with a decrease in concentration ability [47]. The discontinuation of therapy due to these side effects typically occurs between 3 and 6 months from the commencement of the medication. Mood dysfunction was also observed in the form of depression [47]. Scheper et al., describe two patients, without a previous psychiatric history, with severe depression as an adverse effect of their cART containing Dolutegravir [48]. The first patient began to feel "gloomy" and down 1 week after beginning therapy, progressing to paranoia and short-tempered outbursts. These symptoms resolved once the patient was switched to a therapy that did not contain dolutegravir. In the second case, the male patient was started on dolutegravir after experiencing insomnia with efavirenz containing cART. Two months after treatment, he endorsed mild depressive symptoms, which progressed to suicidal thoughts and ideations after 4 months of cART. As a result, his cART was modified to remove dolutegravir; the symptoms, including the depressed mood, resolved within 2 weeks. Quercia et al., meticulously analyzed data from four clinical trials involving 2634 patients with 1315 treated with dolutegravir to ascertain the commonality of mood dysfunctions such as depression, sleep changes, and anxiety [49]. The rates of mood dysfunctions were noticeably lower when directly compared to efavirenz-based treatment in the SINGLE study. In the SINGLE study, the percentage of those treated with dolutegravir who experienced insomnia was 17%, anxiety was 7%, depression was 8%, suicidal ideation was <1%, and nightmares/abnormal dreams was 10%. The number of efavirenz patients in that same study was as follows: insomnia was 12%, anxiety was 7%, depression was 11%, suicidal ideation was 2%, and nightmares/abnormal dreams were 21%. As previously mentioned, it is important to note the overall occurrence of mood symptoms was lower with dolutegravir cART.

Another integrase inhibitor, raltegravir, was also found to have similar side effects, although the incidence rate was lower than that of dolutegravir [49]. Raltegravir had a higher incidence of patients ceasing medication therapy due to depression, a total of 6 overall patients, compared to 1 patient on dolutegravir and seven on efavirenz [49]. Insomnia and abnormal dream activity was found to be the cause for the medical therapy cessation of 4 patients on dolutegravir and 10 patients on efavirenz. None of the raltegravir patients dropped the medication regimen due to insomnia. Neither raltegravir nor dolutegravir had patients cessation therapy due to anxiety, whereas efavirenz had four patients stop therapy [50].

Furthermore, the FDA performed a meta-analysis of five phase 3 trials to investigate further the reported link between these integrase inhibitors and neuropsychiatric symptoms [51]. The exact numbers for these symptoms regarding integrase inhibitor therapy were as follows: depression was 7.1%, sleep disorders were 0.4%, suicidal ideations were 0.6%, and memory dysfunction was 0.5%. The study subsequently found an overall decrease in the incidence of neuropsychiatric dysfunctions involving integrase inhibitors when directly compared to efavirenz-based cART. Of note, Harris et al. reported that raltegravir-based therapy was found to exacerbate depression in four cases [52]. Overall, incidences of depression brought on due to raltegravir therapy are more difficult to find. Dolutegravir research is still being conducted to fully understand the side effect profile of this drug along with the side effects of the other integrase inhibitors [53].

4 Conclusion

As previously mentioned, recent data shows the virus to directly damage the central and peripheral nervous systems, leading to a high prevalence of neuropsychiatric and neurocognitive disorders (including depression, anxiety, and mania), also termed NeuroAIDS. Of such conditions, mood disorders are among the most common mental health conditions in the general population suggesting a strong association between HIV neuropathology. Additionally, the use of cART medications in this population has been associated with various neuropsychiatric symptoms covered here. Of note, cART medications have also been associated with other psychiatric and neuropsychiatric symptoms not covered in this article; our focus was mood dysfunction–related therapy. A more complete review is pending, including other HIV-related neuropsychiatric conditions and cART CNS effects such as the previously mentioned psychosis, ototoxicity, optic neuropathy, and other neurotoxicity [54, 55]. The appearance of these symptoms may be a culmination of multifactorial causes such as an increase of pro-inflammatory cytokines, direct CNS toxicity, the alteration of astrocytes that line the blood-brain-barrier, mitochondrial dysfunction, oxidative stress, or other neuropathologic processes not well understood [56]. Interestingly, mitochondrial dysfunction and increased pro-inflammatory cytokines (such as IL-6), both of which are seen with cART, may result in decreased GABA adrenergic tone. This loss of tone is similar to what is seen in patients diagnosed with schizophrenia [57]. Further research on the pathophysiology of HIV neuropsychiatric complications and on the utilization of cART in the treatment of HIV-related psychiatric symptoms is warranted.

Conflicts of Interest The authors declare no conflicts of interest.

References

1. Ritchie H and Roser M. (2019) - "Mental Health". Published online at Ourworlddata.org Retrieved from https://ourworldindata.org/mental-health'.
2. HERT, M., CORRELL, C. U., BOBES, J., et al. Physical illness in patients with severe mental disorders. I. Prevalence, impact of medications and disparities in health care. World Psychiatry 2011 (10) 52–77
3. Parks J, Svendsen D, Singer P. et al. Morbidity and mortality in people with serious mental illness. Alexandria: National Association of State Mental Health Program Directors (NASMHPD) Medical Directors Council, 2006
4. McCombe JA, Noorbakhsh F, Buchholz C, Trew M, Power C. NeuroAIDS: a watershed for mental health and nervous system disorders. J Psychiatry Neurosci. 2009;34(2):83–85
5. Merrill JE, Chen IS. HIV-1, macrophages, glial cells, and cytokines in AIDS nervous system disease. FASEB J. 1991 Jul;5(10):2391–7
6. Kilbourne A.M., A.C Justice, L Rabeneck, M Rodriguez-Barradas, S Weissman, General medical and psychiatric comorbidity among HIV-infected veterans in the post-HAART era, Journal of Clinical Epidemiology, Volume 54, Issue 12, Supplement 1, 2001, Pages S22–S28
7. Prince M, Patel V, Saxena S, Maj M, Maselko J, Phillips MR, Rahman A. Lancet. 2007 Sep 8; 370(9590):859–77
8. American Psychiatric Association. Diagnostic and Statistical Manual of Mental Disorders. American Psychiatric Publishing; 2000. Text Revision
9. Symptoms of posttraumatic stress and death anxiety in persons with HIV and medication adherence difficulties. Safren SA, Gershuny BS, Hendriksen E. AIDS Patient Care STDS. 2003 Dec; 17(12):657–64.
10. Sherr L, Clucas C, Harding R, Sibley E, Catalan J. HIV and depression - a systematic review of interventions. Psychol Health Med. 2011 Oct; 16(5):493–527.
11. Fuchs D., A. Forsman, L. Hagberg, M. Larsson, G. Norkrans, G. Reibnegger, E.R. Werner, and H. Wachter. Decreased Tryptophan in Patients with HIV-1 Infection. Journal of Interferon Research. Jan 1990. Volume: 10 Issue 6: May 5, 2009
12. Morgan DG, May PC, Finch CE. Dopamine and serotonin systems in human and rodent brain: Effects of age and neurodegenerative disease. Journal of the American Geriatrics Society. 1987;35:334–345
13. Schroecksnadel K.,M. Sarcletti,C. Winkler,B. Mumelter,G. Weiss,D. Fuchs,G. Kemmler,R. Z angerle. Quality of life and immune activation in patients with HIV-infection. Brain, Behavior, and Immunity. Elsevier, August 2008
14. Gostner JM, Becker K, Kurz K and Fuchs D (2015) Disturbed amino acid metabolism in HIV: association with neuropsychiatric symptoms. Front. Psychiatry 6:97.
15. Keegan MR, Chittiprol S, Letendre SL, et al. Tryptophan Metabolism and Its Relationship with Depression and Cognitive Impairment Among HIV-infected Individuals. Int J Tryptophan Res. 2016;9:79–88. Published 2016 Oct 26.

16. Fernandez F, Ruiz P, editors. Psychiatric aspects of HIV/AIDS. Lippincott Williams and Wilkins; Philadelphia, PA: 2006
17. Sewell MC, Goggin KJ, Rabkin JG, Ferrando SJ, McElhiney MC, Evans S. Anxiety syndromes and symptoms among men with AIDS: a longitudinal controlled study. Psychosomatics. 2000 Jul-Aug; 41(4):294–300.
18. Robertson K., Carmen Bayon, Jean-Michel Molina, Patricia McNamara, Christiane Resch, Jose A. Muñoz-Moreno, Ranjababu Kulasegaram, Knud Schewe, Angel Burgos-Ramirez, Cristina De Alvaro, Esther Cabrero, Matthew Guion, Michael Norton & Jean van Wyk (2014) Screening for neurocognitive impairment, depression, and anxiety in HIV-infected patients in Western Europe and Canada, AIDS Care, 26:12, 1555–1561.
19. Etheldreda Nakimuli-Mpungu, M.B.Ch.B., M.Med.(Psych), Seggane Musisi, F.R.C.P.(C), Steven Kiwuwa Mpungu, M.B.Ch.B., M.Sc., and Elly Katabira, M.R.C.P., F.R.C.P. Primary Mania Versus HIV-Related Secondary Mania in Uganda. American Journal of Psychiatry 2006 163:8, 1349–1354
20. Lyketsos CG, Schwartz J, Fishman M, et al: AIDS mania. J Neuropsychiatry Clin Neurosci 1997; 9:277–279.
21. Dubé, B., Benton, T., Cruess, D. G., & Evans, D. L. (2005). Neuropsychiatric manifestations of HIV infection and AIDS. Journal of psychiatry & neuroscience: JPN, 30(4), 237–246
22. Martin, L. & Kagee, A. Lifetime and HIV-Related PTSD Among Persons Recently Diagnosed with HIV. AIDS and Behavior (2011) 15: 125.
23. Machtinger, E.L., Wilson, T.C., Haberer, J.E. et al. Psychological Trauma and PTSD in HIV-Positive Women: A Meta-Analysis. AIDS and Behavior (2012) 16: 2091.
24. Meredith Y Smith, Jennifer Egert, Gary Winkel, Jeffrey Jacobson. The impact of PTSD on pain experience in persons with HIV/AIDS. Pain 2002 (98:1–2), 9–17.
25. Kelly B., Raphael M.D., F. Judd, et al. Posttraumatic stress disorder in response to HIV infection. General Hospital Psychiatry 1998 (20:6), 345–352
26. Breet E., Ashraf Kagee, Soraya Seedat. HIV-related stigma and symptoms of post-traumatic stress disorder and depression in HIV-infected individuals: does social support play a mediating or moderating role?, AIDS Care 2014 (26:8) 947–951
27. Onyebuchi-Iwudibia O., Amy Brown. HIV and depression in Eastern Nigeria: The role of HIV-related stigma, AIDS Care 2014 (26:5) 653–657
28. Akay, C., Cooper, M., Odeleye, A. et al. Antiretroviral drugs induce oxidative stress and neuronal damage in the central nervous system. Journal of Neurovirology 2014 (20: 39) 39–53
29. Lutu, M. R., Nzuza, S., Mato, P. E., et al. DNA polymerase-γ hypothesis in nucleoside reverse transcriptase-induced mitochondrial toxicity revisited: A potentially protective role for citrus fruit-derived naringenin? European Journal of Pharmacology 2019 (852) 159–166.
30. Waller D., Sampson A., Reverse Transcriptase Inhibitors. Reverse Transcriptase Inhibitors. Medical Pharmacology and Therapeutics (Fifth Edition), 2018 (51)581–629
31. Lutu, M. R., Nzuza, S., Mato, P. E., et al. DNA polymerase-γ hypothesis in nucleoside reverse transcriptase-induced mitochondrial toxicity revisited: A potentially protective role for citrus fruit-derived naringenin? European Journal of Pharmacology 2019 (852) 159–166
32. Copeland W. C. The mitochondrial DNA polymerase in health and disease. Sub-cellular biochemistry 2010(50) 211–222
33. Manda, K. R., Banerjee, A., Banks, W. A., & Ercal, N. Highly active antiretroviral therapy drug combination induces oxidative stress and mitochondrial dysfunction in immortalized human blood–brain barrier endothelial cells. Free Radical Biology and Medicine 2011(50:7), 801–810.
34. Brennan, A. T., Davies, M. A., Bor, J. et al. Has the phasing out of stavudine in accordance with changes in WHO guidelines led to a decrease in single-drug substitutions in first-line antiretroviral therapy for HIV in sub-Saharan Africa?. AIDS 2017 (31:1), 147–157
35. Akay, C., Cooper, M., Odeleye, A. et al. Antiretroviral drugs induce oxidative stress and neuronal damage in the central nervous system. J. Neurovirol 2014 (20:1) 39–53

36. Guha, D., Wagner, M., & Ayyavoo, V. Human immunodeficiency virus type 1 (HIV-1)-mediated neuroinflammation dysregulates neurogranin and induces synaptodendritic injury. Journal of neuroinflammation 2018(15:1), 126
37. Campbell S, Macqueen G. The role of the hippocampus in the pathophysiology of major depression. J Psychiatry Neurosci. 2004 (29:6), 417–26.
38. Fumaz, C. R., Gonzalez-Garcia, M., Borras, X. et al. (2012). Psychological stress is associated with high levels of IL-6 in HIV-1 infected individuals on effective combined antiretroviral treatment. Brain, Behavior, and Immunity 2012(26:4), 568–572.
39. Shah, A., Gangwani, M. R., Chaudhari, N. S., et al. Neurotoxicity in the Post-HAART Era: Caution for the Antiretroviral Therapeutics. Neurotoxicity Research 2016, (30:4), 677–697.
40. Yilmaz, A., Watson, V., Dickinson, L., & Back, D. Efavirenz Pharmacokinetics in Cerebrospinal Fluid and Plasma over a 24-Hour Dosing Interval. Antimicrobial Agents and Chemotherapy 2012(56:9), 4583–4585.
41. Guidelines for the use of antiretroviral agents in pediatric HIV infection (2019), http://aidsinfo.nih.gov/guidelines. Accessed 2 May 2019.
42. Wijer, L. V., Mchaile, D. N., Mast, Q. D., et al. Neuropsychiatric symptoms in Tanzanian HIV-infected children receiving long-term efavirenz treatment: A multicentre, cross-sectional, observational study. The Lancet HIV 2019 (6:4).
43. Mugusi, S., Ngaimisi, E., Janabi, M., et al. Neuropsychiatric manifestations among HIV-1 infected African patients receiving efavirenz-based cART with or without tuberculosis treatment containing rifampicin. European journal of clinical pharmacology 2018 (74:11) 1405–1415.
44. Hasse B., H. F. Günthard, G. Bleiber, et al. Efavirenz Intoxication Due to Slow Hepatic Metabolism, Clinical Infectious Diseases Feb 2005 (40:3) e22–e23.
45. Kandel, C. E., & Walmsley, S. L. (2015). Dolutegravir - a review of the pharmacology, efficacy, and safety in the treatment of HIV. Drug design, development and therapy, 9, 3547–3555.
46. Walmsley SL, Antela A, Clumeck N, et al. SINGLE Investigators Dolutegravir plus abacavir-lamivudine for the treatment of HIV-1 infection. N Engl J Med. 2013;369
47. Hoffmann, C., Welz, T., Sabranski, M., Kolb, M., Wolf, E., Stellbrink, H., & Wyen, C. (2016). Higher rates of neuropsychiatric adverse events leading to dolutegravir discontinuation in women and older patients. HIV Medicine, 18(1), 56–63.
48. Scheper, H., van Holten, N., Hovens, J. G. F. M., & de Boer, M. G. J. (2018). Severe depression as a neuropsychiatric side effect induced by dolutegravir. HIV Medicine, 19(4), e58–e59.
49. Quercia R, Roberts J, Murungi A, et al. Psychiatric Adverse Events from the DTG ART-naïve Phase 3 Clinical Trials. HIV Drug Therapy. Glasgow, Abstract; 2016. p. 210, 23–6.
50. Yombi JC. Dolutegravir Neuropsychiatric Adverse Events: Specific Drug Effect or Class Effect. AIDS Rev. 2018;20(1):14–26.
51. Viswanathan P, Baro E, Soon G, et al. Neuropsychiatric Adverse Events As- sociated With Integrase Strand Transfer Inhibitors. Seattle WA (USA): confer- ence on Retroviruses and Opportunistic Infections (CROI); 2017. p. 13–6.
52. Harris, M., Larsen, G., & Montaner, J. S. (2008). Exacerbation of depression associated with starting raltegravir: A report of four cases. Aids, 22(14), 1890–1892.
53. Elzi, L., Erb, S., Furrer, H., Cavassini, M., Calmy, A., Vernazza, P., ... Swiss HIV Cohort Study Group (2017). Adverse events of raltegravir and dolutegravir. AIDS (London, England), 31(13), 1853–1858.
54. Jason Simdon, Dan Watters, Stephen Bartlett, Elizabeth Connick, Ototoxicity Associated with Use of Nucleoside Analog Reverse Transcriptase Inhibitors: A Report of 3 Possible Cases and Review of the Literature, Clinical Infectious Diseases, Volume 32, Issue 11, 1 June 2001, Pages 1623–1627
55. Luzhansky JZ, Pierce AB, Hoy JF, Hall AJ. Leber's hereditary optic neuropathy in the setting of nucleoside analogue toxicity. AIDS. 2001;15(12):1588–9.

56. Cohen, J., D'Agostino, L., Tuzer, F., & Torres, C. (2018). HIV antiretroviral therapy drugs induce premature senescence and altered physiology in HUVECs. Mechanisms of Ageing and Development, 175, 74–82.
57. Brand, S., Moller, M., & Harvey, B. (2015). A Review of Biomarkers in Mood and Psychotic Disorders: A Dissection of Clinical vs. Preclinical Correlates. Current Neuropharmacology, 13(3), 324–368.

Marijuana and HIV

Charurut Somboonwit, David Rutenburg, Asa Oxner, and Lynette Menezes

Abstract Marijuana (cannabis), a naturally occurring herb with two active constituents, D9-tetrahydrocannabinol (THC) and cannabidiol (CBD), has been recognized for its medicinal properties for centuries. Marijuana is the most used of all illicit substances. Marijuana use is more common among people living with HIV/AIDS (PLWH) than the general population. It has been shown to have some therapeutic effects in PLWH, including stimulating appetite, reducing anxiety, providing analgesia, and mitigating neurologic and behavioral health symptoms. However, it also has many negative neurocognitive effects.

Here, we describe how THC/CBD interacts with human metabolic and immunologic pathways and affect the efficiency of acquiring HIV infection, persistence of viral load, and chronic inflammation. We also discuss therapeutic applications of THC/CBD to several syndromes common in HIV patients, including HIV-associated neurologic decline, liver fibrosis and co-infection with viral hepatitis, cardiovascular disease, and behavioral health symptoms. Finally, we review the evidence of marijuana's impact on patient's adherence to their antiretroviral medications and the likelihood of achieving process measures along the HIV care continuum.

Keywords Marijuana · Cannabis · Cannabinoids · D9-tetrahydrocannabinol (THC) · Cannabidiol (CBD) · HIV · People living with HIV/AIDS (PLWH) ·

C. Somboonwit (✉) · L. Menezes
Department of Internal Medicine, Division of Infectious Diseases and International Medicine, University of South Florida, Morsani College of Medicine, Tampa, FL, USA
e-mail: charurut@usf.edu; lmenezes@usf.edu

A. Oxner
University of South Florida, Division of General Internal Medicine, Morsani Center for Healthcare, Tampa, FL, USA
e-mail: aoxner@usf.edu

D. Rutenburg
Department of Internal Medicine, Division of Pulmonary and Critical Care, Rutgers University, Newark, NJ, USA

C. Somboonwit et al. (eds.), *Global Virology IV: Viral Disease Diagnosis and Treatment Delivery in the 21st Century*,
https://doi.org/10.1007/978-3-031-57369-9_9

Inflammation · Immunity · Metabolism · Addiction · HIV-associated neurologic decline · Liver fibrosis and co-infection with viral hepatitis · Cardiovascular disease · Behavioral health symptoms

1 Introduction to Marijuana

Marijuana (cannabis) is one of the most widely used illicit substances globally. In 2021, an estimated 219 million people (4.3%) between the ages of 15 and 64 from 130 countries reported cannabis use at least once in the past year [1] compared to 60 million opioid users [1]. The oldest record of marijuana use dates back to 2727 BC, written by the Chinese Emperor Shen Nung [2], and it has been continually documented in Chinese medical literature for the past 1800 years [3]. Marijuana was introduced into Western medicine in the nineteenth century from the Indian subcontinent, where it was historically popular for its medical benefits as well as inebriation and psychological effects from marijuana and its addictive properties. Marijuana's soaring popularity is due to views of it being less risky than tobacco or alcohol. Its acceptance has influenced a societal trend of considering it relatively benign, changing traditional views on substance use. There is also a notable decline in concerns over marijuana dependency, suggesting a shift in how it is used and perceived regarding addiction, which reflects a considerable cultural change, with marijuana becoming recognized as a more acceptable and less dangerous option within different societies [4]**.** In the United States, marijuana was advertised for its analgesic, anti-inflammatory, anti-spasmodic, and sedative effects and remained recommended in Western medicine pharmacopeia until the 1920s [5]. Because other drugs with similar effects were discovered, marijuana use declined, and advertisements linking it with a Mexican name, marihuana, further discredited its use. By 1970, the United States Food and Drug Administration (FDA) classified it as a Schedule I drug, indicating a substance with a high potential for abuse and no accepted medical use [5].

More than 100 different cannabinoids and chemical ingredients have been identified in the cannabis plant [6]. D9-tetrahydrocannabinol (THC) and cannabidiol (CBD) are the two cannabinoids targeted for overexpression via selective cultivation [7]. Most products have a high concentration of THC, CBD, or both. 'Cannabis is classified' by species (i.e., 'Indica' vs. 'Sativa') and further by strain names like 'Jack Herer' or 'OG Kush'. However, the chemical composition, or "chemotype," is crucial for its effects and varies based on cultivation and processing conditions. Traditionally, smoking dried cannabis flowers was the most common method of use. Smoking remains a popular administration route, but edibles, vaporized concentrates, tongue strips, lingual sprays, and lozenges are all gaining popularity [7–10].

THC has psychoactive properties, while CBD does not. THC is partially agonistic to CB1 cannabinoid receptors and is responsible for dependence with long-term use [6]. THC causes acute cannabis intoxication, manifesting as euphoria, impaired cognition, anxiety or paranoia, and increased appetite [6]. In contrast, CBD lacks

Table 1 Cannabinoid receptors

	CB1 receptors	CB2 receptors
Location	Central nervous system, especially in regions associated with cognition, memory, motor coordination, and emotion Peripheral tissues, to a lesser extent	Peripheral tissues, especially cells associated with the immune system Immune cells like macrophages and microglia Less abundant in the CNS compared to CB1 receptors
Function	Neuromodulate causing neurotransmitter release in the brain Play roles in pain perception, mood regulation, appetite, and memory	Regulate immune responses and functioning Play a role in inflammation and pain management Activation of CB2 receptors does not produce the psychoactive effects associated with CB1 receptor activation
Physiological effects	Activation of CB1 receptors causes psychoactive effects, altered cognition, and changes in motor control	Activation of CB2 receptors influences immune responses, leading to anti-inflammatory effects, and does not produce the psychoactive effects associated with CB1 receptor activation
Therapeutic potential	CB1 receptor agonists or antagonists are explored for treating various CNS disorders, including pain, anxiety, and neurodegenerative diseases	CB2 receptor agonists are being studied for treating inflammatory diseases, pain, and even certain types of cancer due to their modulatory effects on the immune system

Adapted from Mackie K [12]

the psychoactive properties caused by THC. Instead, it binds to non-CB1 receptors, such as GPR55, TRPV1, and 5-HT1A, allowing for a more acceptable risk/benefit profile for its therapeutic effects in inflammation, autism, seizures, and anxiety/posttraumatic stress disorder [6, 11] (Table 1).

The CB1 and CB2 receptors are integral components of the endocannabinoid system but have distinct characteristics and functions. This table shows the differences between CB1 and CB2 receptors, which are crucial for developing targeted therapies that can modulate the endocannabinoid system without unwanted side effects, particularly those associated with the psychoactive properties of CB1 receptor activation.

Cannabis has variable pharmaco-kinetics depending on the intake route. THC is primarily metabolized in the liver through hydroxylation by the cytochrome P450 (CYP 450) enzymes, particularly CYP 450 2C9, 2C19, and 3A4. This process generates a potent psychoactive metabolite, 11-OH-THC. Over 100 THC metabolites are produced, including various hydroxy compounds, ketones, aldehydes, and carboxylic acids. Besides the liver, THC is metabolized in tissues like the brain, intestines, and lungs, each utilizing different hydroxylation pathways. The lung contributes significantly to THC's side-chain hydroxylation. When consumed orally, THC reaches peak plasma concentration after 1 h with a half-life of up to 30 h. It undergoes first-pass metabolism in the liver by the CYP 450 enzyme system, where 11-OH-THC is produced [6]. When inhaled, the cannabinoids are rapidly absorbed into the bloodstream with a peak concentration within minutes and then degraded within 30 min, forming a less psychoactive metabolite. Cannabinoids can interact with the hepatic CYP 450 enzyme system and activate CYP3A4. After repeated doses, some cannabinoids can induce interaction with other drugs metabolized by CYP 450 [5, 13]. Alternatively, in vitro, CBD has shown to be an inhibitor of the CYP 450 enzyme system [6].

2 Epidemiology and Marijuana Use Among People Who Live with HIV (PLWH)

In 2022, 39 million people were living with HIV worldwide [14]. In the United States, strong HIV prevention efforts, including, but not limited to, access to testing, HIV care, condoms, sterile syringes, adult male circumcision, and HIV prevention education programs, have led to a decline in the annual number of new cases [15]. Moreover, a significant number of people living with HIV now experience improved survival and quality of life because of the widespread availability of antiretroviral therapy (ART). However, many challenges, such as poor access to healthcare and low socioeconomic status, continue to impact this vulnerable population now living with a chronic infection. Disparity depends further on when this population has a relationship with substances. The relationship between doctors and patients who use drugs is complex and marked by mutual mistrust. Physicians often feel deceived and face challenges in treating pain and withdrawal in these patients, leading to inconsistent care and poor communication. In turn, it makes patients feel mistreated or biased against, deepening the cycle of mistrust. The problem is worsened by societal views on addiction, affecting both treatment and medical training [16].

Cannabis use is more common among HIV-infected individuals than the general population [17]. Globally, West and Central Africa, as well as sub-Saharan Africa, share a large burden of HIV/AIDS, with 6.1 million people living with HIV [18]. This region also has the highest prevalence of cannabis use at 13.2% (34.3 million users) [1]. In the United States, one study of more than 3000 HIV-infected adults across four US cities found that nearly one in four (24.3%) reported marijuana use in the 3 months prior to the study [17]. In contrast, 5.6% of adults 26 years and older reported cannabis use in the past month in a national prevalence survey of the general population by SAMHSA [19]. In a smaller study among 703 HIV-infected adults from seven community health centers in Florida, cannabis use in the past 3 months was much higher at 33.2% [20]. In Canada, 43% of HIV-infected individuals reported cannabis use [21, 22]. Socioeconomically, cannabis users were more likely to report lower household income and injection drug use and less likely to report "excellent" health [21, 22].

Recent evidence suggests that both recreational and medicinal use of cannabis is on the rise. Across the lifespan of the general population, the prevalence of cannabis use in the past year was highest among adolescents between the ages of 18–25 years (34.8% users) [23]. In a separate analysis of participants from the 2002–2014 National Survey on Drug Use, Mauro, et al. found the prevalence of daily (>300 days/year) and nondaily cannabis adult users steadily increased since 2007 in comparison to a declining trend in daily use among adolescents 12–17 years old [24]. Cannabis use varies by gender and pregnancy status [23]. Men were 1.5 times more likely than women to use cannabis. Pregnant women had lower odds of using cannabis than nonpregnant women. Cannabis use decreased further in the third trimester [23].

Prevalence of marijuana use is commonly found in PLWH. In the United States, 23% of PLWH reported marijuana use, with the major reasons being reducing depression/anxiety and stimulating appetite [25]. In Canada, 43% of PLWH admitted use of marijuana, with 63% of these people using marijuana for health purposes, specifically to relieve HIV-associated symptoms, including nausea, anxiety, pain, and depression [21, 22]. Compared to placebo, marijuana reduced pain caused by medication-induced peripheral neuropathy by 34% [26]. Another reason PLWH use marijuana is to mitigate the side effects of ART, mainly nausea and vomiting, and thus to increase adherence to the medication. In one study, smoking marijuana had the greatest positive effect of 3.3 times more adherence to ART than PLWH, who did not smoke marijuana [27].

3 Marijuana and Substance Use Disorder

Cannabis use disorder (CUD) is a prevalent but underrecognized and undertreated disorder [28]. The Diagnostic and Statistical Manual of Mental Disorders (DSM-5) diagnostic criteria for CUD must include a minimum of 2 of 11 symptoms in a 12-month period, such as craving, need of increased amount for intoxication or desired effect, lack of control, withdrawal, and deleterious effects on social activities and personal and professional responsibilities. CUD is rated on a scale of mild, moderate, or severe based on the number of symptoms [28].

In the 2019 global burden of disease study, the prevalence of CUD cases was 23.8 million, with males (15.63 million) representing twice the prevalence of females (8.61 million). Adolescents between the ages of 20–24 had the highest prevalence of CUD among cannabis users [29]. Recent US data reveal that 19.5% of lifetime cannabis users were diagnosed with CUD [28]. Of these, 23% suffered from severe forms of CUD (meeting six or more diagnostic criteria) [28]. One study comparing users during the periods 2001–2002 and 2012–2013 found nearly a two-fold increase in CUDs from 1.5% to 2.9%, respectively [30]. Adult men, individuals with less than a college education, low-income individuals, and those living in urban areas were more likely to experience CUD [30]. Studies have shown a link between high-potency formulations of cannabis products and a higher risk for CUD [8, 31].

Co-morbid psychiatric conditions (mood disorders, schizophrenia, and anxiety disorders) [32] and other substance use disorders such as alcohol use [33] and tobacco use [30], prescription stimulant use, and cocaine use disorders [34] are common with cannabis use and cannabis use disorder. Other published reports indicate individuals with posttraumatic stress disorder [33], obsessive-compulsive disorder [35], personality disorders [33], and attention deficit hyperactivity disorder (ADHD) [36] are more likely to have used cannabis and have cannabis use disorders than those without these psychiatric conditions. Overall, although cannabis use has been associated with other psychiatric and substance use disorders, previous cross-sectional and prospective studies have not demonstrated a relationship with either serious or chronic medical conditions or death from medical conditions [37, 38].

Although a controversial subject, evidence now suggests that chronic cannabis use can result in dependency. In fact, roughly 1 in 11 who initiate cannabis use will become addicted. The percentage of those addicted increases to 16% if cannabis use is initiated during early adolescence and up to 50% among daily cannabis smokers [39]. Additionally, there is clinical recognition of a "cannabis withdrawal syndrome" presenting with anxiety, irritability, sleep issues, dysphoria, and craving that can result in relapse [40].

4 Marijuana and HIV Immunology

HIV infection results in chronic inflammation, and ongoing immune activation is likely causal to the morbidity of HIV despite treatment with antiretrovirals [41]. THC and CBD have demonstrated numerous actions on the inflammatory cascade via modified cytokine production, induced apoptosis, and reduced immune cell populations, including specific antigen-presenting cells and T cells expressing certain human leukocyte antigens (HLA) [42–44]. Overall, cannabis reduces the likelihood of HIV replication within human HIV reservoirs by limiting generalized immune activation. Cannabis also has direct anti-HIV effects on T cell receptors [45].

Research on mechanisms of cannabinoid modulation of the immune system and inflammatory cascade show mixed pro-inflammation and anti-inflammation effects. For example, THC enhances tumor necrosis factor release and mimics chronic inflammatory states such as HIV infection [46] and has been shown to increase HIV replication in one mouse model [47]. By contrast, in mice models and simian immunodeficiency virus (SIV) infected macaques, THC reduces viral load, disease progression, and tissue inflammation in colonic tissue [48–51]. These authors found associations with persistently inhibited IL-2 [48] or THC-activated microRNA, which exerted an epigenomic regulation over inflammatory cytokines when immune cells were exposed to THC prior to being exposed to HIV [49–52].

Studies have evaluated the effects of marijuana on CD4 population counts, but further immunological studies are needed to investigate any possible effect of cannabis on CD4 cells' functional properties [53, 54]. Abrams et al. evaluated the CD4 count in a randomized, placebo-controlled study. They found no adverse effects on CD4 count in marijuana users compared to nonusers, but their study was limited by a 21-day measurement period [55]. On the contrary, other studies have found lower populations of CD38+ CD4+ cells and CD8+ cells in patients who were heavy cannabis users compared to noncannabis users [43] and decreased populations of pro-inflammatory monocytes displaying CD14 and CD16 HLA antigens in heavy cannabis users [44].

Cannabinoids can also directly inhibit HIV entry into human T cells by interfering with the signal transduction of HIV's CXCR4 receptor [56]. Both THC and CBD have been shown to interfere with HIV cell entry, and the effect seems to be mediated by binding to the CBD2 receptor [52, 56]. Additionally, activation of the CB2 cannabinoid receptors is associated with reduced efficiency of HIV replication within host immune cells, indicating a potential second mechanism by which marijuana could be anti-inflammatory in the setting of HIV [57, 58].

5 Health Effects of Marijuana

Marijuana and other cannabinoids have been used to relieve somatic disease symptoms, including glaucoma, chemotherapy-associated nausea, chronic pain, neuropathic pain in multiple sclerosis, and epilepsy for centuries. Among PLWH, it has been an appetite stimulant for treating AIDS wasting syndrome [59–63]. However, there is limited data on other health effects of cannabis in PLWH.

5.1 Neurocognitive Function and NeuroAIDS

The effects of THC on brain development and cerebral executive functioning are well-studied in non-HIV patients. Exposure of the brain to THC during prenatal development can lead to recalibration of the reward system with increased sensitivity to other drugs and interference with axonal connectivity due to decreased cytoskeletal dynamics [64]. Similarly, marijuana exposure in adolescence is associated with declining IQ due to reduced axonal connectivity. In adolescents, the greatest reductions in axonal synapses occur in the precuneus (a key area for highly integrative thought processing) and fimbria (active in learning and memory) [65–67]. These findings suggest that certain brain regions might be more vulnerable to the long-term effects of marijuana, which is consistent with another study that used imaging in chronic cannabis users vs. nonusers to show reduced activity and volume of neurons in the prefrontal and hippocampal regions [68]. Overall, the effects of THC on neurocognitive function are reduction in executive function, inhibitory controls, and processing due to culling of neuron connectivity.

The neurocognitive consequences of cannabis use among PLWH are less well-known. Studies show marijuana use exacerbates or accelerates HIV-associated cognitive deficits, the greatest being memory impairment, because HIV infiltrates the central nervous system (CNS). The memory system is particularly vulnerable to outside insults, including HIV and cannabis [69, 70]. Due to the additive effects of HIV on the CNS, both past and current cannabis use has been shown to negatively impact scores on complex motor tasks [71]. The effects of HIV appear to be seen in those experiencing symptoms of HIV because they no longer have neuronal reserves that can be relied upon [70]. The CNS HIV Antiretroviral Therapy Effects Research (CHARTER) study found that 52% of HIV-infected persons have a neurocognitive impairment, with the greatest deficits in learning, memory, attention, and executive functioning [72, 73]. Given this existing insult from HIV, one study found poorer neurocognitive performance among HIV patients for each day of marijuana use and years living with HIV [74].

More research is needed to show how marijuana use impacts neurocognitive functioning among PLWH. Specific areas lacking in the current literature include the effects of marijuana on PLWH, who already have neurocognitive decline, and research specific to the CB1 receptors in the setting of HIV, which are known to mediate psychotropic phenotypes of marijuana use. Currently, data about the neurologic effects of marijuana use in PLWH are limited due to sample diversity, high polysubstance abuse confounder effect, and the use of episodic memory assessments [19, 70].

5.2 *Liver Fibrosis and Viral Hepatitis*

Marijuana smoking is common among HIV/hepatitis C virus (HCV) coinfected persons, and most studies of the effects of marijuana on liver fibrosis in PLWH include HIV/HCV coinfected patients. Several studies evaluating the association between marijuana use and the progression of liver fibrosis found no evidence that marijuana promotes fibrosis or cirrhosis in patients with HIV/HCV coinfection [75–77].

Interestingly, the molecular effects of CB1 and CB2 receptors on liver fibrosis are contradictory. THC binding to CB1 receptors in the liver results in a profibrogenic phenotype due to increased expression of transforming growth factor β–1, increased stellate cell production, and decreased apoptosis [78]. Additionally, CB1 receptor activation promotes hepatic fatty acid production and decreases adiponectin (a normally antifibrotic cytokine), mediating fibrosis in nonalcoholic fatty liver disease [79–81]. On the other hand, CBD binding to CB2 receptors in the liver promotes antifibrogenic effects via decreased macrophage oxidative stress, suppressed cytotoxic T cells, and regulation of tumor necrosis factor and interleukin-1 towards an anti-inflammatory phenotype [82–84].

Daily cannabis use may be a protective factor against steatosis in HIV/HCV co-infected patients [85, 86]. A meta-analysis of 5 million patients found a reduction in the prevalence of nonalcoholic steatohepatitis in marijuana users [87]. Studies posit the mechanism for protection against hepatic steatosis includes lower insulin resistance among cannabis users who are HIV/HCV coinfected [86, 87]. This protection of marijuana against steatosis, which is likely mediated by CB2 receptor activation, warrants further research into potential therapeutics for PLWH and those with non-alcoholic fatty liver disease, the leading cause of cirrhosis in the United States.

5.3 Cardiovascular Disease

Heavy marijuana use in PLWH was independently associated with higher cardiovascular events after adjusting for age, tobacco exposure, viral load, and cardiac risk factors [88]. Marijuana use has also been associated with myocardial infarction and cerebrovascular events. The proposed mechanisms are from cannabinoids directly increasing vascular resistance in two different ways. One is decreasing coronary microcirculation by binding to CB1 receptors in arterial blood vessels. The other mechanism is by cannabinoids acting indirectly via effects on vasoactive mediators, such as nitric oxide [89, 90]. In PLWH, independent of marijuana, there is a 1.5–2-fold relative risk increase in coronary and cerebrovascular events, likely due to inflammatory and immune-mediated pathways caused by HIV. Further research must be done to study the potential negative coronary and cerebrovascular outcomes of marijuana use in PLWH.

The cardiac effects of marijuana are not limited to the vascular system. Marijuana can cause electrophysiologic abnormalities leading to conduction problems, namely atrial fibrillation. It can reduce the amplitude of P-waves on the electrocardiogram, decrease the sinoatrial conduction time, shorten the A-H interval time, and lower the mean atrioventricular conduction time. By means of these mechanisms, the risk of atrial fibrillation increases [91].

5.4 Lung Cancer

The effects of long-term marijuana smoking on the risk of lung cancer are unclear. The largest study found no association between marijuana smoking and lung cancer after adjustment for tobacco smoking [92, 93].

5.5 Behavioral Health, Adherence, and Effects on the HIV Care Continuum

Marijuana use can exacerbate symptoms of anxiety and depression, which are common among PLWH [94, 95]. There is a strong association between cannabis and psychoses of various types. In fact, psychotic symptoms in marijuana smokers are more severe among heavy users and those who start smoking at a younger age [96]. This association is so profound that adolescent cannabis use is considered a preventable risk factor for psychosis [93]. The risk of developing schizophrenia after cannabis use is highest in those with a genetic predisposition to schizophrenia but persists after controlling for family history [93]. This relationship is complicated since having a diagnosis of behavioral or mood disorders also predicts marijuana use [97]. Another behavioral health disorder that is common in PLWH is insomnia. However, a recent study found marijuana did not negatively impact sleep in PLWH, although marijuana was associated with poor sleep scores in HIV-negative participants [98]. A systematic review reported the most common side effects of cannabis use to be sleepiness, concentration difficulties, and thirst [99]. More research is needed on the relationships between insomnia, marijuana use, and HIV infection.

There are concerns about a link between substance use, engagement in risky sexual behaviors, and acquisition of HIV. Both marijuana and alcohol have been found to synergistically increase the likelihood of sexual encounters and acceptability of unprotected sex with a main partner [100] and to independently increase the overall likelihood to engage in sexual intercourse [100, 101]. However, studies showing an increased prevalence of sexually transmitted infections, including HIV, in marijuana users versus nonusers are lacking.

Reducing gaps along the HIV care continuum has become important for implementation and public health scientists. The first step in the care continuum is a diagnosis of all infected persons and positivity awareness. A study of young black men who have sex with men found heavy marijuana users were more likely to be unaware of their HIV status [102]. Other studies have investigated the relationship between daily cannabis use and adherence to optimal ART therapy or retention in care measured by missed HIV clinic appointments. The data is mixed, with one study finding no compromise in adherence to ART [103], while other studies show reduced adherence to ART regimens [104, 105] and more missed HIV clinic appointments [106]. The final care continuum step is suppression of viral load, which was studied as part of the Multicenter AIDS Cohort Study (MACS) that found marijuana use did not have any clinically significant adverse effects on CD4 count or viral load [107].

6 Conclusion

The active cannabinoids, THC and CBD, in marijuana have several promising medicinal uses including immunomodulators, with the potential for targeted drug development in the fields of HIV, hepatitis (both viral hepatitis and the immune dysfunction of nonalcoholic steatohepatitis), cardiovascular diseases mediated by inflammation within the microvasculature, and multiple sclerosis.

The lack of psychogenic effects of CBD makes it more appealing for development as a pure cannabinoid therapeutic. Multiple studies have found the side effects of marijuana to be minor, apart from psychogenic effects.

The main limitations in current research include an abundance of short-term study designs, minimal functional studies in vivo, and high confounder potential from other co-exposures, socio-behavioral factors of marijuana users, and drug-disease confounding, such as schizophrenia and marijuana abuse.

In most states in the United States, in the District of Columbia, Guam, Puerto Rico, and the US Virgin Islands, recreational and medicinal marijuana was recently legalized. Additionally, HIV/AIDS is an approved indication for medical marijuana use; this invites concern for the general population health effects of marijuana on HIV control and overall physiologic and psychiatric health. Marijuana may have important therapeutic benefits for individual HIV-infected patients to alleviate symptoms, mitigate their co-morbid diseases, and augment quality of life. However, on a population-wide level, marijuana may increase gaps in the HIV care continuum. Marijuana may have negative neurologic impacts in the areas of memory and executive functioning, which may affect ART adherence and increase high-risk sexual behavior and further viral disease spread. Cardiovascular abnormalities have also been reported. Further research is needed to understand the physiologic and biochemical effects of marijuana in PLWH.

Acknowledgments Brian Giunta, MD, PhD, for additional references.

Authors Contributions Manuscript writing: all authors.

Conflicts of Interest None

References

1. United Nations Office on Drugs and Crime (UNODC), *World Drug Report 2023*. 2023: United Nations publication, 2023.
2. Drug Enforcement Administration (DEA) *Cannabis, Coca, & Poppy: Nature's Addictive Plants*.
3. Brand, E.J. and Z. Zhao, *Cannabis in Chinese Medicine: Are Some Traditional Indications Referenced in Ancient Literature Related to Cannabinoids?* Front Pharmacol, 2017. **8**: p. 108.
4. National Center for Drug Abuse Statistics. *Drug Abuse Statistics*. 2023 [cited 2023 November 28].
5. Abrams, D.I. and M. Guzman, *Cannabis in cancer care*. Clin Pharmacol Ther, 2015. **97**(6): p. 575–86.
6. Huestis, M.A., *Human cannabinoid pharmacokinetics*. Chem Biodivers, 2007. **4**(8): p. 1770–804.
7. ElSohly, M.A., et al., *Changes in Cannabis Potency Over the Last 2 Decades (1995–2014): Analysis of Current Data in the United States*. Biol Psychiatry, 2016. **79**(7): p. 613–9.
8. Spindle, T.R., M.O. Bonn-Miller, and R. Vandrey, *Changing landscape of cannabis: novel products, formulations, and methods of administration*. Curr Opin Psychol, 2019. **30**: p. 98–102.
9. Sexton, M., et al., *Evaluation of Cannabinoid and Terpenoid Content: Cannabis Flower Compared to Supercritical CO2 Concentrate*. Planta Med, 2018. **84**(4): p. 234–241.
10. Russo, E.B., *Taming THC: potential cannabis synergy and phytocannabinoid-terpenoid entourage effects*. Br J Pharmacol, 2011. **163**(7): p. 1344–64.

11. Devinsky, O., et al., *Trial of Cannabidiol for Drug-Resistant Seizures in the Dravet Syndrome.* N Engl J Med, 2017. **376**(21): p. 2011–2020.
12. Mackie, K., *Cannabinoid receptors: where they are and what they do.* J Neuroendocrinol, 2008. **20 Suppl 1**: p. 10–4.
13. Engels, F.K., et al., *Medicinal cannabis does not influence the clinical pharmacokinetics of irinotecan and docetaxel.* Oncologist, 2007. **12**(3): p. 291–300.
14. World Health Organization (WHO). *HIV and AIDS.* 2023 July 13, 2023 September 8, 2023]; Available from: https://www.who.int/news-room/fact-sheets/detail/hiv-aids.
15. Division of HIV Prevention, N.C.f.H., Viral Hepatitis, STD, and TB Prevention, Centers for Disease Control and Prevention. *Effectiveness of Prevention Strategies to Reduce the Risk of Acquiring or Transmitting HIV.* 2022 June 17, 2022 [cited 2023 September 20, 2023]; Available from: https://www.cdc.gov/hiv/risk/estimates/preventionstrategies.html.
16. Merrill, J.O., et al., *Mutual mistrust in the medical care of drug users: the keys to the "narc" cabinet.* J Gen Intern Med, 2002. **17**(5): p. 327–33.
17. Mimiaga, M.J., et al., *Substance use among HIV-infected patients engaged in primary care in the United States: findings from the Centers for AIDS Research Network of Integrated Clinical Systems cohort.* Am J Public Health, 2013. **103**(8): p. 1457–67.
18. AVERT. *HIV and AIDS in West and Central Africa Overview.* 2018 10 December 2018 [cited 2019 May 26]; Available from: https://www.avert.org/hiv-and-aids-west-and-central-africa-overview.
19. Skalski, L.M., et al., *Memory Impairment in HIV-Infected Individuals with Early and Late Initiation of Regular Marijuana Use.* AIDS Behav, 2018. **22**(5): p. 1596–1605.
20. Mannes, Z.L., et al., *The association of therapeutic versus recreational marijuana use and antiretroviral adherence among adults living with HIV in Florida.* Patient Prefer Adherence, 2018. **12**: p. 1363–1372.
21. Cinti, S., *Medical marijuana in HIV-positive patients: what do we know?* J Int Assoc Physicians AIDS Care (Chic), 2009. **8**(6): p. 342–6.
22. Furler, M.D., et al., *Medicinal and recreational marijuana use by patients infected with HIV.* AIDS Patient Care STDS, 2004. **18**(4): p. 215–28.
23. Center for Behavioral Health Statistics and Quality, *2017 National Survey on Drug Use and Health: Detailed Tables.* 2018, Rockville, MD: Substance Abuse and Mental Health Services Administration.
24. Mauro, P.M., et al., *Age Differences in Daily and Nondaily Cannabis Use in the United States, 2002–2014.* J Stud Alcohol Drugs, 2018. **79**(3): p. 423–431.
25. Prentiss, D., et al., *Patterns of marijuana use among patients with HIV/AIDS followed in a public health care setting.* J Acquir Immune Defic Syndr, 2004. **35**(1): p. 38–45.
26. Abrams, D.I., et al., *Cannabis in painful HIV-associated sensory neuropathy: a randomized placebo-controlled trial.* Neurology, 2007. **68**(7): p. 515–21.
27. de Jong, B.C., et al., *Marijuana use and its association with adherence to antiretroviral therapy among HIV-infected persons with moderate to severe nausea.* J Acquir Immune Defic Syndr, 2005. **38**(1): p. 43–6.
28. Hasin, D.S., et al., *Prevalence and Correlates of DSM-5 Cannabis Use Disorder, 2012–2013: Findings from the National Epidemiologic Survey on Alcohol and Related Conditions-III.* Am J Psychiatry, 2016. **173**(6): p. 588–99.
29. Shao, H., et al., *Trends of the Global Burden of Disease Attributable to Cannabis Use Disorder in 204 Countries and Territories, 1990–2019: Results from the Disease Burden Study 2019.* Int J Ment Health Addict, 2023: p. 1–23.
30. Hasin, D.S., et al., *Prevalence of Marijuana Use Disorders in the United States Between 2001–2002 and 2012–2013.* JAMA Psychiatry, 2015. **72**(12): p. 1235–42.
31. van der Pol, P., et al., *Cross-sectional and prospective relation of cannabis potency, dosing and smoking behaviour with cannabis dependence: an ecological study.* Addiction, 2014. **109**(7): p. 1101–9.

32. Martins, S.S. and D.A. Gorelick, *Conditional substance abuse and dependence by diagnosis of mood or anxiety disorder or schizophrenia in the U.S. population.* Drug Alcohol Depend, 2011. **119**(1–2): p. 28–36.
33. Kerridge, B.T., et al., *DSM-5 cannabis use disorder in the National Epidemiologic Survey on Alcohol and Related Conditions-III: Gender-specific profiles.* Addict Behav, 2018. **76**: p. 52–60.
34. Hayley, A.C., C. Stough, and L.A. Downey, *DSM-5 cannabis use disorder, substance use and DSM-5 specific substance-use disorders: Evaluating comorbidity in a population-based sample.* Eur Neuropsychopharmacol, 2017. **27**(8): p. 732–743.
35. Teesson, M., et al., *Prevalence, correlates and comorbidity of DSM-IV Cannabis Use and Cannabis Use Disorders in Australia.* Aust N Z J Psychiatry, 2012. **46**(12): p. 1182–92.
36. De Alwis, D., et al., *Attention-deficit/hyperactivity disorder subtypes and substance use and use disorders in NESARC.* Addict Behav, 2014. **39**(8): p. 1278–85.
37. Degenhardt, L., et al., *The global epidemiology and contribution of cannabis use and dependence to the global burden of disease: results from the GBD 2010 study.* PLoS One, 2013. **8**(10): p. e76635.
38. Calabria, B., et al., *Does cannabis use increase the risk of death? Systematic review of epidemiological evidence on adverse effects of cannabis use.* Drug Alcohol Rev, 2010. **29**(3): p. 318–30.
39. Hall, W. and L. Degenhardt, *Adverse health effects of non-medical cannabis use.* Lancet, 2009. **374**(9698): p. 1383–91.
40. Gorelick, D.A., et al., *Diagnostic criteria for cannabis withdrawal syndrome.* Drug Alcohol Depend, 2012. **123**(1–3): p. 141–7.
41. Costiniuk, C.T., et al., *Oral cannabinoids in people living with HIV on effective antiretroviral therapy: CTN PT028-study protocol for a pilot randomised trial to assess safety, tolerability and effect on immune activation.* BMJ Open, 2019. **9**(1): p. e024793.
42. Nagarkatti, P., et al., *Cannabinoids as novel anti-inflammatory drugs.* Future Med Chem, 2009. **1**(7): p. 1333–49.
43. Manuzak, J.A., et al., *Heavy Cannabis Use Associated With Reduction in Activated and Inflammatory Immune Cell Frequencies in Antiretroviral Therapy-Treated Human Immunodeficiency Virus-Infected Individuals.* Clin Infect Dis, 2018. **66**(12): p. 1872–1882.
44. Rizzo, M.D., et al., *HIV-infected cannabis users have lower circulating CD16+ monocytes and IFN-gamma-inducible protein 10 levels compared with nonusing HIV patients.* Aids, 2018. **32**(4): p. 419–429.
45. Henriquez, J.E., et al., *Interferon-α-Mediated Activation of T Cells from Healthy and HIV-Infected Individuals Is Suppressed by Δ(9)-Tetrahydrocannabinol.* J Pharmacol Exp Ther, 2018. **367**(1): p. 49–58.
46. Williams, J.C., et al., *Delta(9)-Tetrahydrocannabinol (THC) enhances lipopolysaccharide-stimulated tissue factor in human monocytes and monocyte-derived microvesicles.* J Inflamm (Lond), 2015. **12**: p. 39.
47. Roth, M.D., et al., *Tetrahydrocannabinol suppresses immune function and enhances HIV replication in the huPBL-SCID mouse.* Life Sci, 2005. **77**(14): p. 1711–22.
48. Massi, P., et al., *Immune function alterations in mice tolerant to delta9-tetrahydrocannabinol: functional and biochemical parameters.* J Neuroimmunol, 1998. **92**(1–2): p. 60–6.
49. Molina, P.E., et al., *Cannabinoid neuroimmune modulation of SIV disease.* J Neuroimmune Pharmacol, 2011. **6**(4): p. 516–27.
50. Molina, P.E., et al., *Cannabinoid administration attenuates the progression of simian immunodeficiency virus.* AIDS Res Hum Retroviruses, 2011. **27**(6): p. 585–92.
51. Chandra, L.C., et al., *Chronic administration of Delta9-tetrahydrocannabinol induces intestinal anti-inflammatory microRNA expression during acute simian immunodeficiency virus infection of rhesus macaques.* J Virol, 2015. **89**(2): p. 1168–81.

52. Williams, J.C., et al., *Delta(9)-Tetrahydrocannabinol treatment during human monocyte differentiation reduces macrophage susceptibility to HIV-1 infection.* J Neuroimmune Pharmacol, 2014. **9**(3): p. 369–79.
53. Eisenstein, T.K. and J.J. Meissler, *Effects of Cannabinoids on T-cell Function and Resistance to Infection.* J Neuroimmune Pharmacol, 2015. **10**(2): p. 204–16.
54. Marcellin, F., et al., *No significant effect of cannabis use on the count and percentage of circulating CD4 T-cells in HIV-HCV co-infected patients (ANRS CO13-HEPAVIH French cohort).* Drug Alcohol Rev, 2017. **36**(2): p. 227–238.
55. Abrams, D.I., et al., *Short-term effects of cannabinoids in patients with HIV-1 infection: a randomized, placebo-controlled clinical trial.* Ann Intern Med, 2003. **139**(4): p. 258–66.
56. Costantino, C.M., et al., *Cannabinoid receptor 2-mediated attenuation of CXCR4-tropic HIV infection in primary CD4+ T cells.* PLoS One, 2012. **7**(3): p. e33961.
57. Ramirez, S.H., et al., *Attenuation of HIV-1 replication in macrophages by cannabinoid receptor 2 agonists.* J Leukoc Biol, 2013. **93**(5): p. 801–10.
58. Milloy, M.J., et al., *High-intensity cannabis use associated with lower plasma human immunodeficiency virus-1 RNA viral load among recently infected people who use injection drugs.* Drug Alcohol Rev, 2015. **34**(2): p. 135–40.
59. Watson, S.J., J.A. Benson, Jr., and J.E. Joy, *Marijuana and medicine: assessing the science base: a summary of the 1999 Institute of Medicine report.* Arch Gen Psychiatry, 2000. **57**(6): p. 547–52.
60. Wilsey, B., et al., *A randomized, placebo-controlled, crossover trial of cannabis cigarettes in neuropathic pain.* J Pain, 2008. **9**(6): p. 506–21.
61. Wallace, M., et al., *Dose-dependent effects of smoked cannabis on capsaicin-induced pain and hyperalgesia in healthy volunteers.* Anesthesiology, 2007. **107**(5): p. 785–96.
62. Collin, C., et al., *Randomized controlled trial of cannabis-based medicine in spasticity caused by multiple sclerosis.* Eur J Neurol, 2007. **14**(3): p. 290–6.
63. Hill, T.D., et al., *Cannabidivarin-rich cannabis extracts are anticonvulsant in mouse and rat via a CB1 receptor-independent mechanism.* Br J Pharmacol, 2013. **170**(3): p. 679–92.
64. Dinieri, J.A. and Y.L. Hurd, *Rat models of prenatal and adolescent cannabis exposure.* Methods Mol Biol, 2012. **829**: p. 231–42.
65. Filbey, F. and U. Yezhuvath, *Functional connectivity in inhibitory control networks and severity of cannabis use disorder.* Am J Drug Alcohol Abuse, 2013. **39**(6): p. 382–91.
66. Meier, M.H., et al., *Persistent cannabis users show neuropsychological decline from childhood to midlife.* Proc Natl Acad Sci U S A, 2012. **109**(40): p. E2657–64.
67. Zalesky, A., et al., *Effect of long-term cannabis use on axonal fibre connectivity.* Brain, 2012. **135**(Pt 7): p. 2245–55.
68. Batalla, A., et al., *Structural and functional imaging studies in chronic cannabis users: a systematic review of adolescent and adult findings.* PLoS One, 2013. **8**(2): p. e55821.
69. Cristiani, S.A., N.D. Pukay-Martin, and R.A. Bornstein, *Marijuana use and cognitive function in HIV-infected people.* J Neuropsychiatry Clin Neurosci, 2004. **16**(3): p. 330–5.
70. Skalski, L.M., et al., *The Impact of Marijuana Use on Memory in HIV-Infected Patients: A Comprehensive Review of the HIV and Marijuana Literatures.* Curr Drug Abuse Rev, 2016. **9**(2): p. 126–141.
71. Gonzalez, R., et al., *Impact of HIV and a history of marijuana dependence on procedural learning among individuals with a history of substance dependence.* J Clin Exp Neuropsychol, 2011. **33**(7): p. 735–52.
72. Heaton, R.K., et al., *HIV-associated neurocognitive disorders persist in the era of potent antiretroviral therapy: CHARTER Study.* Neurology, 2010. **75**(23): p. 2087–96.
73. Heaton, R.K., et al., *HIV-associated neurocognitive disorders before and during the era of combination antiretroviral therapy: differences in rates, nature, and predictors.* J Neurovirol, 2011. **17**(1): p. 3–16.

74. Attonito, J.M., et al., *Exploring Substance Use and HIV Treatment Factors Associated with Neurocognitive Impairment among People Living with HIV/AIDS.* Front Public Health, 2014. **2**: p. 105.
75. Marcellin, F., et al., *No influence of cannabis use on liver stiffness in HIV-HCV co-infected patients (ANRS CO13 HEPAVIH cohort study).* Eur J Gastroenterol Hepatol, 2019. **31**(2): p. 274–276.
76. Kelly, E.M., et al., *Marijuana Use Is Not Associated With Progression to Advanced Liver Fibrosis in HIV/Hepatitis C Virus-coinfected Women.* Clin Infect Dis, 2016. **63**(4): p. 512–8.
77. Liu, T., et al., *Marijuana use in hepatitis C infection does not affect liver biopsy histology or treatment outcomes.* Can J Gastroenterol Hepatol, 2014. **28**(7): p. 381–4.
78. Teixeira-Clerc, F., et al., *CB1 cannabinoid receptor antagonism: a new strategy for the treatment of liver fibrosis.* Nat Med, 2006. **12**(6): p. 671–6.
79. Marcellin, F., et al., *Cannabinoids and reduced risk of hepatic steatosis in HIV-HCV co-infection: paving the way for future clinical research.* Expert Rev Anti Infect Ther, 2018. **16**(5): p. 377–380.
80. Osei-Hyiaman, D., et al., *Endocannabinoid activation at hepatic CB1 receptors stimulates fatty acid synthesis and contributes to diet-induced obesity.* J Clin Invest, 2005. **115**(5): p. 1298–305.
81. Kamada, Y., et al., *Enhanced carbon tetrachloride-induced liver fibrosis in mice lacking adiponectin.* Gastroenterology, 2003. **125**(6): p. 1796–807.
82. Julien, B., et al., *Antifibrogenic role of the cannabinoid receptor CB2 in the liver.* Gastroenterology, 2005. **128**(3): p. 742–55.
83. Mukhopadhyay, P., et al., *Cannabidiol protects against hepatic ischemia/reperfusion injury by attenuating inflammatory signaling and response, oxidative/nitrative stress, and cell death.* Free Radic Biol Med, 2011. **50**(10): p. 1368–81.
84. Croxford, J.L. and T. Yamamura, *Cannabinoids and the immune system: potential for the treatment of inflammatory diseases?* J Neuroimmunol, 2005. **166**(1–2): p. 3–18.
85. Nordmann, S., et al., *Daily cannabis and reduced risk of steatosis in human immunodeficiency virus and hepatitis C virus-co-infected patients (ANRS CO13-HEPAVIH).* J Viral Hepat, 2018. **25**(2): p. 171–179.
86. Carrieri, M.P., et al., *Cannabis Use and Reduced Risk of Insulin Resistance in HIV-HCV Infected Patients: A Longitudinal Analysis (ANRS CO13 HEPAVIH).* Clin Infect Dis, 2015. **61**(1): p. 40–8.
87. Farooqui, M.T., et al., *Marijuana is not associated with progression of hepatic fibrosis in liver disease: a systematic review and meta-analysis.* Eur J Gastroenterol Hepatol, 2019. **31**(2): p. 149–156.
88. Lorenz, D.R., et al., *Marijuana Use Impacts Midlife Cardiovascular Events in HIV-Infected Men.* Clin Infect Dis, 2017. **65**(4): p. 626–635.
89. Thomas, G., R.A. Kloner, and S. Rezkalla, *Adverse cardiovascular, cerebrovascular, and peripheral vascular effects of marijuana inhalation: what cardiologists need to know.* Am J Cardiol, 2014. **113**(1): p. 187–90.
90. Stanley, C. and S.E. O'Sullivan, *Vascular targets for cannabinoids: animal and human studies.* Br J Pharmacol, 2014. **171**(6): p. 1361–78.
91. Rezkalla, S. and R.A. Kloner, *Cardiovascular effects of marijuana.* Trends in cardiovascular medicine, 2019. **29**(7): p. 403–407.
92. Hashibe, M., et al., *Marijuana use and the risk of lung and upper aerodigestive tract cancers: results of a population-based case-control study.* Cancer Epidemiol Biomarkers Prev, 2006. **15**(10): p. 1829–34.
93. Volkow, N.D., et al., *Adverse health effects of marijuana use.* N Engl J Med, 2014. **370**(23): p. 2219–27.
94. Patton, G.C., et al., *Cannabis use and mental health in young people: cohort study.* Bmj, 2002. **325**(7374): p. 1195–8.

95. Di Forti, M., et al., *Daily use, especially of high-potency cannabis, drives the earlier onset of psychosis in cannabis users.* Schizophr Bull, 2014. **40**(6): p. 1509–17.
96. Fergusson, D.M. and J.M. Boden, *Cannabis use and later life outcomes.* Addiction, 2008. **103**(6): p. 969–76; discussion 977–8.
97. Elkington, K.S., et al., *Marijuana Use and Psychiatric Disorders in Perinatally HIV-Exposed Youth: Does HIV Matter?* J Pediatr Psychol, 2016. **41**(3): p. 277–86.
98. Lim, A.C. and A.D. Thames, *Differential relationships between cannabis consumption and sleep health as a function of HIV status.* Drug Alcohol Depend, 2018. **192**: p. 233–237.
99. Lutge, E.E., A. Gray, and N. Siegfried, *The medical use of cannabis for reducing morbidity and mortality in patients with HIV/AIDS.* Cochrane Database Syst Rev, 2013(4): p. CD005175.
100. Metrik, J., et al., *Sexual Risk Behavior and Heavy Drinking Among Weekly Marijuana Users.* J Stud Alcohol Drugs, 2016. **77**(1): p. 104–12.
101. Kaljee, L., et al., *Cross-sectional data on alcohol and marijuana use and sexual behavior among male and female secondary school students in New Providence, The Bahamas.* Int J Adolesc Med Health, 2016. **28**(2): p. 133–40.
102. Morgan, E., et al., *Marijuana Use Among Young Black Men Who Have Sex With Men and the HIV Care Continuum: Findings From the uConnect Cohort.* Subst Use Misuse, 2016. **51**(13): p. 1751–9.
103. Slawson, G., et al., *High-intensity cannabis use and adherence to antiretroviral therapy among people who use illicit drugs in a Canadian setting.* AIDS Behav, 2015. **19**(1): p. 120–7.
104. Zhang, M., et al., *CB2 receptor activation attenuates microcirculatory dysfunction during cerebral ischemic/reperfusion injury.* Microvasc Res, 2009. **78**(1): p. 86–94.
105. Thompson, A.B., et al., *Regular Marijuana Use is Associated with Poor Viral Suppression in HIV-Infected Adolescents and Young Adults.* AIDS Behav, 2018. **22**(4): p. 1363–1372.
106. Kipp, A.M., et al., *Daily Marijuana Use is Associated with Missed Clinic Appointments Among HIV-Infected Persons Engaged in HIV Care.* AIDS Behav, 2017. **21**(7): p. 1996–2004.
107. Okafor, C.N., et al., *Prevalence and correlates of marijuana use among HIV-seropositive and seronegative men in the Multicenter AIDS Cohort Study (MACS), 1984–2013.* Am J Drug Alcohol Abuse, 2017. **43**(5): p. 556–566.

Mpox: An Emerging Global Orthopoxvirus Outbreak

Hanan Ibrahim and Ambika Eranki

Abstract Mpox (monkeypox) is a viral infection caused by the Monkeypox virus (MPXV) — an Orthopoxvirus. MPXV was discovered over five decades ago, and the first human case was reported in 1970. The disease has now gained notoriety since a global outbreak occurred in 2022–2023 (Monkeypox. https://www.who.int/health-topics/monkeypox. Accessed 10 Sept 2023). Mpox can spread by direct and zoonotic transmission. The reemergence of MPXV was noted after the end of the smallpox vaccination era in 1980 (Mpox (monkeypox). https://www.who.int/news-room/fact-sheets/detail/monkeypox. Accessed 10 Sept 2023). Given its propensity to cause severe disease in immunocompromised individuals, as well as the challenges associated with diagnosis, management, and prevention, this is an important disease for clinicians to be aware of, diagnose promptly, and manage appropriately.

Keywords Mpox · Monkeypox · Orthopoxviridae · Zoonosis · Global outbreak · Monkeys · Chimpanzees · Rodents · Sexual transmission risk · Clades · Strains · West Africa · CDC · NIH · WHO

1 History and Background

Monkeypox virus (MPXV) was first described in 1958 in two shipments of monkeys arriving in Denmark from Singapore. There were two cases of this pox-like disease among the monkeys, which were both nonfatal. The exact source of this virus was unknown; however, it is speculated that the virus spread could have

H. Ibrahim
University of South Florida, Morsani College of Medicine, Tampa, FL, USA

A. Eranki (✉)
Division of Infectious Diseases, Department of Internal Medicine, University of South Florida, Morsani College of Medicine, Tampa, FL, USA
e-mail: aeranki@usf.edu

C. Somboonwit et al. (eds.), *Global Virology IV: Viral Disease Diagnosis and Treatment Delivery in the 21st Century*,
https://doi.org/10.1007/978-3-031-57369-9_10

occurred from a monkey that was a companion to the two affected monkeys in that shipment or could have also been from one or both of the affected monkeys. Possibly, the infection spread from either a recent infection in the respiratory tract or a past asymptomatic infection. However, the virus was present in the tissues of the companion monkey [1]. African rodents may be the source of the disease MPXV infections, which have occurred in squirrels, rats, mice, monkeys, and prairie dogs and then spread to humans [2].

MPXV was determined to be zoonotic in 1970, and the first human cases appeared in West Africa: in the Democratic Republic of Congo, Liberia, and Sierra Leone. There were six cases, all nonfatal and all in smallpox-unvaccinated patients. These were the first human cases of monkeypox, and due to their clinical similarity to the smallpox virus, they were initially misdiagnosed as smallpox. However, after samples were analyzed from the patients in several laboratories, the identification of MPVX was made. The first case was diagnosed in a 9-month-old boy from the Democratic Republic of Congo. The four cases following were in Liberia that involved children ranging from 4 to 9 years of age. One was an isolated case, while the other three were in contact with each other. The final case that year was in Sierra Leone; the patient was a 24-year-old man who reported killing and removing the stomach and intestines from a monkey a couple of weeks prior to the illness. In all these cases, no other member of the patients' households had contracted the virus. However, in all these areas, there were large numbers of monkeys and apes (gorillas and chimpanzees). Some of these monkeys or apes had either low antibody levels against MPXV (low hemagglutination inhibition antibodies) or were positive for the virus. Since these monkeys or apes were often killed and eaten by those living in overlapping habitats, this could have been a possible avenue for the disease to transfer between the monkeys or apes to humans [1]. In late 2022, the World Health Organization (WHO) endorsed and started using the term "mpox" in lieu of "monkeypox"—this is based on the WHO best practice guidelines for "neutral" naming of emerging diseases to mitigate any discriminatory or pejorative language based on racial/ethnic or other stereotypes [3].

2 Clinical Presentation

2.1 Background

Mpox is a zoonotic viral infection belonging to the genus *Orthopoxviridae* (family: *Poxviridae*) [4]. The family of poxviruses is large; however, the most notorious member of this family is the smallpox virus. The first human case of MPXV originated in the Democratic Republic of Congo (DRC) in 1970. Since then, the virus has been endemic in Central and West Africa including Cameroon, the Central African Republic, DRC, Gabon, Ghana, Cote D'Ivoire, Liberia, Nigeria, the Republic of the Congo, and Sierra Leone [5]. There are two major clades reported (Central and West Africa or the Congo Basin [CB] or Clade I and West Africa [WA] or Clade II)[6]. Of these, the Congo Basin clade has a higher case fatality ratio (CFR). Upon analyzing both strain's genomes, it was found that the West African strain had deletions and fragmentations in open reading frames, likely causing reduced virulence. However, the Central African strain suppresses the patient's T-cell responses and inhibits the complement enzymes, allowing the virus to develop in the body further and, hence, impact the host negatively [4]. Regardless of clade, the mortality rate is higher in immunocompromised individuals and in young children. Of note, people living with HIV, especially those advanced with suppressed CD4 lymphocyte counts, have also been noted to have higher rates of mpox infection and more severe illness. Per the Centers for Disease Control (CDC), the ongoing outbreak in the United States is caused by clade IIb [West African] of the mpox virus [7].

2.2 *Typical Risk Factors*

The hypothesized risk factor for zoonotic transfer of MPXV is through hunting and eating bushmeat. These practices are frequent in endemic areas. Another possible point of zoonotic transmission of this disease is through contact with African rodents, as they are theorized to be the original host of this disease [8]. Since the exact mode of transmission has not been fully elucidated yet, the consensus is that the virus can be acquired via bites/scratches from infected animals, consumption of infected animal meat (for zoonotic transmission), respiratory droplets, direct contact with infectious lesions or bodily fluids, sexual contact, and fomite transmission [4].

2.3 *Clinical Features of Mpox*

The overall clinical syndrome of mpox is similar to smallpox; however, it is much less lethal. The clinical course of mpox involves a stepwise progression. There is the incubation period, the prodrome stage, and finally, the rash stage. Each of these stages varies in symptoms and transmissibility. The overall severity of the illness depends on the mode of acquisition as well as the immune status and comorbidities of the host [9].

The incubation period of this virus ranges from 3 to17 days on average [9]. This is when the virus establishes infection in the host; therefore, the infected individual is asymptomatic until the prodromal stage begins. The individual is not contagious during the incubation period. Current guidance from the Centers for Disease Control (CDC) recommends monitoring persons for 3 weeks if they have been exposed to mpox. After this period, the illness will develop and subsequently enter the symptomatic phase, typically lasting 2–4 weeks [9].

The prodrome stage marks the early stage of symptoms. Prodromal symptoms may include malaise, myalgias, fatigue, fevers, sore throat, cough, and lymphadenopathy [2]. Notably, lymphadenopathy is unique to mpox (like human cowpox virus infection). Lymph nodes in the cervical chains, axillae, and inguinal areas can be involved, which can be unilateral or bilateral [9]. During this stage of the illness, persons are contagious due to the potential to spread infectious virions via bodily fluids/secretions [10]. It is recommended that they isolate if they start developing symptoms [11]. This stage, however, can occur either before or after the rash or may not occur at all [9].

The third stage of the illness is the rash stage. However, it has been reported that patients have presented with rash without a prodrome during this mpox outbreak, that is, only presenting with localized lesions. This is the most contagious phase of the infection. The rash progresses via various stages of evolution over 2–4 weeks. The initial rash is macular, that is, flat lesions, lasting about 1–2 days. Following this, the rash evolves into the papular stage, meaning the lesions look more raised and are palpable. After this, the lesions convert into the "vesicular" form, which refers to small fluid-filled blisters (vesicles are usually less than 5 mm in size). This phase also lasts for about 2 days. The last phase in the evolution of the rash is the pustular phase, which lasts longer, up to a week: The lesions are filled with opaque fluid, and these pustules have been described as round, firm, and rubbery in consistency. All these phases are highly infectious on direct or indirect contact. The rash eventually scabs over, and the scabs can last for 7–14 days before falling off spontaneously. Once all the scabs fall off a person, they are no longer contagious, and new skin will form [9].

The dispersion of the lesions around the patient's body varies from person to person, and some patients may experience few skin lesions. In contrast, for others, lesions may cover the affected areas of the body with hundreds or more. These lesions can appear on the face as well as in areas such as the mouth and throat, hand palms and soles of the feet, and the genital areas, including the anus [1]. The rash can also coalesce into larger areas of involvement in more severe disease. Bacterial superinfection like staphylococcal skin and soft tissue infections can also occur in severe disease [12]. The mucosal/genital lesions of mpox can be misdiagnosed as other STIs such as herpes simplex virus (HSV), syphilis, and disseminated gonorrhoea [13].

Severe illness with MPXV is characterized by numerous lesions that can coalesce and become necrotic or hemorrhagic, severe lymphadenopathy resulting in the obstruction of airway passages or the GI tract, multisystem involvement with nodular lesions, superimposed bacterial infections with systemic inflammatory response, eye involvement, and CNS involvement [14].

Another concern with severe disease is when there is involvement of areas such as the pharynx, for example, that causes severe morbidity due to complications like strictures or lumen obstruction. Some of these can also require surgical management. Severe illness can also be seen in other immunocompromised populations such as infants, pregnant women, hematopoietic or solid organ transplant recipients, people on long-term immunosuppressive agents, persons who live with HIV, and others [14].

2.4 *Transmission Among Humans*

The transmission of mpox among humans can occur through direct contact with the lesions on the body of an infected person, which can be in the mouth or the genitals. Viral particles are present in the skin lesions and can infect contacts through skin breaks and open mucus membranes. Therefore, this would include skin-to-skin touching or sex (vaginal, oral, and anal), kissing mouth-to-mouth or the skin directly, additionally including any respiratory droplets that are exchanged during prolonged close contact. In that case, it is believed that the virus can transmit via respiratory droplets to the exposed person and enter via mucosal membranes such as the respiratory or genital tract [1].

3 Diagnosis

There are several laboratory modalities to detect and diagnose mpox. This is essential to treat and prevent virus spread to other people. Molecular diagnosis with polymerase chain reaction (PCR) testing for mpox is generally highly sensitive and specific and recommended by several organizations, including the WHO and the CDC. Several real-time PCR assays have been developed, targeting different regions in the mpox virus genome. While real-time PCR is highly reliable, it is expensive and requires specialized equipment and conditions to be done correctly. Isothermal methods like loop-mediated isothermal amplification (LAMP) are an alternative method for rapid mpox detection. LAMP is a simpler, cost-effective technique that amplifies nucleic acids at a constant temperature and eliminates the need for expensive equipment, making LAMP suitable for diagnostics in regions with limited laboratory resources and supplies [15].

4 Prevention and Challenges: Prior to the 2022 Outbreak

There were several challenges during the initial mpox outbreak, especially in endemic countries, a notable one being contact tracing. Another challenge is the need for more resources to diagnose the mpox cases. Many countries in which the virus is endemic do not have the funding to increase testing and materials and train medical personnel. Yet another challenge is that the mpox virus is prone to mutation and has mutated before as it changed from a zoonotic-focused disease to a sexually transmitted disease, creating a new transmittable route [16].

Preventative methods have focused on extensive laboratory testing from researchers and healthcare professionals, the use of smallpox vaccines and treatment, and educating people who are at high risk of contracting the virus, such as animal owners, healthcare providers, veterinarians, and other professionals who handle animals. For example, the outbreak in 2003 where cases of mpox were found in the United States was due to the importation of prairie dogs, which are native rodents to West Africa. Spread of the virus during those case investigations revealed contact with a sick animal either through touching it, getting scratched and breaking the skin, getting bit, cleaning the animal's cage, or touching the bedding of a sick animal. With all the cases, person-to-person contact was not a factor in spreading the virus. Therefore, preventative measures focused heavily on awareness of interacting with certain species of animals [17]. Tracking potentially infected animals and if they were linked to any humans or had contact with them is another form of prevention. There is also a need for health education initiatives to teach the public how to prevent the acquisition and transmission of MPXV by spreading awareness of what it is. Finally, to stop international spread, it has been necessary to create and enforce laws that prohibit the transportation and importation of certain species of rodents, specifically from Africa, including prairie dogs, as they are a possible host of the disease [17].

5 Mpox Outbreak During the COVID-19 Pandemic

Since its discovery, several outbreaks of mpox have occurred sporadically from 2003 to 2017/2018 in Nigeria, Singapore, Israel and the US, United Kingdom (UK) and other European countries in 2022 [18, 19]. Mpox resurged in May 2022, when two cases of mpox were reported in the United Kingdom (UK). It originated from a British resident who had traveled to Nigeria (where this virus is endemic), and while there was presenting symptoms of mpox for a couple of days. When they returned to the United Kingdom, they were still contagious, spreading the virus to close contacts [20]. Since discovering these patients having mpox, there has been a surge in cases present in countries across Europe, North America, Latin America, and the Middle East [21].

Among these patients, disease characteristics were different from the traditional presentations of mpox. The majority of people infected were 20–50-year-old men that were gay or bisexual or other men who have sex with men (MSM). This led to investigating sexual transmission as a form of spreading mpox [21]. As it was found, most of the infections during the outbreak had been transmitted through sexual contact with symptomatic people. Although most infections that were spread sexually were through male-to-male sexual contact, heterosexual sexual transmission, skin-to-skin contact with children, contaminated needles, and lack of PPE when treating mpox patients were also reported as transmission events of the virus [21] (Table 1 and Fig. 1).

The lack of awareness and knowledge of this virus is another challenge, as communities that are not informed on how to stay safe and protected are more susceptible to getting the virus and spreading it [16]. Since this outbreak occurred and was declared a "public health emergency of international concern" by the World Health Organization (WHO) while the Coronavirus Disease 2019 (COVID-19) pandemic

Table 1 Comparison of typical presentations of Mpox vs. presentations in 2022

Prior to 2022 outbreak	2022 outbreak
Skin rashes appear after the prodromal phases [22]	Half of patients report having a rash as their first symptom [22]
Common rash locations: head and neck, trunk, upper limbs, and lower limbs. This also included the palmar and plantar in the majority of patients [23]	Common rash locations: genitalia, trunk, upper limb, and anal/perianal. The genitalia involvement was much more common [23]
The percentage of men was only slightly higher than that of women (55.7% vs 44.3%) [24]	99% of the patients in the included studies were cases of men [24]
The highest percentage of cases belonged to the 0–9 years age group (38.9%), followed by 10–19 (27.8%), 20–29 (17.2%), 30–39 (10.1%), and older than 40 years of age (6.0%) [24]	The majority of patients in the 2022 outbreak were 30–39 years of age (46.7%), followed by 40–49 (23.2%), 20–29 (21.7%), 50–59 (6.5%), older than 60 (1.1%), and younger than 20 years (0.8%) [24]
Majority of cases (61.64%) were due to a history of animal exposure and 38.35% of cases were due to human-to-human transmission [23]	Almost all cases were due to human-to-human transmission. Most cases reported skin-to-skin contact with symptomatic or asymptomatic patients, mainly through sexual contact or attending mass gatherings or contact with contaminated objects in places where Mpox was prevalent [23]
Majority of cases were reported in endemic countries or due to travel in endemic countries [23]	Majority of cases are from non-endemic countries (America, Europe, and Asia) [23]
Diagnostic testing with a reverse transcriptase polymerase chain reaction (RT-PCR) test from the skin lesions [23]	Diagnostic test from an RT-PCR test through anal, and oropharyngeal swabs [23]
Most patients had moderate skin rash severity (25–99 skin lesions) [23]	Majority of patients had mild skin rash severity (<25 skin lesions) [23]
11.4% of confirmed positive patients had HIV [23]	41.1% of confirmed positive patients had HIV [23]
Of reported cases (majority being unknown), the rate of mortality was 4.2% [23]	The mortality rate of cases was 0.2% [23]

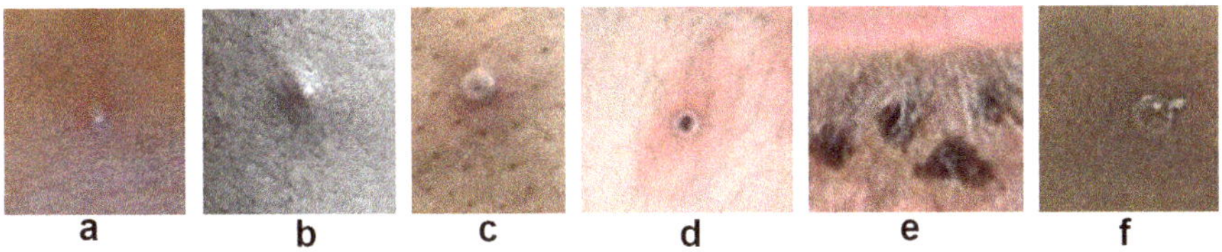

Fig. 1 The images below show the progression of skin lesions with mpox patients. Figure (**a**) shows the early pustule. Figure (**b**) is a small pustule. Figure (**c**) is the umbilicated pustule. Figure (**d**) is the ulcerated lesion. Figure (**e**) is crusting of mature lesions. Figure (**f**) is the partially removed scab. (Mpox rash photos credited from UK Health Security Agency. Images contain public sector information licensed from the Open Government License v3.0 [25])

was ongoing, there were unique challenges associated with already overburdened public health systems and dearth of appropriate resources to combat the outbreak [26].

6 Treatment and Prevention: How to Mitigate the Outbreak Post 2022

6.1 *Managing and Treating Mpox*

Patients with mild symptoms may not require medication, but for those with severe symptoms, antihistamines, analgesics, emollients, and topical steroids can help manage physical skin lesions. Antiviral medications like tecovirimat, which inhibits viral protein, and cidofovir, an inhibitor of viral DNA synthesis, may also be used for treatment. Tecovirimat stands out as the preferred primary treatment choice and can be used alongside antiretroviral therapy in cases of HIV co-infection. Monitoring for nephrotoxicity is crucial when administering cidofovir, and brincidofovir, another viral DNA synthesis inhibitor could be considered in specific scenarios, such as hepatic impairment or during pregnancy [21]. The current guidance recommends supportive care and conservative management for most mpox infections unless the patient is immunocompromised with severe disease or the lesions are in an anatomic area, which is associated with a high risk of complications/sequelae [14].

The current agents available to treat mpox include the following:

(A) Tecovirimat (TPOXX, ST-246) is approved by the Food and Drug Administration (FDA) for the treatment of smallpox but has been shown to treat other orthopoxviruses in animal models. This is the recommended first-line agent at the time of writing this.

(B) Brincidofovir, a Cidofovir relative, is also FDA-approved for smallpox treatment, but no current human study has demonstrated its effectiveness in mpox. There is in vitro and animal data showing that it also works for other orthopoxvirus infections. It is available through an e-IND ("FDA authorized single patient emergency use") via the CDC when patients test positive for mpox and have severe disease or are at high risk of developing severe disease and meet certain criteria set out by the CDC [14].

6.2 *Vaccinations and Preventative Measures*

High-risk groups can benefit from vaccination, with the live, nonreplicating modified vaccinia Ankara (MVA) vaccine (JYNNEOS) being the primary choice. However, in cases where MVA is unavailable, the replication-competent smallpox vaccine (ACAM2000) can be considered, although with a higher risk of adverse reactions. MVA is safe for immunocompromised individuals and is administered in two doses, 4 weeks apart [21]. These vaccines can be used as pre-exposure prophylaxis to prevent high-risk individuals from contracting the virus or as post-exposure prophylaxis for individuals who have already been exposed to prevent the virus from developing [12]. There are two methods to administering the mpox vaccine to individuals, either through subcutaneous or intradermal administration. The subcutaneous is the standard method and has an injection volume of 0.5 mL. The intradermal administration requires a lower volume of the injection, only 0.1 mL, and can be used under an Emergency Use Authorization (EUA) [12]. Due to the lower volume of injections and the limited supply of the vaccine during the 2022 outbreak, the intradermal vaccine has become the predominant one [27].

6.3 Medications Mechanism of Action

Tecovirimat effectively inhibits the replication and spread of the virus by targeting the F13L protein preventing the release of the infectious virus from infected cells, thus limiting the virus's ability to spread from cell to cell [28].

There is an ongoing multicenter, phase III clinical trial looking at the efficacy of Tecovirimat in mpox, which is funded by the National Institutes of Allergy and Infectious Diseases (NIAID), called the STOMP (Study of Tecovirimat for Mpox); it is currently enrolling trial participants across multiple US sites [26].

7 Conclusion

Since its outbreak in 2022, mpox has become an important infection in the public health realm, especially in the area of sexually transmitted diseases. Due to its nature of transmissibility through sexual behaviors, there is now a need for a multipronged approach to prevent mpox to at-risk populations. An example of this would be administering the vaccine to certain high-risk groups, such as men who have sex with men, transgender/non-binary individuals with new STIs or multiple sexual partners, people who have had sex at commercial venues or those attending large public events, those with diagnosed HIV, and partners of these high-risk groups [21]. Clinicians including primary care providers need to be well-versed in recognizing the clinical syndrome of mpox and knowing how to distinguish it from other infectious diseases, especially in challenging situations, such as when there are isolated genital lesions, which can present a diagnostic conundrum. Having a high index of suspicion is important to pick up diagnostic clues to initiate prompt isolation measures and treatment. By adopting several preventative measures, like vaccination to at-risk communities, we can collectively work toward mitigating the impact of mpox within our communities and stopping its further spread.

Conflicts of Interest The authors declare no conflicts of interest.

References

1. Cho, C. T.; Wenner, H. A. Monkeypox Virus. *Bacteriol. Rev.* **1973**, *37* (1), 1–18. https://doi.org/10.1128/br.37.1.1-18.1973.
2. Moyo, E.; Musuka, G.; Murewanhema, G.; Moyo, P.; Dzinamarira, T. Monkeypox Outbreak: A Perspective on Africa's Diagnostic and Containment Capacity. *Int. J. Infect. Dis.* **2022**, *123*, 127–130. https://doi.org/10.1016/j.ijid.2022.08.016.
3. *WHO recommends new name for monkeypox disease.* https://www.who.int/news/item/28-11-2022-who-recommends-new-name-for-monkeypox-disease (accessed 2023-09-10).
4. Kaler, J.; Hussain, A.; Flores, G.; Kheiri, S.; Desrosiers, D. Monkeypox: A Comprehensive Review of Transmission, Pathogenesis, and Manifestation. *Cureus 14* (7), e26531. https://doi.org/10.7759/cureus.26531.
5. *Mpox (formerly Monkeypox) | NIH: National Institute of Allergy and Infectious Diseases.* https://www.niaid.nih.gov/diseases-conditions/Mpox (accessed 2023-09-10).
6. *Monkeypox.* https://www.who.int/health-topics/monkeypox (accessed 2023-09-10).
7. CDC. *Mpox and HIV.* Centers for Disease Control and Prevention. https://www.cdc.gov/poxvirus/mpox/prevention/hiv.html (accessed 2023-10-19).
8. Quiner, C. A.; Moses, C.; Monroe, B. P.; Nakazawa, Y.; Doty, J. B.; Hughes, C. M.; McCollum, A. M.; Ibata, S.; Malekani, J.; Okitolonda, E.; Carroll, D. S.; Reynolds, M. G. Presumptive Risk Factors for Monkeypox in Rural Communities in The Democratic Republic of the Congo. *PLOS ONE* **2017**, *12* (2), e0168664. https://doi.org/10.1371/journal.pone.0168664.
9. *Clinical Recognition | Mpox | Poxvirus | CDC.* https://www.cdc.gov/poxvirus/Mpox/clinicians/clinical-recognition.html (accessed 2023-09-10).
10. *Mpox (formerly Monkeypox): What You Need To Know.* Yale Medicine. https://www.yalemedicine.org/news/monkeypox-Mpox-symptoms-treatment (accessed 2023-09-13).
11. *Information For Healthcare Professionals | Mpox | Poxvirus | CDC.* https://www.cdc.gov/poxvirus/Mpox/clinicians/index.html (accessed 2023-09-10).
12. *Interim Clinical Considerations for Use of JYNNEOS and ACAM2000 Vaccines during the 2022 U.S. Mpox Outbreak | Mpox | Poxvirus | CDC.* https://www.cdc.gov/poxvirus/mpox/clinicians/vaccines/vaccine-considerations.html (accessed 2023-09-30).
13. Low, N.; Bachmann, L. H.; Ogoina, D.; McDonald, R.; Ipekci, A. M.; Quilter, L. A. S.; Cevik, M. Mpox Virus and Transmission through Sexual Contact: Defining the Research Agenda. *PLOS Med.* **2023**, *20* (1), e1004163. https://doi.org/10.1371/journal.pmed.1004163.
14. *Treatment Information for Healthcare Professionals | Mpox | Poxvirus | CDC.* https://www.cdc.gov/poxvirus/Mpox/clinicians/treatment.html (accessed 2023-09-13).
15. Silva, S. J. R. da; Kohl, A.; Pena, L.; Pardee, K. Clinical and Laboratory Diagnosis of Monkeypox (Mpox): Current Status and Future Directions. *iScience* **2023**, *26* (6), 106759. https://doi.org/10.1016/j.isci.2023.106759.
16. Nyame, J.; Punniyakotti, S.; Khera, K.; Pal, R. S.; Varadarajan, N.; Sharma, P. Challenges in the Treatment and Prevention of Monkeypox Infection; A Comprehensive Review. *Acta Trop.* **2023**, *245*, 106960. https://doi.org/10.1016/j.actatropica.2023.106960.
17. Past *U.S. Cases and Outbreaks | Mpox | Poxvirus | CDC.* https://www.cdc.gov/poxvirus/Mpox/outbreak/us-outbreaks.html (accessed 2023-09-10).
18. *Mpox (monkeypox).* https://www.who.int/news-room/fact-sheets/detail/monkeypox (accessed 2023-09-10).
19. *Multi-country monkeypox outbreak in non-endemic countries.* https://www.who.int/emergencies/disease-outbreak-news/item/2022-DON385 (accessed 2023-10-18).

20. Hraib, M.; Jouni, S.; Albitar, M. M.; Alaidi, S.; Alshehabi, Z. The Outbreak of Monkeypox 2022: An Overview. *Ann. Med. Surg.* **2022**, *79*, 104069. https://doi.org/10.1016/j.amsu.2022.104069.
21. Chaudhari, S.; Treffeisen, L.; Virk, J.; Parikh, T.; Gopalakrishnan Ravikumar, N. P.; Goti, A. M.; Goyal, L.; Yashi, K. The 2022 Monkeypox Epidemic and What Has Led to the Current State of the Disease in the US: A Systematic Review. *Cureus* **2023**. https://doi.org/10.7759/cureus.33515.
22. CDC. *Detection & Transmission of Mpox Virus During the 2022 Clade IIb Out.* Centers for Disease Control and Prevention. https://www.cdc.gov/poxvirus/Mpox/about/science-behind-transmission.html (accessed 2023-09-10).
23. Hatami, H.; Jamshidi, P.; Arbabi, M.; Safavi-Naini, S. A. A.; Farokh, P.; Izadi-Jorshari, G.; Mohammadzadeh, B.; Nasiri, M. J.; Zandi, M.; Nayebzade, A.; Sechi, L. A. Demographic, Epidemiologic, and Clinical Characteristics of Human Monkeypox Disease Pre- and Post-2022 Outbreaks: A Systematic Review and Meta-Analysis. *Biomedicines* **2023**, *11* (3), 957. https://doi.org/10.3390/biomedicines11030957.
24. Li, P.; Li, J.; Ayada, I.; Avan, A.; Zheng, Q.; Peppelenbosch, M. P.; de Vries, A. C.; Pan, Q. Clinical Features, Antiviral Treatment, and Patient Outcomes: A Systematic Review and Comparative Analysis of the Previous and the 2022 Mpox Outbreaks. *J. Infect. Dis.* **2023**, *228* (4), 391–401. https://doi.org/10.1093/infdis/jiad034.
25. *Mpox (monkeypox): background information*. GOV.UK. https://www.gov.uk/guidance/monkeypox (accessed 2023-09-13).
26. *STOMP*. https://www.stomptpoxx.org/main (accessed 2023-09-13).
27. Owens, L. E.; Currie, D. W.; Kramarow, E. A.; Siddique, S.; Swanson, M.; Carter, R. J.; Kriss, J. L.; Boersma, P. M.; Lee, F. C.; Spicknall, I.; Hurley, E.; Zlotorzynska, M.; Gundlapalli, A. V. JYNNEOS Vaccination Coverage Among Persons at Risk for Mpox — United States, May 22, 2022–January 31, 2023. *Morb. Mortal. Wkly. Rep.* **2023**, *72* (13), 342–347. https://doi.org/10.15585/mmwr.mm7213a4.
28. Merchlinsky, M.; Albright, A.; Olson, V.; Schiltz, H.; Merkeley, T.; Hughes, C.; Petersen, B.; Challberg, M. The Development and Approval of Tecoviromat (TPOXX®), the First Antiviral against Smallpox. *Antiviral Res.* **2019**, *168*, 168–174. https://doi.org/10.1016/j.antiviral.2019.06.005.

Community-Acquired Respiratory Infections in HCT and Immunosuppressed Patients

Aliyah Baluch and Olga Klinkova

Abstract Community-acquired respiratory infections in the immunosuppressed patient population continue to be highly frequent, with negligible treatment options. The best approach is preventing infection followed by timely microbiologically based polymerase chain reaction (PCR) testing to assess which patients require droplet isolation. In terms of therapy, influenza is the only respiratory infection with an FDA-approved therapy, oseltamivir.

Keywords Influenza A and B · Parainfluenza 1 through 4 · Respiratory Syncytial Virus · Human Metapneumovirus · Rhinovirus/Enterovirus · Bocavirus · Ki · Wu · Adenovirus · Immunosuppressed lower respiratory infections · Multiplex PCRs · Multi-virus specific T cell therapy

1 Introduction

Respiratory infections continue to be a great source of morbidity and mortality in the population of immunosuppressed persons, such as recipients of solid organ transplants and hematopoietic stem cell transplants (HSCT) [1, 2]. This chapter will review which organisms have a high level of pathogenicity, potential therapeutic options, and other interventions that are in the pipeline for therapy. Due to the recent pandemic of SARS-CoV 2 (COVID-19), this subject will be in a separate chapter. Immunosuppressed patients often have impaired T-cell immunity related to their primary diseases and the resultant therapies utilized, especially during the conditioning phase of an HSCT. The lack of T-cell immunity increases a patient's risk of acquiring respiratory infections (compared to a nonimmunosuppressed person), as

A. Baluch (✉) · O. Klinkova
Division of Infectious Diseases, Department of IHM/Internal Medicine, Moffitt Cancer Center, Tampa, FL, USA
e-mail: aliyah.baluch@moffitt.org

C. Somboonwit et al. (eds.), *Global Virology IV: Viral Disease Diagnosis and Treatment Delivery in the 21st Century*,
https://doi.org/10.1007/978-3-031-57369-9_11

well as increasing the risk of progression from an uncomplicated upper respiratory infection to a lower respiratory tract infection (LRTI) with higher mortality and morbidity [3]. Risk-stratifying interventions are critical when assessing patients, especially for patients on systemic corticosteroid use or low lymphocyte counts as factors for progression to LRTI [4].

2 Epidemiology

Influenza, parainfluenza (types 1 through 4, but especially type 3), respiratory syncytial virus (RSV), human metapneumovirus (hMPV), and adenovirus are traditionally the respiratory viruses known to have the highest level of morbidity and mortality in transplant recipients. RSV, as an example, has been stated to infect approximately 2–17% of HSCT recipients. Others, such as bocavirus, increase a patient's morbidity due to cough but are not necessarily linked to an increased mortality risk. Rhinovirus/enterovirus, one of the etiologies of the common cold, also continues to have the highest rates of respiratory infections in the immunosuppressed patient population [5] but with little risk of LRTI. Coronavirus, even before COVID-19, had a variety of seasonal variants that more often led to URIs than LRTIs. KI and Wu, like bocavirus, have been found in the immunosuppressed but with unknown ramifications [6]. Adenovirus can present as many different types of infections, but most importantly, it can present as both a URTI and an LRTI infection with a very high mortality rate [7].

Despite the known seasonality of most respiratory infections, in the immunosuppressed population, they can occur at any point in the year. Immunosuppressed patients, such as HSCT recipients, are at risk during their inpatient stays when they are exposed to their visitors and/or hospital staff if they are unmasked. The respiratory infectious risk continues to escalate once they are discharged and out of the community, especially if they are not compliant with mask-wearing policies [2]. Prolonged shedding of respiratory infections can also occur in the immunosuppressed, especially HSCT recipients, leading to added needs for appropriate isolation per infection control practices, especially in the clinic setting. The median duration of respiratory virus shedding is 3 weeks, which should be considered when doing follow-up multiplex PCR testing [8].

3 Challenges to Prevention and Treatment

In terms of prevention, where applicable, vaccination post-HSCT is utilized. Influenza vaccination is the main respiratory infection that has a potential annual vaccine. As of 2023, no respiratory infections other than COVID-19 have widely available vaccine-preventative options. Masking and good hand hygiene become the cornerstone for preventing community-acquired respiratory infections. As seen with the implementation of nearly universal masking during the start of the COVID-19 pandemic, the incidence of community-acquired respiratory infections fell to near nil before rebounding again due to mask-wearing fatigue. In terms of antiviral treatment, it is based specifically on the infection. Influenza A and B are commonly treated with oral oseltamivir 75 mg twice daily for 5 days, ideally within 48 h of the onset of symptoms.

Prolonged courses of oseltamivir can be considered in immunosuppressed populations such as HSCT (up to 10 days) [9]. There needs to be more data available for baloxavir and other options for influenza in immunosuppressed patients. Palivizumab is a monoclonal therapy FDA-approved for prevention in the pediatric population, though some are currently assessing it as a potential co-therapy [10]. It is not currently considered a mainstream option for preventing RSV in the adult immunosuppressed population. Patients infected with RSV can be assessed for the use of oral ribavirin as well as immunoglobulin supplementation. Dosing of oral ribavirin does need to be appropriately adjusted for acute or chronic kidney dysfunction. Aerosolized ribavirin and oral ribavirin have similar outcomes, but it is significantly easier to access and give to the patient promptly the oral preparation [11, 12]. Parainfluenza 1 through 4 and human metapneumovirus have less strong data to support any therapeutic interventions other than supportive care and the consideration for immunoglobulin supplementation if the IgG level is low (traditionally less than 400 mg/dl) [13, 14]. Adenovirus infection, especially if lung infections are part of a disseminated case, should be considered for treatment with cidofovir in all immunosuppressed patients [7, 15].

4 Newer Approaches

4.1 Microbiology

Microbiological testing is one of the newest changes in the field of respiratory infections. Viral culture is rarely done in either academic or community hospitals. Viral antigen or multiplex PCRs have become the norm for workups in walk-in clinics, emergency rooms, and inpatient services. This allows for a rapid turnaround for diagnosis and good antimicrobial stewardship leading to less use of inappropriate antimicrobials for viral infections.

4.2 Treatment

In the pipeline of antivirals circa 2023, alloVIR® is being assessed as an allogenic, off-the-shelf, multi-virus specific T cell (VST) therapy targeting respiratory infections and/or disseminated viral infection. This is an example of the types of products being assessed for those with T-cell deficiencies that are at high risk of progression of infection from URTI to LRTI (https://www.allovir.com/products/alvr106 last accessed on 2/18/2023) [16]. However, VST therapy is limited to matching HLA alleles in the patient to what is available in the available banks, having the appropriate expansion of the T cells specific for the infection when needed, and access to the patient [17]. VST infusions appear to be a safe alternative option for treatment with an ongoing need for monitoring for de novo cases of GVHD. It is also advantageous that VSTs appear to have multispecificity for infections, leading to a broad antiviral applicability, which is appealing as often HCT recipients can be infected with more than one infection at a time [18]. There are ongoing trials assessing which respiratory infections can be optimally treated with VSTs.

5 Conclusion

Respiratory infections continue to be an issue for immunosuppressed patients, especially HCT recipients. As recent infectious outbreaks have shown, a simple infection via droplets can spread rapidly and create not just an upper respiratory infection but a lower respiratory infection as well. In this patient population, LRTIs can lead to diffuse alveolar hemorrhage, lung graft versus host disease, and a number of life-threatening issues. The highest mortality rates appear to be from influenza A and adenovirus [19], but as seen with the COVID-19 pandemic of 2020 onward, it can quickly change. It is best to prevent these groups of infections with masks and good hand hygiene as there are still very few therapies that are robust and freely available for HCT recipients in a timely manner. The use of antigens and even more so the use of multiplex PCRs has greatly increased the ability to diagnose URTI and LRTI infections.

Coming full circle in terms of treatment of respiratory infections in the immunosuppressed and particularly in the HCT population, the discussion should always include the possibility of decreasing the patient's immunosuppression allowing the innate immune system to rebound and attempt to clear the infection. If the net state of immunosuppressed was better balanced, then ideally, the infections would have a higher chance of being self-limited. Unfortunately, especially if a patient is still neutropenic related to recent chemotherapy or status post recent monoclonal antibody therapy (example for graft versus host disease), these are not therapies that can be reversed. These then become difficult respiratory infectious cases with minimal options other than supportive care with oxygen.

Conflicts of Interest The author reports no potential conflicts of interest.

References

1. Fishman JA, Practice ASTIDCo. Introduction: infection in solid organ transplant recipients. Am J Transplant. 2009;9 Suppl 4:S3–6.
2. Tomblyn M, Chiller T, Einsele H, Gress R, Sepkowitz K, Storek J, et al. Guidelines for preventing infectious complications among hematopoietic cell transplantation recipients: a global perspective. Biol Blood Marrow Transplant. 2009;15(10):1143–238.
3. Soldatou A, Davies EG. Respiratory virus infections in the immunocompromised host. Paediatr Respir Rev. 2003;4(3):193–204.
4. Seo S, Gooley TA, Kuypers JM, Stednick Z, Jerome KR, Englund JA, et al. Human Metapneumovirus Infections Following Hematopoietic Cell Transplantation: Factors Associated With Disease Progression. Clin Infect Dis. 2016;63(2):178–85.
5. Ison MG. Respiratory viral infections in transplant recipients. Antivir Ther. 2007;12(4 Pt B):627–38.
6. Kumar D. Emerging viruses in transplantation. Curr Opin Infect Dis. 2010;23(4):374–8.
7. Florescu DF, Schaenman JM, Practice ASTIDCo. Adenovirus in solid organ transplant recipients: Guidelines from the American Society of Transplantation Infectious Diseases Community of Practice. Clin Transplant. 2019;33(9):e13527.
8. Milano F, Campbell AP, Guthrie KA, Kuypers J, Englund JA, Corey L, et al. Human rhinovirus and coronavirus detection among allogeneic hematopoietic stem cell transplantation recipients. Blood. 2010;115(10):2088–94.
9. Chemaly RF, Shah DP, Boeckh MJ. Management of respiratory viral infections in hematopoietic cell transplant recipients and patients with hematologic malignancies. Clin Infect Dis. 2014;59 Suppl 5(Suppl 5):S344–51.
10. Shah JN, Chemaly RF. Management of RSV infections in adult recipients of hematopoietic stem cell transplantation. Blood. 2011;117(10):2755–63.
11. Foolad F, Aitken SL, Shigle TL, Prayag A, Ghantoji S, Ariza-Heredia E, et al. Oral Versus Aerosolized Ribavirin for the Treatment of Respiratory Syncytial Virus Infections in Hematopoietic Cell Transplant Recipients. Clin Infect Dis. 2019;68(10):1641–9.
12. Khanna N, Widmer AF, Decker M, Steffen I, Halter J, Heim D, et al. Respiratory syncytial virus infection in patients with hematological diseases: single-center study and review of the literature. Clin Infect Dis. 2008;46(3):402–12.
13. Englund JA. Diagnosis and epidemiology of community-acquired respiratory virus infections in the immunocompromised host. Biol Blood Marrow Transplant. 2001;7 Suppl:2S–4S.
14. Shah DP, Shah PK, Azzi JM, El Chaer F, Chemaly RF. Human metapneumovirus infections in hematopoietic cell transplant recipients and hematologic malignancy patients: A systematic review. Cancer Lett. 2016;379(1):100–6.
15. Ison MG, Hayden RT. Adenovirus. Microbiol Spectr. 2016;4(4).
16. ALVR106: An investigational allogeneic, off-the-shelf, multi-virus specific T cell (VST) therapy targeting respiratory viruses. Available from: https://www.allovir.com/products/alvr106.
17. O'Reilly RJ, Prockop S, Hasan AN, Koehne G, Doubrovina E. Virus-specific T-cell banks for 'off the shelf' adoptive therapy of refractory infections. Bone Marrow Transplant. 2016;51(9):1163–72.
18. Tzannou I, Papadopoulou A, Naik S, Leung K, Martinez CA, Ramos CA, et al. Off-the-Shelf Virus-Specific T Cells to Treat BK Virus, Human Herpesvirus 6, Cytomegalovirus, Epstein-Barr Virus, and Adenovirus Infections After Allogeneic Hematopoietic Stem-Cell Transplantation. J Clin Oncol. 2017;35(31):3547–57.
19. Raboni SM, Nogueira MB, Tsuchiya LR, Takahashi GA, Pereira LA, Pasquini R, et al. Respiratory tract viral infections in bone marrow transplant patients. Transplantation. 2003;76(1):142–6.

COVID-19 in Cancer and Hematopoietic Stem Cell Transplant Recipients

Aliyah Baluch

Abstract COVID-19 is a novel coronavirus infection that rapidly spread throughout the world, creating unprecedented issues on all fronts that led to interruptions in the care of cancer patients. It presents in a bimodal fashion with an initial upper respiratory infection, which can escalate with hyperinflammation in a subpopulation of patients. Risk factors for severe COVID-19 are age and a cancer diagnosis. Within the population of cancer patients are some of the most immunosuppressed: autologous and allogeneic hematopoietic cell transplant (HCT) patients and chimeric antigen receptor therapy (CAR T) patients. When available, preventative strategies, as well as ever-evolving treatment options, were utilized to care for these patient populations.

Keywords COVID-19 · SARs CO-V · nCoV · Coronavirus · Cancer · Immunosuppressed-related COVID · Hematopoietic Cell Transplant-Related COVID19

1 Introduction

In January 2020, the World Health Organization (WHO) announced there was an outbreak in Wuhan, China, caused by the 2019 novel coronavirus (nCoV or SARS-CoV-2) leading to the disease state of COVID-19. After the initial infection, it created a wide range of symptoms, from a mild upper respiratory infection to acute respiratory distress syndrome (ARDS) and, ultimately, death [1]. After exposure, the median time to symptom development with the initial wave of infection was 5 days (range of 2–14 days) [2]. Also, during the initial wave of infection in China,

A. Baluch (✉)
Division of Infectious Diseases, Department of IHM/Internal Medicine, Moffitt Cancer Center, Tampa, FL, USA
e-mail: aliyah.baluch@moffitt.org

C. Somboonwit et al. (eds.), *Global Virology IV: Viral Disease Diagnosis and Treatment Delivery in the 21st Century*,
https://doi.org/10.1007/978-3-031-57369-9_12

approximately 80% of people infected would go on to clear the infection and improve [3]. The remaining patients would often progress with multiple organ failure in addition to hypoxic respiratory failure prior to their demise [4]. The pathogenesis hypothesis of the multi-organ failure was an inflammatory cascade akin to a cytokine storm related to inflammatory dysregulation secondary to the coronavirus infection [5].

It rapidly adjusted the way cancer care was being accessed by patients, including significant delays in appropriate care [6]. Initially, all questions had to be asked and then answered in terms of prevalence, presentation, and if the outcomes were different in cancer patients if they were infected with COVID-19. When the epidemic was announced, healthcare systems had to leverage their access to antigen or PCR testing for COVID-19. Due to the strain on the world's resources, there were often shortages of reagents for testing, leading to the need to send samples to reference labs, creating a further delay in diagnosis, appropriate isolation, and later therapy. Pharmaceutical companies worked hard to create preventative and curative therapies for COVID-19. Unfortunately, as the coronavirus continues to evolve, it has succeeded in evading certain interventions. As of 3/1/2023, there are no longer any preventative monoclonals on the market, even for immunosuppressed populations. There are vaccines available with the ongoing consideration of an annual booster against COVID-19 for all people, including the immunosuppressed [7].

2 Epidemiology

In Western countries, most patients infected with nCoV presented with non-specific symptoms such as fever, shortness of breath, and concomitant chest discomfort. Labs were consistent with low oxygen saturation, and imaging showed ground-glass opacities on a chest x-ray or CT scan. Risk factors in the general population for severe COVID-19 have been found to be advanced age, diabetes, hypertension, chronic kidney disease, obesity, cancer diagnosis, and immunosuppression. These risk factors increased the risk of progression from a simple upper respiratory tract infection to a lower respiratory tract infection with increased mortality [1]. A meta-analysis as of 2021 showed a pooled in-hospital mortality risk in patients with cancer and COVID-19 to be 14.1% [8]. Active cancer appears to have a higher mortality rate than cancer in remission. It is still unclear as to the etiology of the increased mortality: directly related to the cancer, the therapy for the cancer, or that by having cancer, the inherent increased risk of death is increased [9]. Regarding the type of cancer and the risk of death from COVID-19, leukemias have higher mortality risks than solid tumors [10]. There seemed to be a trend toward higher mortality due to respiratory failure in COVID-19-infected hematological patients when neutropenia was resolved or reversed, though it is unclear if it is related to G-CSF or not [11, 12].

To assess COVID-19 in HCT recipients, the Center for International Blood and Marrow Transplant Research (CIBMTR) implemented a special form for

COVID-19-related data capture on March 27, 2020, and collected data until Aug 12, 2020.

There were 318 HCT recipients in this review, showing that both autologous and allogeneic HCT recipients who got COVID-19 had a poorer outcome. The median time to infection was 23 months for autologous HCT recepients and 17 months for allogeneic HCT recipients, i.e., most patients were out in the community when infected. Overall survival one-month post-infection was 68% (95% CI 58–77) for recipients of allogeneic HCT and 67% (55–78) for recipients of autologous HCT, though this is still lower than the general population [13].

3 Challenges to Prevention and Treatment

The biggest challenge during the COVID-19 is the moving target of the actual coronavirus. As it mutates, it evades the current interventions of the day (preventative therapies, vaccine interventions, and treatments). In the current era of electronic medical records, when nCoV initially became a pandemic information technology, hospitals had to build plans for testing and treatment. The microbiology lab had to create assays and up-scale them to allow for the screening of all cancer patients before therapy, especially chemotherapy, and immunotherapy. National bodies overseeing standards, such as the American Society for Transplantation and Cellular Therapy (ASTCT), created guidelines that strongly recommended that during the epidemic of COVID-19, all donor cells should be frozen before initiating recipient conditioning therapy to help minimize any potential delays due to infection. These added steps greatly strained hospital systems, especially the transplant groups caring for the immunosuppressed. Access to appropriate protective clothing and masks for patients and hospital providers became remarkably difficult, especially during the first waves of COVID-19.

As with all respiratory infections, good hand hygiene and appropriate isolation of infectious people are important. For clearance before starting conditioning therapy for an autologous HCT, allogeneic HCT, or CAR T recipient, it was recommended that the patient be tested and proven to be PCR-negative for COVID-19. If a patient was found to be positive for COVID-19, delaying therapy was deemed to be the best course of action for the patient due to the risk of progression of the infection. In contrast, the patient is severely lymphopenic, neutropenic, and immunosuppressed. While being treated for active COVID-19 and waiting afterward for the virus to clear, there was an ongoing concern that the patient in question could relapse with his/her primary disease (and die). This probably has been the greatest challenge during the epidemic for transplanters, attempting to mitigate the risk of the infection versus the risk of losing the optimal time to transplant [2].

In cases of nCoV-infected HCT recipients, the recommendation was to isolate the PCR-positive cases away from those that were not infected. COVID-19 cases were not to be kept on laminar air flow and positive pressure floors due to the risk of nosocomial infection to the other patients on the floor. Traditionally, respiratory infections in the HCT population shed 3 weeks, but it has been found that nCoV had a median duration of shedding of 7.7 weeks (range 2 to 18.7). There were even 2 outliers s/p CAR T that had persistent shedding >5 months post-infection [14, 15]. The remaining uninfected patients undergoing transplant were kept on their transplant floors with minimal interactions with the outside world to decrease their potential exposure to nCoV. This intervention was specifically challenging for the patients as they are often on the unit for multiple weeks at a time, leading to the potential for decreased mental wellness due to a lack of social interactions [2].

4 Newer Approaches

Vaccine, used to induce protective antibodies, recommendations continue to be updated, especially in the immunosuppressed population. Pfizer-BioNTech (bivalent) and Moderna (bivalent) are the main vaccine options recommended by the CDC, with updated boosters to protect against the original COVID-19 and the Omicron BA.4 and BA.5 variants [16].

Treatment for COVID-19 in cancer, and especially in HCT patients, has become very streamlined as of 2023. The PINETREE Study showed efficacy for using outpatient remdesivir for 3 days. In contrast, for severe COVID-19, it is still recommended that the patient be admitted for supplemental oxygen and remdesivir for 5 to 10 days [17, 18]. Cancer patients with COVID-19 continue to have a higher pretest probability of requiring supplemental oxygen therapy [19]. Baricitinib is a Janus kinase-STAT signaling inhibitor (JAK-STAT) that has been shown to have antiviral and anti-inflammatory abilities, leading to potentially lower cytokine levels. This has been shown to be advantageous in combination with remdesivir, leading to reduced recovery time in hospitalized patients with COVID-19 pneumonia [20, 21].

Oral outpatient options have arrived for the treatment of COVID-19, leading to a decrease in morbidity and mortality. Molnupiravir is best used within 2 to 3 days of the onset of symptoms. No significant drug-drug interactions are making it advantageous for immunosuppressed populations that may still be on calcineurin or mTOR inhibitors. Paxlovid (nirmatrelvir/ritonavir), on the other hand, is also an oral outpatient option to treat nCoV, but as it inhibits CYP3A, significant drug-drug interactions are making it disadvantageous for the use in immunosuppressed patients on active therapy [22, 23].

5 Conclusion

The pandemic related to COVID-19 required innovative thinking by healthcare systems across the world, including cancer physicians. At every potential entry point of the infection, interventions had to be assessed. Then, they implemented measures to protect the most vulnerable: auto- and allogenic HCT recipients as well as CAR T patients. Hospitals collaborated to share information on how best to test, treat, and protect their cancer patients promptly. The issue of COVID-19 is less dire in 2023, but there are still clusters of infections leading to poor outcomes. But as the virus continues to mutate and infect, it is losing pathogenicity. Whereas one of the common symptoms of the initial wave of COVID-19 in Italy and then in the US was anosmia, now, in 2023, most COVID-19 symptoms are related to rhinorrhea, cough, or fever. In the HCT field of work, things are nearing back to normal with the loosening of many "COVID-19 era" restrictions. COVID-19 PCR screening is still required prior to HCT or CAR T prior to the start of conditioning chemotherapy, but visitation rules are back to pre-2020 regulations. Hospital epidemiologists continue to monitor the trends of all respiratory infections in this vulnerable population but hopefully without the fear of another imminent high-fatality-related infection on the horizon.

Conflicts of Interest The author reports no potential conflicts of interest.

References

1. Goldman JD, Robinson PC, Uldrick TS, Ljungman P. COVID-19 in immunocompromised populations: implications for prognosis and repurposing of immunotherapies. J Immunother Cancer. 2021;9(6).
2. Ljungman P, Mikulska M, de la Camara R, Basak GW, Chabannon C, Corbacioglu S, et al. The challenge of COVID-19 and hematopoietic cell transplantation; EBMT recommendations for management of hematopoietic cell transplant recipients, their donors, and patients undergoing CAR T-cell therapy. Bone Marrow Transplant. 2020;55(11):2071–6.
3. Wu Z, McGoogan JM. Characteristics of and Important Lessons From the Coronavirus Disease 2019 (COVID-19) Outbreak in China: Summary of a Report of 72 314 Cases From the Chinese Center for Disease Control and Prevention. JAMA. 2020;323(13):1239–42.
4. Siddiqi HK, Mehra MR. COVID-19 illness in native and immunosuppressed states: A clinical-therapeutic staging proposal. J Heart Lung Transplant. 2020;39(5):405–7.
5. Del Valle DM, Kim-Schulze S, Huang HH, Beckmann ND, Nirenberg S, Wang B, et al. An inflammatory cytokine signature predicts COVID-19 severity and survival. Nat Med. 2020;26(10):1636–43.
6. Collaborative CO. Effect of COVID-19 pandemic lockdowns on planned cancer surgery for 15 tumour types in 61 countries: an international, prospective, cohort study. Lancet Oncol. 2021;22(11):1507–17.
7. Coronavirus (COVID-19) Update: FDA Limits Use of Certain Monoclonal Antibodies to Treat COVID-19 Due to the Omicron Variant 2022. Available from: https://www.fda.gov/news-events/press-announcements/coronavirus-covid-19-update-fda-limits-use-certain-monoclonal-antibodies-treat-covid-19-due-omicron.
8. Zarifkar P, Kamath A, Robinson C, Morgulchik N, Shah SFH, Cheng TKM, et al. Clinical Characteristics and Outcomes in Patients with COVID-19 and Cancer: a Systematic Review and Meta-analysis. Clin Oncol (R Coll Radiol). 2021;33(3):e180–e91.
9. Kuderer NM, Choueiri TK, Shah DP, Shyr Y, Rubinstein SM, Rivera DR, et al. Clinical impact of COVID-19 on patients with cancer (CCC19): a cohort study. Lancet. 2020;395(10241):1907–18.

10. Lee LYW, Cazier JB, Starkey T, Briggs SEW, Arnold R, Bisht V, et al. COVID-19 prevalence and mortality in patients with cancer and the effect of primary tumour subtype and patient demographics: a prospective cohort study. Lancet Oncol. 2020;21(10):1309–16.
11. Kuri-Cervantes L, Pampena MB, Meng W, Rosenfeld AM, Ittner CAG, Weisman AR, et al. Comprehensive mapping of immune perturbations associated with severe COVID-19. Sci Immunol. 2020;5(49).
12. Dentone C, Vena A, Loconte M, Grillo F, Brunetti I, Barisione E, et al. Bronchoalveolar lavage fluid characteristics and outcomes of invasively mechanically ventilated patients with COVID-19 pneumonia in Genoa, Italy. BMC Infect Dis. 2021;21(1):353.
13. Sharma A, Bhatt NS, St Martin A, Abid MB, Bloomquist J, Chemaly RF, et al. Clinical characteristics and outcomes of COVID-19 in haematopoietic stem-cell transplantation recipients: an observational cohort study. Lancet Haematol. 2021;8(3):e185–e93.
14. Mushtaq MU, Shahzad M, Chaudhary SG, Luder M, Ahmed N, Abdelhakim H, et al. Impact of SARS-CoV-2 in Hematopoietic Stem Cell Transplantation and Chimeric Antigen Receptor T Cell Therapy Recipients. Transplant Cell Ther. 2021;27(9):796 e1– e7.
15. Milano F, Campbell AP, Guthrie KA, Kuypers J, Englund JA, Corey L, et al. Human rhinovirus and coronavirus detection among allogeneic hematopoietic stem cell transplantation recipients. Blood. 2010;115(10):2088–94.
16. Available from: https://www.cdc.gov/coronavirus/2019-ncov/vaccines/stay-up-to-date.html.
17. Gottlieb RL, Vaca CE, Paredes R, Mera J, Webb BJ, Perez G, et al. Early Remdesivir to Prevent Progression to Severe Covid-19 in Outpatients. N Engl J Med. 2022;386(4):305–15.
18. Beigel JH, Tomashek KM, Dodd LE, Mehta AK, Zingman BS, Kalil AC, et al. Remdesivir for the Treatment of Covid-19 - Final Report. N Engl J Med. 2020;383(19):1813–26.
19. Liu C, Zhao Y, Okwan-Duodu D, Basho R, Cui X. COVID-19 in cancer patients: risk, clinical features, and management. Cancer Biol Med. 2020;17(3):519–27.
20. Kalil AC, Patterson TF, Mehta AK, Tomashek KM, Wolfe CR, Ghazaryan V, et al. Baricitinib plus Remdesivir for Hospitalized Adults with Covid-19. N Engl J Med. 2021;384(9):795–807.
21. Magro G. COVID-19: Review on latest available drugs and therapies against SARS-CoV-2. Coagulation and inflammation cross-talking. Virus Res. 2020;286:198070.
22. Wen W, Chen C, Tang J, Wang C, Zhou M, Cheng Y, et al. Efficacy and safety of three new oral antiviral treatment (molnupiravir, fluvoxamine and Paxlovid) for COVID-19:a meta-analysis. Ann Med. 2022;54(1):516–23.
23. Marzolini C, Kuritzkes DR, Marra F, Boyle A, Gibbons S, Flexner C, et al. Recommendations for the Management of Drug-Drug Interactions Between the COVID-19 Antiviral Nirmatrelvir/Ritonavir (Paxlovid) and Comedications. Clin Pharmacol Ther. 2022;112(6):1191–200.

Epstein-Barr Virus: Acute Infection, Oncogenesis, Prevention and Pitfalls

Guy Handley

Abstract Primary infection by the Epstein-Barr virus (EBV) often occurs without symptoms, and by adulthood seroprevalence stands at approximately 95%. When symptomatic, it presents with acute mononucleosis after which most patients recover without specific therapy and without sequelae. Uncommon but potentially life-threatening complications of acute infection are discussed. Latent infection and immortalization occur, and through oncogenic pathways EBV-associated malignancies may develop. Patients receiving solid organ or hematopoietic stem cell transplants carry additional risks for EBV-associated post-transplant lymphoproliferative disorders through loss or reduction of T-cell surveillance but have differing epidemiology and presentations. These are reviewed along with treatment strategies based on consensus guidelines. Testing platforms carry specific pitfalls as detection of EBV DNA after acute infection is often episomal or non-encapsulated DNA rather than virion and varies based on the specimen tested. While no durable vaccine exists, potential targets and past trials are reviewed.

Keywords Epstein-Barr Virus · Mononucleosis · Post-transplant Lymphoproliferative Disease · Malignancy · Vaccine, Hemophagocytic Lymphohistiocytosis

---

G. Handley (✉)
Division of Infectious Disease and International Medicine, Department of Internal Medicine, Morsani College of Medicine, University of South Florida, Tampa, FL, USA
e-mail: Guyhandley@usf.edu

C. Somboonwit et al. (eds.), *Global Virology IV: Viral Disease Diagnosis and Treatment Delivery in the 21st Century*,
https://doi.org/10.1007/978-3-031-57369-9_13

1 Introduction

The Epstein-Barr virus (EBV), or human herpesvirus 4, was initially discovered in a patient with Burkitt's lymphoma and is associated with multiple malignancies [1, 2]. Most infections are thought to be asymptomatic [3, 4]. Clinical cases present with infectious mononucleosis after several weeks of incubation. This syndrome is characterized by the triad of fever, tonsillar pharyngitis, and lymphadenopathy, with cases reported as early as 1915 [5–9]. At least one early research technician studying the virus developed mononucleosis, with subsequent production of virus specific antibodies suggesting a link [5]. Prospective and cohort studies among university-age students have shown variable incidences of symptomatic infection occurring in 25–74% of cases [10–13]. Seroprevalence increases with age and stands at over 90–95% by adulthood [8, 10, 14].

Two genomic sub-types exist, EBV-1 and EBV-2, with notable differences in the expression of latent infection cycle EBV nuclear antigen (EBNA) genes and EBV encoded small RNAs (EBER) [3, 15]. EBV-1 is the dominant strain in the United States and Europe, but both may be more evenly distributed in other regions worldwide [3]. Initial infections are thought to occur in oropharyngeal epithelial cells after contact with infected saliva, followed by lytic viral replication [16–18]. The virus then secondarily infects B lymphocytes, leading to a life-long infection and a viral reservoir of 1–50 B-cells per million [18–20]. This delay potentially explains the prolonged incubation period, as symptoms occur only after the development of cytotoxic T-cells directed against infected B lymphocytes [7, 21]. The process is still not fully understood. After initial infection, asymptomatic shedding may occur from periodic reactivation in a significant proportion of healthy individuals, increasing in times of stress or critical illness [22, 23]. Certain populations, such as solid organ transplant recipients or patients with hematologic malignancy, may exhibit shedding in up to 50–92% of patients [23, 24]. Interestingly virus produced from epithelial cells may be more effective at infecting B-cells and conversely virus produced by B-cells may be more effective at infecting epithelial cells through altered tropism in in vitro models [16]. Viral latency in B-cell populations occurs through complex interactions and the production of EBNA, EBER and latent infection membrane proteins (LMP) without the production of true viral particles while typically maintaining multiple episomal genetic copies [2, 25, 26]. Although not discussed extensively in this chapter and not fully understood, these products assist in establishing B-cell proliferation, inhibit apoptosis, and can lead to cell line immortalization and oncogenesis [2, 25, 27]. NK and T-cells may also be infected and may be associated with similar lymphoproliferative or malignant disorders [28]. Immortalization leads to a population of infected lymphocytes that may produce viral-specific antigens but not undergo lytic viral replication. Assays that measure quantitative EBV DNA levels may colloquially refer to an EBV "viral load." In many of these patients, particularly after the acute initial EBV infection or in EBV-associated malignancies, the EBV DNA detected is typically cell-associated, episomal or non-encapsulated naked DNA present due to cellular death and apoptosis rather than the true lytic phase virions seen in acute mononucleosis [28, 29]. As such, this chapter will refer to EBV reactivation or viremia as EBV DNA-emia unless the distinction is made or a particular study defines the term. Prior EBV infection may be associated with autoimmune diseases, including systemic lupus erythematosus, multiple sclerosis or rheumatoid arthritis, by serving as a possible environmental trigger or through antigen mimicry, but this is not fully understood and has not been clearly defined [30].

Though most of these latently infected B-cells are cleared by host immune responses, some express EBV gene products that inhibit lytic cycle viral replication and may assist in immune system evasion and the development of lymphoproliferative disorders [25, 26, 31, 32]. EBV gene products are thought to be implicated in about 1.5% of cancer cases worldwide [4]. Different patterns of gene expression are associated with specific EBV-associated malignancies [27, 33]. These may provide targets for therapeutics, preventative treatments or vaccines to prevent EBV-related complications. This chapter focuses on the clinical aspects of EBV infection and potential subsequent complications in the adolescent and adult populations and then discusses available therapeutics as well as potential future targets for treatment or prevention.

2 Acute EBV Infection and Complications

2.1 Infectious Mononucleosis

As noted, primary EBV infection, when symptomatic, presents with acute mononucleosis. While classically defined by fever, tonsillar pharyngitis and lymphadenopathy, infectious mononucleosis may also present with a prolonged illness that can include fatigue, headaches, myalgias, respiratory symptoms or abdominal pain with a range of 3–54 days of symptoms [8]. A generalized rash that may be petechial, maculopapular or urticarial has been described after administration of antibiotics, classically amoxicillin, in patients with infectious mononucleosis and is thought to occur through a delayed hypersensitivity reaction [34, 35]. Patients typically have no reactions after future administration. Atypical lymphocytosis is common but not specific to EBV and may be related to the increase in cytotoxic CD8+ T-cells, which correlates with disease severity [36, 37].

Heterophile antibodies, which have had a known association with infectious mononucleosis as early as 1932, are typically positive, and rapid testing platforms provide 85% sensitivity with near 100% specificity in symptomatic patients but may be absent in young children or early in infection [38, 39]. It is important to note that positive tests may occur with other conditions such as human immunodeficiency virus, lymphoma, systemic lupus erythematosus, and other viral infections [32]. When the diagnosis is unclear or a heterophile antibody is negative, specific antibodies against viral capsid antigen, nuclear antigen or early antigen may be obtained with commercial assays and can suggest acute infection. In particular, the absence of the nuclear antigen IgG, which is not typically seen until at least 4 weeks after symptom onset, may suggest acute infection [39, 40]. As EBV establishes latency in B lymphocytes, it is not uncommon to have detectable EBV specific antibodies, particularly IgM, for months to years after initial infection due to other viral infections as well as non-infectious etiologies that may drive B lymphocytes activation [40, 41]. Serologic testing should only be obtained in the appropriate clinical context, as detection of antibodies does not necessarily imply reactivation, persistent acute infection or complications related to EBV itself. EBV polymerase chain reaction (PCR) is rarely needed in acute infections and may be negative early in the clinical course [7]. In acute infection, blood viral load may be significantly higher in more symptomatic patients, but oral viral load has not been shown to correlate with symptoms [36]. Most patients recover and no specific treatment is indicated or has been shown to improve outcomes; however, in rare instances, death has occurred with complications that are further described [32].

2.2 *Complications of Acute EBV Infection*

Complications from acute infection can involve nearly any organ system and are too numerous to review here [2]. It is worth noting those that may be life-threatening. Splenomegaly is common in at least half of cases and can be life-threatening if a splenic rupture develops [42]. The estimated rate of this complication is 1–2 per 1000 cases, and of those, half occur spontaneously, although it may develop after trauma [43]. Critical airway obstruction from massive lymphadenopathy or edema may occur and corticosteroids have been utilized in such cases [8, 32]. Potentially life-threatening neurologic complications include Guillain-Barré syndrome, meningoencephalitis, transverse myelitis and cerebritis [44, 45]. A potential, but poorly understood, association with Lemierre's syndrome and *Fusobacterium necrophorum* infection may exist [46, 47]. Fatal complications from myocarditis, cardiac arrhythmia, liver failure, and hematologic complications have been reported [8, 32]. Multiple instances of organ failure due to infiltration of tissues by activated or EBV-infected lymphocytes, rather than infection of tissue specific cells by EBV virus, have been described [48–50]. Overall, such complications are exceedingly rare and without specific proven treatments, but they may be encountered in clinical practice. It is important to note that rare but defined primary immunodeficiencies exist, notably the X-linked Duncan disease, which may predispose to fulminant or fatal acute infectious mononucleosis [51, 52].

2.3 *Chronic Active EBV Infection*

Chronic active EBV infection (CAEBV) conversely is an entity characterized by persistent mononucleosis-like symptoms or hepatosplenomegaly that may progress to a potential life-threatening lymphoproliferative disorder characterized by

T-cell or NK-cell infiltration of vital organs [53, 54]. It is thought this may occur after EBV infection of T or NK-cells rather than B-cells, as patients often develop loss of B-cells, and they may progress to T, B or NK-cell lymphomas [54, 55]. Additional complications include multi-organ failure, opportunistic infection or hemophagocytosis [56]. Proposed guidelines define CAEBV not only with detectable EBV DNA-emia and serologies but also with proven tissue infiltration by EBV infected lymphocytes [54, 56]. Multiple therapeutic modalities have been attempted, including acyclovir, azathioprine, corticosteroids, cyclosporine, interferons, intravenous immunoglobulin (IVIG), rituximab, cytotoxic chemotherapy or EBV-specific cytotoxic T lymphocytes; however, all have a transient or minimal effect on the disease [54]. Without stem cell transplant, outcomes are poor, with death occurring approximately 10–15 years after diagnosis [54, 57–59]. Some experts recommend combination therapies with ganciclovir and either histone deacetylase inhibitors or bortezomib based on limited clinical data as a bridge to stem cell transplant [59].

Acute infectious mononucleosis often requires several weeks before recovery. Some patients have reported a chronic fatigue syndrome attributed to EBV; however, no data or evidence has been shown to prove a causal link to such a syndrome [60, 61]. Patients may present with a wide range of symptoms, including myalgias, headaches or sore throat in the absence of objective laboratory or clinical findings attributable to EBV [60, 61]. As noted, EBV-specific antibodies or detection of EBV DNA in blood are variably present in asymptomatic patients and are not reliable to diagnose a chronic active infection. Multiple studies have found no significant difference in EBV specific antibody titers in such patients compared to those without chronic fatigue symptoms [61–64]. This makes it difficult to attribute a chronic fatigue syndrome to the EBV virus itself, likely owing to the 90–95% seroprevalence by adulthood, and as discussed later, anti-virals have not been shown to have any effect due to how the latent EBV genome is maintained.

2.4 Hemophagocytic Lymphohistiocytosis

Hemophagocytic lymphohistiocytosis (HLH) has been described as an acute EBV infection. It is a clinical syndrome of pathogenic lymphocyte and histiocyte activation leading to infiltration of multiple sites, including bone marrow, hepatic tissue, the spleen and lymph nodes [65–67]. Immune deficiencies, such as *GATA2* deficiency, may be associated with the development of HLH [68]. Though rare, HLH primarily complicates acute EBV infection in pediatric patient populations and is almost exclusively driven by a monoclonal population of EBV infected T-cells, leading to extensive cytokine overactivity [69–71]. Though far more rare, HLH after acute EBV infection driven by monoclonal populations of NK and B-cells has also been described [71].

HLH associated with EBV outside of primary infection is driven by EBV-associated malignancies such as lymphoma [72]. While lymphomas of T/NK-cell origin may have a higher predisposition to HLH, particularly in the Western hemisphere, they may also be associated with B-cell lymphomas [72, 73]. A particular challenge comes from the utilization of EBV DNA quantitative PCR in whole blood or plasma to diagnose HLH in patients with lymphoproliferative disorders or lymphomas. Detection of EBV DNA, which may be more common in patients with malignancy and HLH, has been used to suggest EBV as a possible trigger [74]. However, as noted, the EBV DNA detected is usually cell-associated, episomal or non-encapsulated naked DNA rather than virion [28, 29]. Detection of EBV by in-situ hybridization on lymphoma tissue may be able to make a distinction when elevated EBV DNA levels are measured in blood [72]. This still cannot differentiate whether EBV or the EBV-related malignancy itself is the etiology behind HLH in these settings, and further study is needed to understand the disease process.

Typical treatments include corticosteroids or IVIG in milder cases, or a combination of dexamethasone and etoposide in more severe cases using the regimen recommended by the HLH-2004 study [71, 75]. The HLH-2004 study demonstrated benefit with the dexamethasone and etoposide combination; however, the authors concluded that cyclosporine, which had been utilized historically, may not significantly impact outcomes [75]. Rituximab may have a role and has been employed as salvage therapy with mixed results [71, 76–78]. Future studies are needed to demonstrate benefit, and notably, those suggesting a benefit typically involve patients with HLH as a complication of acute EBV infection in pediatric and adolescent populations rather than HLH associated with malignancy [77, 78]. Ultimately, stem cell transplantation may be required in refractory cases [71].

3 Lymphoproliferative Disorders

EBV-associated lymphoproliferative disorders refer to the unchecked growth of subsets of EBV-infected lymphocytes. Disease severity ranges from benign polyclonal hyperplasia to polyclonal or monoclonal lymphomas; over time, the incidence has continued to increase [79, 80]. Loss or reduction of active T-cell surveillance is thought to underlie the disease process, as suggested with mouse models depleted of T-cells [32, 81]. Multiple genetic disorders are associated with EBV-associated lymphoproliferative disorders but are not discussed here [82]. Though most cases are EBV-related, a significant proportion of post-transplant lymphoproliferative disease (PTLD) is EBV negative [83]. The World Health Organization classification includes four subtypes: nondestructive PTLD, polymorphic PTLD, monomorphic PTLD and Hodgkin's lymphoma [83]. This section will focus primarily on EBV-related PTLD noting specific differences in each population.

3.1 Post-transplant Lymphoproliferative Disease in Solid Organ Transplant

EBV-associated PTLD is primarily a host-derived process in solid organ transplantation (SOT) but can originate from donors as well [84]. The rate varies by type of transplant among renal (0.8–2.5%), pancreatic (0.5–5%), liver (1–5.5%), heart (2–8%) and lung (3–10%) allograft recipients, but can be significantly higher in intestinal or multi-organ transplant recipients [80, 85]. EBV and CMV mismatch between donor and recipient and negative recipient EBV serostatus are associated with higher rates of PTLD; moreover, particularly in pediatric populations, acute primary EBV infection may be a trigger to develop disease in the post-transplant setting [80, 85–87]. Caucasian race and age at the time of transplant may be risk factors specifically for the development of post-transplant lymphoma [88]. Lymphocyte depleting regimens, such as anti-thymocyte globulin or anti-CD3 antibodies, or lymphocyte depleting prophylaxis, have been associated with higher rates of PTLD and EBV-related malignancy after SOT [85, 87, 89–91].

Notably, PTLD originating from the recipient may present with more disseminated disease, while donor-derived PTLD may be more restricted to the allograft [92]. EBV-positive PTLD occurs more frequently within the first year after transplant, whereas the proportion of EBV negative PTLD increases farther from transplant [85]. While EBV may be the predominant etiology for most, it is more often seen in non-germinal center B-cell subtypes [83, 85]. EBER has been detected in allograft tissues prior to the onset of PTLD in much greater frequency than control groups and may be useful to predict future disease development [93]. More frequent detection of EBV DNA in the blood or plasma with higher levels or rapid increases may suggest increased risks [94, 95]. Depending on the tumor sub-type, cell-free plasma EBV DNA may have higher specificity than DNA detected from mononuclear cells, but there are issues with a lack of standardization and laboratory techniques that can significantly alter measured DNA levels. The optimal method of monitoring EBV DNA has not been established; moreover, EBV DNA-emia is not uncommon and, at lower levels, has not been clearly associated with PTLD development [28, 95, 96].

The American Society of Transplantation (AST) provides extensive guidance regarding PTLD after SOT [96]. Notable EBV-specific preventative or therapeutic strategies are included, but the disease requires a multidisciplinary approach with extensive therapeutic options that cannot all be presented here. In larger studies and meta-analyses, anti-viral prophylaxis has not been shown to reduce the incidence of PTLD, though induction regimens containing rituximab may have lower long-term risk [96–98]. When higher levels of EBV DNA are detected, smaller studies have suggested a potential benefit for pre-emptive rituximab; however, guidelines conclude there is not sufficient evidence to support this approach at present in part due to difficulties in evaluating EBV DNA and inability to define a "set-point" or threshold to indicate need for therapy [96, 99, 100]. Reduction in immunosuppression when possible remains the primary recommendation for prevention in high-risk populations or initial treatment when PTLD develops [96, 101]. Surgical resection or radiotherapy may be an option for localized disease [80, 96]. There is no convincing evidence to support a benefit for anti-viral agents; however, in IgM positive subtypes of PTLD, some lytic phase gene expression has been found [102]. Despite this, a therapeutic benefit for anti-virals has not been found. Perhaps the most data support a rituximab-based approach in CD 20+ EBV-associated PTLD with multiple trials showing variable benefit, with some patients achieving remission with rituximab and reduction of immunosuppression alone [85, 96, 103–105]. Cytotoxic chemotherapy with CHOP (cyclophosphamide, doxorubicin, Oncovin or generic name vincristine, and prednisone) in combination with rituximab has also been utilized, with potential increased survival [96, 106, 107]. A rituximab-based approach for CD20+ PTLD followed by consideration of consolidation by risk-stratified sequential treatment (RSST) trial protocols is strongly recommended in adult populations by AST guidelines [96, 107].

Additional agents under investigation or with potential benefit include Bruton's tyrosine kinase inhibition, proteasome inhibition, checkpoint inhibitors, anti-CD30 therapy, or mTOR inhibition [85, 96]. In recent mouse models, however, programmed death receptor (PD)-1 blockade leads toward the progression of EBV-positive PTLD [108]. Adoptive cellular therapy with cytotoxic T-lymphocytes (CTLs) has more evidence in PTLD following hematopoietic cell transplant (HCT), but has been utilized in cases of refractory PTLD after SOT [96, 109–111]. It has been proposed that the host origin of most PTLD in solid organ transplant, rather than donor-derived in HCT, may reduce the efficacy of this modality, and third-party donor CTLs with high rates of HLA match may provide the most benefit [96, 109, 110, 112]. CTLs can be considered for refractory EBV-positive PTLD, and in at least some case reports, they have been used as secondary prophylaxis in high-risk patients [113, 114]. EBV-related PTLD presents a challenge for clinicians. Anti-virals have not shown significant benefit, but primary therapy typically involves a reduction in immunosuppression, followed by rituximab-based therapy in CD20+ disease. Adoptive cellular therapies may be considered for refractory cases.

3.2 *Post-transplant Lymphoproliferative Disease in Hematopoietic Cell Transplant*

EBV-associated PTLD is thought to occur due to a similar loss or reduction of active T-cell surveillance following allogeneic HCT, but risk factors and presentations may differ. While extremely rare, cases have been described after autologous HCT [115]. This section will focus primarily on allogeneic HCT. Nearly all cases are donor-derived following HCT and typically occur within 1 year of transplant, with most occurring within the first 6 months [80, 116]. Larger studies report an overall incidence of 0.2%–3.2% of transplants, though the rate varies significantly based on a number of risk factors [80, 116, 117]. In a large study, rates differed among donor types, including matched family donors (1.2%), mismatched family donors, including haploidentical transplants (2.8%), matched unrelated donors (4%) and mismatched unrelated donors (11.2%) [117]. Cord blood transplantation may carry around a 4.2% incidence [118]. In this study, patients were prospectively monitored for EBV DNA-emia with a cumulative detection rate of 5.0% at 3 years though none of those evaluated had detectable EBER or LMP1, suggesting a more latent process [118]. Among the most significant risk factors are T-cell depleting conditioning regimens. Overall rates may be lower if lymphocyte depletion is extensive, including B-cells and T-cells, rather than T-cell selective regimens [94, 116, 117, 119]. Interestingly, in cord blood transplantation, myeloablative regimens may carry a lower risk of PTLD compared to non-myeloablative regimens, particularly those that include anti-thymocyte globulin (ATG), with a reported difference of 2% vs. 21%; however, this requires further study as DNA-emia was included with PTLD as an EBV-related complication in the reported outcome [120]. As noted, cases occur after haploidentical transplantation, but a large study found no cases in 396 haploidentical transplantation patients who had received post-transplantation cyclophosphamide for graft-versus-host disease (GVHD) prophylaxis during the 12-month follow-up period [121]. Less established but reported risk factors include age over 50, prolonged maintenance immunosuppression for GVHD, prior splenectomy, second HCT and donor/recipient EBV serologic mismatch [80, 116, 122, 123]. Protective factors that may be associated with lower rates of PTLD include rituximab exposure within 6 months after HCT, sirolimus use, GVHD prophylaxis with post-transplant cyclophosphamide and early CD4+ T-cell reconstitution [122].

While EBV DNA-emia may occur with PTLD, there are several pitfalls that must be considered with DNA monitoring, pre-emptive and prophylactic strategies. EBV DNA-emia is common after HCT in high-risk populations and does not necessarily imply progression to PTLD. The European Conference on Infections in Leukemia (ECIL) provides substantial guidance regarding the management of EBV and PTLD after HCT [123]. Initial peak levels of EBV DNA appear similar between T-cell depleting and non-T-cell depleting conditioning regimens, but recurrent DNA-emia appears to be more frequent after T-cell depletion, and in at least one study, PTLD only occurred in patients with EBV DNA-emia and T-cell depleting conditioning [94]. In these high-risk patients, DNA-emia occurred in 54% of patients following T-cell depleting regimens and 28% of those with non T-cell depleting regimens; however, only 10 of those 46 patients (21.7%) with EBV DNA-emia and T-cell depletion developed PTLD [94]. Those with EBV-associated PTLD had a median DNA level of 110,000 copies/mL EBV DNA (range: 1800–790,000 copies/mL) at the time of diagnosis. This suggests that the majority of patients with DNA-emia will not progress to PTLD, even in the higher risk populations, and may not require therapy, though EBV DNA-emia may be a risk factor. Notably no patients with plasma EBV DNA levels less than 1000 copies/mL developed PTLD even in this highest risk population [94]. In this and other similar studies measuring EBV DNA levels, potential confounders include a lack of standardization and variability between specimens, such as plasma versus whole blood.

Despite this limitation, EBV DNA-emia is likely a risk factor for the development of PTLD, and thus prophylactic strategies in high-risk patients would seem to have a potential benefit. A retrospective analysis of 55 high-risk patients receiving ATG conditioning were given rituximab 200 mg (fixed dose) on day +5 compared to a historical control group of 68 patients found to have a lower rate of EBV DNA-emia (56% vs. 85%; $p = 0.0004$) in whole blood samples [124]. There was no significant difference in survival at 180 days or rates of PTLD; however, only one case of EBV-associated PTLD was diagnosed [124]. Rituximab therapy carries its own risks, which should be carefully considered prior to administration. In a study of patients receiving rituximab early following HCT, most developed significant and potentially life-threatening cytopenias up to 12 months following administration, along with the potential for increased invasive infections [125]. Alternatively, pre-emptive strategies have been evaluated but face similar risks. In a study evaluating pre-emptive rituximab therapy, 49 HCT patients with T-cell depleting conditioning regimens were prospectively monitored and 17 (35%) developed an EBV DNA-emia threshold of 1000 copies/mL in plasma ($n = 17$; 35%) [126]. Of these 15 were given rituximab 375 mg/m^2 as a pre-emptive intervention, while two had lymphadenopathy at time of screening and were found to already have PTLD on pathologic specimens [126]. Compared to a historical cohort, the authors concluded there was a lower rate of PTLD incidence and mortality with the pre-emptive strategy [126]. It should be noted that this conclusion is confounded by including two patients with PTLD at the time of detection of EBV DNA >1000 copies/mL and would require a better controlled study to prove a benefit. There was not a statistically significant difference in incidence, though a reduced trend existed, of PTLD. Additionally, a bias toward estimated survival may exist as patients were screened at a set timepoint and then compared to a historical control group with more lead time to measure survival. This was also a study in one of the highest risk patient populations with substantial T-cell depleting conditioning regimens. A subsequent study of 78 allogeneic HCT patients receiving rituximab pre-emptive therapy in which specific EBV DNA levels were not reported, found high rates of bacterial (64%), viral (59%) and fungal (17%) infections in the 36-month follow-up period [127]. Of these, 52 patients were compared to a historic control group who had not received rituximab, and there was a statistically higher rate of bacterial (55% vs. 35%; $p = 0.048$) infections [127]. Non-relapse mortality over 36 months also trended higher in the rituximab group as compared to controls (35% vs. 15%; $p = 0.16$) [127]. Another study of 219 high-risk HCT patients monitored weekly for EBV DNA in whole blood samples and given rituximab 375 mg/m^2 weekly until resolution of EBV DNA-emia if they developed EBV reactivation, defined as two consecutive weekly measurements of greater than 5000 UI/mL EBV DNA [128]. Of these, 117 developed EBV reactivation as defined by the authors and received rituximab. The patients were compared to a control group of 112 patients without EBV reactivation. No differences in relapse incidence, progression-free survival or overall survival at 24 months were found between the groups, though patients receiving rituximab had a statistically significant higher rate of receipt of intravenous immunoglobulin (IVIG) and lower overall immunoglobulin levels [128]. Though not statistically significant, there was a trend toward worse survival at 24 months in patients receiving rituximab (59% vs. 82%; $p = 0.23$) though multiple confounders are potentially present and a causal relationship cannot be established.

Based on these limited studies, the ECIL guideline recommends EBV DNA monitoring by quantitative PCR, at least weekly, from within the first month after allogeneic HCT until at least 4 months afterward, and potentially longer in high-risk patients [123]. Whole blood, serum and plasma are all considered acceptable sources. Prophylactic rituximab should only be considered in the highest risk populations as efficacy data is lacking, while anti-virals, IVIG and interferon are not recommended [123]. Significant EBV DNA-emia is noted as a potential indication for pre-emptive therapy with rituximab; however, the guideline notes that the threshold for what constitutes significant or high-level EBV DNA-emia has not been established or defined, and the rate of increase in EBV DNA levels is likely significant as well [123]. Reduction in immunosuppression is recommended. Third-party EBV CTLs are potential options to be considered for prophylaxis and pre-emptive therapies [123, 129]. Once EBV-associated PTLD is diagnosed, weekly rituximab 375 mg/m^2 with reduction of immunosuppression is recommended as first-line therapy [123]. Second-line options include donor lymphocyte infusion (DLI), though at a cost of higher risk of GVHD, the presence of which may be a contraindication to DLI itself, and EBV CTLs [123, 130–132]. No specific therapy exists for central nervous system EBV-associated PTLD, but the guideline discusses potential options [123].

In high-risk patients, EBV DNA-emia is not uncommon after HCT; however, evidence for prophylactic or pre-emptive therapy is not definitive. As noted while guidelines and studies discuss and have differing definitions of EBV reactivation, it may be a misnomer as the majority of EBV DNA detected in these populations is episomal or naked non-encapsulated DNA reflecting its latent nature rather than lytic virion [28, 29, 118]. A specific threshold for initiation of pre-emptive therapy with rituximab has not been established, and significant potential complications of rituximab therapy exist. Though this threshold has not been established, in at least one series of HCT patients with >50,000 copies/mL of EBV DNA in serum had 100% mortality [133]. Testing platforms and the type of specimen used to measure EBV DNA are not often standardized. It is likely that the persistence of detectable EBV DNA as well as rate of increase over time are as important as a threshold of EBV DNA in determining which patients may benefit from intervention.

4 Immortalization and EBV-Associated Malignancies

EBV has been associated with numerable malignancies, which utilize latent EBV gene products for oncogenesis. After primary infection, EBV immortalization in B-cell populations occurs through variable expression rates of EBERs, EBNAs, LMPs and micro-RNA expression by way of complex pathways and receptor mimicry, preventing typical activated B-cell apoptosis and leading to a static peripheral resting memory B-cell state where infectious virion is not produced [25, 26, 31, 134]. By this process, the EBV genome becomes episomal, with a reduced but lifelong expression of gene products [2]. Conversely oral hairy leukoplakia, a nonmalignant lesion of unchecked lytic replication in immunosuppressed patients, may maintain the linear infectious virion [135]. Additionally, by deactivation and suppression of genes that promote growth during the initial infection, infected B-cells may persist in a benign memory resting state, evading immune system surveillance [26]. Such immortalization and lifelong latency require a successful transition through this process. Multiple pathways to oncogenesis exist whereby an infected cell may continue to express genes leading to proliferation, likely more frequent in immunosuppressed patients who may not clear such populations, or through persistent expression of LMPs and growth promoting EBNAs in both infected B-cell and non-B-cell populations [25, 26]. Malignancies, including Burkitt's lymphoma, Hodgkin's lymphoma, various B and T-cell lymphomas, and epithelial cell tumors, including nasopharyngeal and gastric carcinomas, may arise at many steps through this transformative process with differing degrees of latent gene product expression [25, 33]. Typically, such malignancies maintain multiple episomal copies of the EBV viral genome [25]. In such malignancies, these episomes or naked non-encapsulated DNA products may be detected by commercial testing platforms but do not usually represent infectious virion [28, 29]. Interestingly, while the majority of cell free DNA detected in tumors of epithelial origin, such as nasopharyngeal carcinoma, is still episomal or naked non-encapsulated DNA products, there may be some proportion that is encapsulated DNA, suggesting possible virion [28, 136]. This may in part be due to the inability of epithelial cells to transition to the resting memory state [26]. Each of these malignancies lead to distinct clinical presentations with differing treatment regimens which are not covered here. Analyses of in vitro cell lines for some malignancies have shown chemotherapy-induced lytic EBV replication and susceptibility to anti-virals; however, evidence of clinical utility for anti-virals has not been demonstrated [2, 137–140]. As such, anti-virals have no clear utility in the treatment of such patients. In part, this is likely due to the fact oncogenesis is driven by latent gene expression rather than lytic viral replication. In some instances, EBV-specific cytotoxic T-lymphocytes may play a role along with traditional chemotherapeutic regimens [134, 141–143].

5 Prevention and Vaccine Strategies

Owing to the rare but potentially catastrophic acute and long-term complications of EBV infection, primary prevention and vaccination could provide substantial benefit. Annually, there are about 200,000 new cases of EBV-associated malignancy in the United States [144]. An ideal vaccine should reduce oncogenesis and overall EBV-related malignancy burden. Highly conserved targets across genomic subtypes provide the most promise. The glycoprotein gp350 provides such a target in both EBV-1 and 2 genomic sub-types [145]. It is highly expressed and interacts with the

CD21 cellular receptor [146, 147]. While acute infection and latent immortalization occur even in the absence of gp350, its presence enhances infection of B-cells in vitro [148]. One of the first vaccine trials utilized a vaccinia strain expressing gp350 and successfully prevented primary infection in six of nine patients compared to none of the control patients exposed to EBV during the 16-month follow-up period [149]. A later double-blinded, randomized control trial with a recombinant gp350 vaccine demonstrated a 78% efficacy in reducing incidence of infectious mononucleosis; however, there was not a significant difference in the rates of asymptomatic EBV infection [150]. To date, a reliable vaccination strategy has not been demonstrated for primary prevention. In part, patients have been shown to have multiple EBV viral strains, which may provide opportunities to bypass vaccine targets [151, 152]. Another epitope target or vaccines that induce a response to multiple epitopes may be required.

While vaccines for primary prevention are lacking, other potential targets involve oncogenic pathways that can lead to EBV-related malignancies. Strategies include targeting EBV-lytic and latent proteins by inducing a T-cell response through vaccination [153–155]. A phase I trial of such targets seemed to demonstrate promise to potentially reduce symptomatic infection [156]. Such a T-cell response may have a role in preventing oncogenesis in high-risk patients, such as solid organ or stem cell transplant recipients particularly until broader immune reconstitution has occurred. While this concept has been evaluated in HCT through the use of adoptive transfer of EBV-specific CTLs, a potential vaccine strategy would likely be more cost-effective and have broader utility [129]. Due to the widespread burden of symptomatic primary infection with rare, but potentially severe, complications and long-term risks of EBV-associated malignancy, an EBV vaccine will continue to be an area of future investigation.

References

1. Epstein MA, Achong BG, Pope JH. Virus in cultured lymphoblasts from a New Guinea Burkitt lymphoma. BMJ. 1967 Apr 29;2(5547):290–1.
2. Longnecker RM, Kieff E, Cohen JI. Epstein-barr virus. In Fields Virology: Sixth Edition. Vol. 1. Wolters Kluwer Health Adis (ESP). 2013
3. Correia S, Bridges R, Wegner F, Venturini C, Palser A, Middeldorp JM, et al. Sequence Variation of Epstein-Barr Virus: Viral Types, Geography, Codon Usage, and Diseases. Longnecker RM, editor. Journal of Virology. 2018 Nov 15;92(22):e01132–18.
4. Farrell PJ. Epstein-Barr Virus and Cancer. Annual Review of Pathology. 2019 Jan 24;14:29–53.
5. Henle G, Henle W, Diehl V. Relation of Burkitt's tumor-associated herpes-type virus to infectious mononucleosis. Proceedings of the National Academy of Sciences. 1968 Jan 1;59(1):94–101.
6. Hall AJ. A Case Resembling Acute Lymphatic Leukæmia, Ending in Complete Recovery. Proceedings of the Royal Society of Medicine. 1915 Apr;8(Med_Sect):15–9.
7. Dunmire SK, Grimm JM, Schmeling DO, Balfour HH, Hogquist KA. The Incubation Period of Primary Epstein-Barr Virus Infection: Viral Dynamics and Immunologic Events. PLoS pathogens. 2015;11(12):e1005286.

8. Dunmire SK, Verghese PS, Balfour HH. Primary Epstein-Barr virus infection. Journal of Clinical Virology. 2018 May;102:84–92.
9. Evans AS, Neiderman JC, Cenabre LC, West B, Richards VA. A Prospective Evaluation of Heterophile and Epstein-Barr Virus-Specific IgM Antibody Tests in Clinical and Subclinical Infectious Mononucleosis: Specificity and Sensitivity of the Tests and Persistence of Antibody. Journal of Infectious Diseases. 1975 Nov 1;132(5):546–54.
10. Balfour HH, Sifakis F, Sliman JA, Knight JA, Schmeling DO, Thomas W. Age-Specific Prevalence of Epstein–Barr Virus Infection Among Individuals Aged 6–19 Years in the United States and Factors Affecting Its Acquisition. The Journal of Infectious Diseases. 2013 Jul 18;208(8):1286–93.
11. Grimm JM, Schmeling DO, Dunmire SK, Knight JA, Mullan BD, Ed JA, et al. Prospective studies of infectious mononucleosis in university students. Clinical & Translational Immunology. 2016 Aug 12;5(8):e94.
12. Crawford DH, Macsween KF, Higgins CD, Thomas R, McAulay K, Williams H, et al. A cohort study among university students: identification of risk factors for Epstein-Barr virus seroconversion and infectious mononucleosis. Clinical Infectious Diseases. 2006;43(3):276–82.
13. Infectious Mononucleosis and its Relationship to EB Virus Antibody: A JOINT INVESTIGATION BY UNIVERSITY HEALTH PHYSICIANS AND P.H.L.S. LABORATORIES. BMJ. 1971 Dec 11;4(5788):643–6.
14. Pereira MS, Blake JM, Macrae AD. EB virus antibody at different ages. BMJ. 1969 Nov 29;4(5682):526–7.
15. Arrand JR, Young LS, Tugwood JD. Two families of sequences in the small RNA-encoding region of Epstein-Barr virus (EBV) correlate with EBV types A and B. Journal of Virology. 1989 Feb;63(2):983–6.
16. Borza CM, Hutt-Fletcher LM. Alternate replication in B cells and epithelial cells switches tropism of Epstein–Barr virus. Nature Medicine. 2002 Jun;8(6):594–9.
17. Li QX, Young LS, Niedobitek G, Dawson CW, Birkenbach M, Wang F, et al. Epstein–Barr virus infection and replication in a human epithelial cell system. Nature. 1992 Mar 26;356(6367):347–50.
18. Niedobitek G, Agathanggelou A, Herbst H, Whitehead L, Wright DH, Young LS. Epstein-Barr virus (EBV) infection in infectious mononucleosis: virus latency, replication and phenotype of EBV-infected cells. The Journal of Pathology. 1997 Jun;182(2):151–9.
19. Allday Martin J, Crawford Dorothy H. Role of epithelium in EBV persistence and pathogenesis of B-cell tumours. The Lancet. 1988 Apr 16;331(8590):855–7.
20. Wagner HJ, Bein G, Bitsch A, Kirchner H. Detection and quantification of latently infected B lymphocytes in Epstein-Barr virus-seropositive, healthy individuals by polymerase chain reaction. Journal of Clinical Microbiology. 1992 Nov;30(11):2826–9.
21. Rickinson AB, Moss DJ. Human cytotoxic T lymphocyte responses to Epstein-Barr virus infection. Annual Review of Immunology. 1997 Apr;15(1):405–31.
22. Gerber P, Lucas S, Nonoyama M, Perlin E, Goldstein L. Oral excretion of Epstein-Barr virus by healthy subjects and patients with infectious mononucleosis. The Lancet. 1972 Nov;2(7785):988–9.
23. Chang R, Lewis J, Reynolds R. Prevalence of oropharyngeal excreters of leukocyte transforming agents among a human population. New England Journal of Medicine. 1973 Dec 20;289(25):1325–9.
24. Chang R, Lewis J, Reynolds R, Sullivan M, Neuman J. Oropharyngeal Excretion of Epstein-Barr Virus by Patients with Lymphoproliferative Disorders and by Recipients of Renal Homografts. Annals of Internal Medicine. 1978 Jan;88(1):34–40.
25. Young LS, Murray PG. Epstein–Barr virus and oncogenesis: from latent genes to tumours. Oncogene. 2003 Aug 11;22(33):5108–21.
26. Thorley-Lawson DA. Epstein-Barr virus: exploiting the immune system. Nature Reviews Immunology. 2001;1(1):75–82.

27. Cohen JI. Epstein–Barr Virus Infection. New England Journal of Medicine. 2000 Aug 17;343(7):481–92.
28. Kimura H, Ito Y, Suzuki R, Nishiyama Y. Measuring Epstein-Barr virus (EBV) load: the significance and application for each EBV-associated disease. Reviews in Medical Virology. 2008 Sep;18(5):305–19.
29. Ryan JL, Fan H, Swinnen LJ, Schichman SA, Raab-Traub N, Covington M, et al. Epstein-Barr Virus (EBV) DNA in Plasma Is Not Encapsidated in Patients With EBV-Related Malignancies. Diagnostic Molecular Pathology. 2004 Jun;13(2):61–8.
30. Lossius A, Johansen J, Torkildsen Ø, Vartdal F, Holmøy T. Epstein-Barr Virus in Systemic Lupus Erythematosus, Rheumatoid Arthritis and Multiple Sclerosis—Association and Causation. Viruses. 2012 Dec 13;4(12):3701–30.
31. Miller CL, Lee JH, Kieff E, Longnecker R. An integral membrane protein (LMP2) blocks reactivation of Epstein-Barr virus from latency following surface immunoglobulin crosslinking. Proceedings of the National Academy of Sciences. 1994 Jan 18;91(2):772–6.
32. Macsween KF, Crawford DH. Epstein-Barr virus—recent advances. The Lancet Infectious Diseases. 2003 Mar;3(3):131–40.
33. Küppers R. B cells under influence: transformation of B cells by Epstein–Barr virus. Nature Reviews Immunology. 2003 Oct;3(10):801–12.
34. Balfour HH, Forte FA, Simpson RB, Zolov DM. Penicillin-related Exanthems in Infectious Mononucleosis Identical to Those Associated with Ampicillin. Clinical Pediatrics. 1972 Jul;11(7):417–21.
35. Thompson DF, Ramos CL. Antibiotic-Induced Rash in Patients With Infectious Mononucleosis. Annals of Pharmacotherapy. 2017 Feb;51(2):154–62.
36. Balfour HH, Odumade OA, Schmeling DO, Mullan BD, Ed JA, Knight JA, et al. Behavioral, Virologic, and Immunologic Factors Associated With Acquisition and Severity of Primary Epstein–Barr Virus Infection in University Students. The Journal of Infectious Diseases. 2012 Oct 24;207(1):80–8.
37. Brigden M, Au S, Thompson S, Brigden S, Doyle P, Tsaparas Y. Infectious mononucleosis in an outpatient population: diagnostic utility of 2 automated hematology analyzers and the sensitivity and specificity of Hoagland's criteria in heterophile-positive patients. Archives of Pathology & Laboratory Medicine. 1999 Oct;123(10):875–81.
38. Paul JR, Bunnell WW. The Presence of Heterophile Antibodies in Infectious Mononucleosis. The American Journal of the Medical Sciences. 1974 Mar;267(3):178
39. Linderholm M, Boman J, Juto P, Linde A. Comparative evaluation of nine kits for rapid diagnosis of infectious mononucleosis and Epstein-Barr virus-specific serology. Journal of Clinical Microbiology. 1994 Jan;32(1):259–61.
40. De Paschale M, Clerici P. Serological diagnosis of Epstein-Barr virus infection: Problems and solutions. World Journal of Virology. 2012 Feb 12;1(1):31–43.
41. Obel N, Høier-Madsen M, Kangro H. Serological and clinical findings in patients with serological evidence of reactivated Epstein-Barr virus infection. APMIS. 1996 Jun;104(6):424–8.
42. Dommerby H, Stangerup S-E., Stangerup M, Hancke S. Hepatosplenomegaly in infectious mononucleosis, assessed by ultrasonic scanning. The Journal of Laryngology & Otology. 1986 May;100(5):573–9.
43. Gayer G, Zandman-Goddard G, Kosych E, Apter S. Spontaneous rupture of the spleen detected on CT as the initial manifestation of infectious mononucleosis. Emergency Radiology. 2003 Apr;10(1):51–2.
44. Grose C, Henle W, Henle G, Feorino PM. Primary Epstein–Barr-Virus Infections in Acute Neurologic Diseases. New England Journal of Medicine. 1975 Feb 20;292(8):392–5.
45. Gilbert JW, Culebras A. Cerebellitis in Infectious Mononucleosis. JAMA: The Journal of the American Medical Association. 1972 May 1;220(5):727.
46. Chacko EM, Krilov LR, Patten W, Lee PJ. Lemierre's and Lemierre's-like syndromes in association with infectious mononucleosis. The Journal of Laryngology & Otology. 2010 Dec;124(12):1257–62.

47. Riordan T. Human Infection with Fusobacterium necrophorum (Necrobacillosis), with a Focus on Lemierre's Syndrome. Clinical Microbiology Reviews. 2007 Oct;20(4):622–59.
48. Mayer HB, Wanke CA, Williams M, Crosson AW, Federman M, Hammer SM. Epstein-Barr Virus-Induced Infectious Mononucleosis Complicated by Acute Renal Failure: Case Report and Review. Clinical Infectious Diseases. 1996 Jun;22(6):1009–18.
49. Papatheodoridis GV, Delladetsima JK, Kavallierou L, Kapranos N, Tassopoulos NC. Fulminant hepatitis due to Epstein-Barr virus infection. Journal of Hepatology. 1995 Sep;23(3):348–50.
50. Kimura H, Nagasaka T, Hoshino Y, Hayashi N, Tanaka N, Lan Xu J, et al. Severe hepatitis caused by Epstein-Barr virus without infection of hepatocytes. Human Pathology. 2001 Jul;32(7):757–62.
51. Purtilo D, Cassel C, Yang J, Harper R. X-linked recessive progressive combined variable immunodeficiency (Duncan's disease). The Lancet. 1975 Apr 26;1(7913):935–40.
52. Tangye SG, Palendira U, Edwards E. Human immunity against EBV—lessons from the clinic. Journal of Experimental Medicine. 2017 Feb;214(2):269–83.
53. Straus SE. The Chronic Mononucleosis Syndrome. Journal of Infectious Diseases. 1988 Mar;157(3):405–12.
54. Cohen JI, Jaffe ES, Dale JK, Pittaluga S, Heslop HE, Rooney CM, et al. Characterization and treatment of chronic active Epstein-Barr virus disease: a 28-year experience in the United States. Blood. 2011 Jun 2;117(22):5835–49.
55. Jones JF, Shurin S, Abramowsky C, Tubbs RR, Sciotto CG, Wahl R, et al. T-Cell Lymphomas Containing Epstein–Barr Viral DNA in Patients with Chronic Epstein–Barr Virus Infections. New England Journal of Medicine. 1988 Mar 24;318(12):733–41.
56. Okano M, Kawa K, Kimura H, Yachie A, Wakiguchi H, Maeda A, et al. Proposed guidelines for diagnosing chronic active Epstein-Barr virus infection. American Journal of Hematology. 2005 Sep;80(1):64–9.
57. Gotoh K, Ito Y, Shibata-Watanabe Y, Kawada J, Takahashi Y, Yagasaki H, et al. Clinical and Virological Characteristics of 15 Patients with Chronic Active Epstein-Barr Virus Infection Treated with Hematopoietic Stem Cell Transplantation. Clinical Infectious Diseases. 2008 May 15;46(10):1525–34.
58. Kawa K, Sawada A, Sato M, Okamura T, Sakata N, Kondo O, et al. Excellent outcome of allogeneic hematopoietic SCT with reduced-intensity conditioning for the treatment of chronic active EBV infection. Bone Marrow Transplantation. 2010 May 24;46(1):77–83.
59. Bollard CM, Cohen JI. How I treat T-cell chronic active Epstein-Barr virus disease. Blood. 2018 Jun 28;131(26):2899–905.
60. Horwitz CA, Henle W, Henle G, Rudnick H, Latts E. Long-Term Serological Follow-Up of Patients for Epstein-Barr Virus After Recovery from Infectious Mononucleosis. Journal of Infectious Diseases. 1985 Jun;151(6):1150–3.
61. Buchwald D, Sullivan J, Komaroff A. Frequency of "Chronic Active Epstein-Barr Virus Infection" in a General Medical Practice. JAMA: The Journal of the American Medical Association. 1987 May 1;257(17):2303–7.
62. Hellinger WC, Smith T, Van Scoy R, Spitzer P, Forgacs P, Edson R. Chronic Fatigue Syndrome and the Diagnostic Utility of Antibody to Epstein-Barr Virus Early Antigen. JAMA: The Journal of the American Medical Association. 1988 Aug 19;260(7):971–3.
63. Buchwald D, Goldenberg DL, Sullivan JL, Komaroff AL. The "chronic, active epstein-barr virus infection" syndrome and primary fibromyalgia. Arthritis & Rheumatism. 1987
64. Matthews DA, Lane TJ, Manu P. Antibodies to Epstein-Barr virus in patients with chronic fatigue. Southern Medical Journal. 1991 Jul;84(7):832–40.
65. Janka G, Imashuku S, Elinder G, Schneider M, Henter J. Infection- and malignancy-associated hemophagocytic syndromes. Secondary hemophagocytic lymphohistiocytosis. Hematology/Oncology Clinics of North America. 1998 Apr;12(2):435–44.

66. Sullivan J, Woda B, Herrod H, Koh G, Rivara F, Mulder C. Epstein-Barr virus-associated hemophagocytic syndrome: virological and immunopathological studies. Blood. 1985 May;65(5):1097–104.
67. Mroczek E, Weisenburger D, Grierson H, Markin R, Purtilo D. Fatal infectious mononucleosis and virus-associated hemophagocytic syndrome. Archives of Pathology & Laboratory Medicine. 1987 Jun;111(6):530–5.
68. Cohen JI, Dropulic L, Hsu AP, Zerbe CS, Krogmann T, Dowdell K, et al. Association of GATA2 Deficiency With Severe Primary Epstein-Barr Virus (EBV) Infection and EBV-associated Cancers. Clinical Infectious Diseases. 2016 Jun 1;63(1):41–7.
69. Noma T, Kou K, Yoshizawa I, Kawano Y, Miyashita T, Mizutani S, et al. Monoclonal proliferation of Epstein-Barr virus-infected T-cells in a patient with virus-associated haemophagocytic syndrome. European Journal of Pediatrics. 1994 Oct;153(10):734–8.
70. Kawaguchi H, Miyashita T, Herbst H, Niedobitek G, Asada M, Tsuchida M, et al. Epstein-Barr virus-infected T lymphocytes in Epstein-Barr virus-associated hemophagocytic syndrome. Journal of Clinical Investigation. 1993 Sep;92(3):1444–50.
71. Imashuku S, Morimoto A, Ishii E. Virus-triggered secondary hemophagocytic lymphohistiocytosis. Acta Paediatrica. 2021 Jun 22;110(10):2729–36.
72. Ishii E, Ohga S, Imashuku S, Yasukawa M, Tsuda H, Miura I, et al. Nationwide Survey of Hemophagocytic Lymphohistiocytosis in Japan. International Journal of Hematology. 2007 Jul;86(1):58–65.
73. Shimazaki C, Inaba T, Shimura K, Okamoto A, Takahashi R, Hirai H, et al. B-cell lymphoma associated with haemophagocytic syndrome: a clinical, immunological and cytogenetic study. British Journal of Haematology. 1999 Mar;104(4):672–9.
74. Strenger V, Merth G, Lackner H, Aberle SW, Kessler HH, Seidel MG, et al. Malignancy and chemotherapy induced haemophagocytic lymphohistiocytosis in children and adolescents—a single centre experience of 20 years. Annals of Hematology. 2018 Jun;97(6):989–98.
75. Bergsten E, Horne A, Aricó M, Astigarraga I, Egeler RM, Filipovich AH, et al. Confirmed efficacy of etoposide and dexamethasone in HLH treatment: long-term results of the cooperative HLH-2004 study. Blood. 2017 Dec 21;130(25):2728–38.
76. Lai W, Wang Y, Wang J, Wu L, Jin Z, Wang Z. Epstein-Barr virus-associated hemophagocytic lymphohistiocytosis in adults and adolescents—a life-threatening disease: analysis of 133 cases from a single center. Hematology. 2018 Jun 29;23(10):810–6.
77. Chellapandian D, Das R, Zelley K, Wiener SJ, Zhao H, Teachey DT, et al. Treatment of Epstein Barr virus-induced haemophagocytic lymphohistiocytosis with rituximab-containing chemo-immunotherapeutic regimens. British Journal of Haematology. 2013 May 21;162(3):376–82.
78. Stefanou C, Tzortzi C, Georgiou F, Timiliotou C. Combining an antiviral with rituximab in EBV-related haemophagocytic lymphohistiocytosis led to rapid viral clearance; and a comprehensive review: Table 1. BMJ Case Reports. 2016 Dec 9;2016:bcr2016216488.
79. Hanto DW. Classification of Epstein-Barr virus-associated posttransplant lymphoproliferative diseases: implications for understanding their pathogenesis and developing rational treatment strategies. Annual Review of Medicine. 1995 Feb;46(1):381–94.
80. Dierickx D, Habermann TM. Post-Transplantation Lymphoproliferative Disorders in Adults. Longo DL, editor. New England Journal of Medicine. 2018 Feb 8;378(6):549–62.
81. Zhang B, Kracker S, Yasuda T, Casola S, Vanneman M, Hömig-Hölzel C, et al. Immune Surveillance and Therapy of Lymphomas Driven by Epstein-Barr Virus Protein LMP1 in a Mouse Model. Cell. 2012 Feb 17;148(4):739–51.
82. Latour S, Fischer A. Signaling pathways involved in the T-cell-mediated immunity against Epstein-Barr virus: Lessons from genetic diseases. Immunological Reviews. 2019 Sep;291(1):174–89.
83. Swerdlow SH, Webber SA, Chadburn A, Ferry JA. Post-transplant lymphoproliferative disorders. In: WHO classification of tumours of haematopoietic and lymphoid tissues: Fourth Edition. IARC Press. 2017

84. Weissmann DJ, Ferry JA, Harris NL, Louis DN, Delmonico F, Spiro I. Posttransplantation Lymphoproliferative Disorders in Solid Organ Recipients Are Predominantly Aggressive Tumors of Host Origin. American Journal of Clinical Pathology. 1995 Jun;103(6):748–55.
85. Peters AC, Akinwumi MS, Cervera C, Mabilangan C, Ghosh S, Lai R, et al. The Changing Epidemiology of Posttransplant Lymphoproliferative Disorder in Adult Solid Organ Transplant Recipients Over 30 Years. Transplantation. 2018 Sep;102(9):1553–62.
86. McDonald RA, Smith JM, Ho M, Lindblad R, Ikle D, Grimm P, et al. Incidence of PTLD in Pediatric Renal Transplant Recipients Receiving Basiliximab, Calcineurin Inhibitor, Sirolimus and Steroids. American Journal of Transplantation. 2008 May;8(5):984–9.
87. Walker RC, Marshall WF, Strickler JG, Wiesner RH, Velosa JA, Habermann TM, et al. Pretransplantation Assessment of the Risk of Lymphoproliferative Disorder. Clinical Infectious Diseases. 1995 May;20(5):1346–53.
88. Smith JM, Rudser K, Gillen D, Kestenbaum B, Seliger S, Weiss N, et al. Risk of Lymphoma after Renal Transplantation Varies with Time: An Analysis of the United States Renal Data System. Transplantation. 2006 Jan;81(2):175–80.
89. Opelz G, Henderson R. Incidence of non-hodgkin lymphoma in kidney and heart transplant recipients. The Lancet. 1993 Dec;342(8886–8887):1514–6.
90. Cockfield SM, Prieksaitis JK, Jewell LD, Parfrey N. Post-transplant lymphoproliferative disorder in renal allograft recipients. Clinical experience and risk factor analysis in a single center. Transplantation. 1993 Jul;56(1):88–96.
91. Opelz G, Naujokat C, Daniel V, Terness P, Döhler B. Disassociation Between Risk of Graft Loss and Risk of Non-Hodgkin Lymphoma With Induction Agents in Renal Transplant Recipients. Transplantation. 2006 May 15;81(9):1227–33.
92. Petit B, Le Meur Y, Jaccard A, Paraf F, Robert CL, Bordessoule D, et al. Influence of host-recipient origin on clinical aspects of posttransplantation lymphoproliferative disorders in kidney transplantation. Transplantation. 2002 Jan 27;73(2):265–71.
93. Randhawa PS, Jaffe R, Demetris AJ, Nalesnik M, Starzl TE, Chen YY, et al. Expression of Epstein–Barr Virus–Encoded Small RNA (by the EBER-1 Gene) in Liver Specimens from Transplant Recipients with Post-Transplantation Lymphoproliferative Disease. New England Journal of Medicine. 1992 Dec 10;327(24):1710–4.
94. van Esser JWJ, van der Holt B, Meijer E, Niesters HGM, Trenschel R, Thijsen SFT, et al. Epstein-Barr virus (EBV) reactivation is a frequent event after allogeneic stem cell transplantation (SCT) and quantitatively predicts EBV-lymphoproliferative disease following T-cell–depleted SCT. Blood. 2001 Aug 15;98(4):972–8.
95. Tsai DE, Douglas L, Andreadis C, Vogl DT, Arnoldi S, Kotloff R, et al. EBV PCR in the Diagnosis and Monitoring of Posttransplant Lymphoproliferative Disorder: Results of a Two-Arm Prospective Trial. American Journal of Transplantation. 2008 May;8(5):1016–24.
96. Allen UD, Preiksaitis JK. Post-transplant lymphoproliferative disorders, Epstein-Barr virus infection, and disease in solid organ transplantation: Guidelines from the American Society of Transplantation Infectious Diseases Community of Practice. Clinical Transplantation. 2019 Jul 23;33(9):e13652.
97. AlDabbagh MA, Gitman MR, Kumar D, Humar A, Rotstein C, Husain S. The Role of Antiviral Prophylaxis for the Prevention of Epstein-Barr Virus-Associated Posttransplant Lymphoproliferative Disease in Solid Organ Transplant Recipients: A Systematic Review. American Journal of Transplantation. 2017 Mar;17(3):770–81.
98. Walti LN, Mugglin C, Sidler D, Mombelli M, Manuel O, Hirsch HH, et al. Association of antiviral prophylaxis and rituximab use with posttransplant lymphoproliferative disorders (PTLDs): A nationwide cohort study. American Journal of Transplantation. 2020 Dec 22;21(7):2532–42.
99. Choquet S, Varnous S, Deback C, Golmard JL, Leblond V. Adapted Treatment of Epstein-Barr Virus Infection to Prevent Posttransplant Lymphoproliferative Disorder After Heart Transplantation. American Journal of Transplantation. 2014 Apr;14(4):857–66.

100. Martin SI, Dodson B, Wheeler C, Davis J, Pesavento T, Bumgardner GL. Monitoring Infection with Epstein-Barr Virus among Seromismatch Adult Renal Transplant Recipients. American Journal of Transplantation. 2011 May;11(5):1058–63.
101. San-Juan R, Manuel O, Hirsch HH, Fernández-Ruiz M, López-Medrano F, Comoli P, et al. Current preventive strategies and management of Epstein–Barr virus-related post-transplant lymphoproliferative disease in solid organ transplantation in Europe. Results of the ESGICH Questionnaire-based Cross-sectional Survey. Clinical Microbiology and Infection. 2015 Jun;21(6):604.e1–9.
102. Morscio J, Finalet Ferreiro J, Vander Borght S, Bittoun E, Gheysens O, Dierickx D, et al. Identification of distinct subgroups of EBV-positive post-transplant diffuse large B-cell lymphoma. Modern Pathology. 2017 Mar;30(3):370–81.
103. Choquet S, Trappe R, Leblond V, Jager U, Davi F, Oertel S. CHOP-21 for the treatment of post-transplant lymphoproliferative disorders following solid organ transplantation. Haematologica. 2007 Feb;92(2):273–4.
104. Evens AM, David KA, Helenowski I, Nelson B, Kaufman D, Kircher SM, et al. Multicenter Analysis of 80 Solid Organ Transplantation Recipients With Post-Transplantation Lymphoproliferative Disease: Outcomes and Prognostic Factors in the Modern Era. Journal of Clinical Oncology. 2010 Feb 20;28(6):1038–46.
105. Gonzalez-Barca E, Domingo-Domenech E, Capote FJ, Gomez-Codina J, Salar A, Bailen A, et al. Prospective phase II trial of extended treatment with rituximab in patients with B-cell post-transplant lymphoproliferative disease. Haematologica. 2007 Nov;92(11):1489–94.
106. Trappe R, Oertel S, Leblond V, Mollee P, Sender M, Reinke P, et al. Sequential treatment with rituximab followed by CHOP chemotherapy in adult B-cell post-transplant lymphoproliferative disorder (PTLD): the prospective international multicentre phase 2 PTLD-1 trial. The Lancet Oncology. 2012 Feb;13(2):196–206.
107. Trappe RU, Dierickx D, Zimmermann H, Morschhauser F, Mollee P, Zaucha JM, et al. Response to Rituximab Induction Is a Predictive Marker in B-Cell Post-Transplant Lymphoproliferative Disorder and Allows Successful Stratification Into Rituximab or R-CHOP Consolidation in an International, Prospective, Multicenter Phase II Trial. Journal of Clinical Oncology. 2017 Feb 10;35(5):536–43.
108. Volk V, Theobald SJ, Danisch S, Khailaie S, Kalbarczyk M, Schneider A, et al. PD-1 Blockade Aggravates Epstein–Barr Virus+ Post-Transplant Lymphoproliferative Disorder in Humanized Mice Resulting in Central Nervous System Involvement and CD4+ T Cell Dysregulations. Frontiers in Oncology. 2021 Jan 12;10:614876.
109. Haque T, Wilkie GM, Jones MM, Higgins CD, Urquhart G, Wingate P, et al. Allogeneic cytotoxic T-cell therapy for EBV-positive posttransplantation lymphoproliferative disease: results of a phase 2 multicenter clinical trial. Blood. 2007 Aug 15;110(4):1123–31.
110. Chiou FK, Beath SV, Wilkie GM, Vickers MA, Morland B, Gupte GL. Cytotoxic T-lymphocyte therapy for post-transplant lymphoproliferative disorder after solid organ transplantation in children. Pediatric Transplantation. 2018 Mar;22(2):e13133.
111. Sun Q, Burton R, Reddy V, Lucas KG. Safety of allogeneic Epstein-Barr virus (EBV)-specific cytotoxic T lymphocytes for patients with refractory EBV-related lymphoma. British Journal of Haematology. 2002 Sep;118(3):799–808.
112. Prockop S, Doubrovina E, Suser S, Heller G, Barker J, Dahi P, et al. Off-the-shelf EBV-specific T cell immunotherapy for rituximab-refractory EBV-associated lymphoma following transplantation. Journal of Clinical Investigation. 2020 Feb 3;130(2):733–47.
113. Liu J, Zhang J, Zhan H, Sun L, Wei L. EBV-specific cytotoxic T lymphocytes for refractory EBV-associated post-transplant lymphoproliferative disorder in solid organ transplant recipients: a systematic review. Transplant International. 2021 Dec;34(12):2483–93.
114. Kim N, Sohn H-J, Oh JH, Jeon Y-W, Lee H-J, Cho H-I, et al. Infusions of Epstein–Barr virus-specific cytotoxic T lymphocytes as post-remission therapy in high-risk post-transplant lymphoproliferative disorder patients: report of two cases. International Journal of Hematology. 2018 May;107(5):596–603.

115. Eckrich MJ, Frangoul H, Knight J, Mosse C, Domm J. A case of pediatric PTLD following autologous stem cell transplantation and review of the literature. Pediatric Transplantation. 2012 Feb;16(1):E15–8.
116. Landgren O, Gilbert ES, Rizzo JD, Socié G, Banks PM, Sobocinski KA, et al. Risk factors for lymphoproliferative disorders after allogeneic hematopoietic cell transplantation. Blood. 2009 May 14;113(20):4992–5001.
117. Styczynski J, Gil L, Tridello G, Ljungman P, Donnelly JP, van der Velden W, et al. Response to Rituximab-Based Therapy and Risk Factor Analysis in Epstein Barr Virus–Related Lymphoproliferative Disorder After Hematopoietic Stem Cell Transplant in Children and Adults: A Study From the Infectious Diseases Working Party of the European Group for Blood and Marrow Transplantation. Clinical Infectious Diseases. 2013 Jun 13;57(6):794–802.
118. Sanz J, Arango M, Senent L, Jarque I, Montesinos P, Sempere A, et al. EBV-associated post-transplant lymphoproliferative disorder after umbilical cord blood transplantation in adults with hematological diseases. Bone Marrow Transplantation. 2014 Mar;49(3):397–402.
119. Mautner J, Bornkamm GW. The role of virus-specific CD4+ T cells in the control of Epstein-Barr virus infection. European Journal of Cell Biology. 2012 Jan;91(1):31–5.
120. Brunstein CG, Weisdorf DJ, DeFor T, Barker JN, Tolar J, van Burik J-AH, et al. Marked increased risk of Epstein-Barr virus-related complications with the addition of antithymocyte globulin to a nonmyeloablative conditioning prior to unrelated umbilical cord blood transplantation. Blood. 2006 Oct 15;108(8):2874–80.
121. Kanakry JA, Kasamon YL, Bolaños-Meade J, Borrello IM, Brodsky RA, Fuchs EJ, et al. Absence of Post-Transplantation Lymphoproliferative Disorder after Allogeneic Blood or Marrow Transplantation Using Post-Transplantation Cyclophosphamide as Graft-versus-Host Disease Prophylaxis. Biology of Blood and Marrow Transplantation. 2013 Oct;19(10):1514–7.
122. Lindsay J, Yong MK, Greenwood M, Kong DCM, Chen SCA, Rawlinson W, et al. Epstein-Barr virus related post-transplant lymphoproliferative disorder prevention strategies in allogeneic hematopoietic stem cell transplantation. Reviews in Medical Virology. 2020 Jul;30(4):e2108.
123. Styczynski J, van der Velden W, Fox CP, Engelhard D, de la Camara R, Cordonnier C, et al. Management of Epstein-Barr Virus infections and post-transplant lymphoproliferative disorders in patients after allogeneic hematopoietic stem cell transplantation: Sixth European Conference on Infections in Leukemia (ECIL-6) guidelines. Haematologica. 2016 Jul;101(7):803–11.
124. Dominietto A, Tedone E, Soracco M, Bruno B, Raiola AM, Van Lint MT, et al. In vivo B-cell depletion with rituximab for alternative donor hemopoietic SCT. Bone Marrow Transplantation. 2012 Jan;47(1):101–6.
125. McIver Z, Stephens N, Grim A, Barrett AJ. Rituximab Administration within 6 Months of T Cell-Depleted Allogeneic SCT is Associated with Prolonged Life-Threatening Cytopenias. Biology of Blood and Marrow Transplantation. 2010 Nov;16(11):1549–56.
126. van Esser JWJ, Niesters HGM, van der Holt B, Meijer E, Osterhaus ADME, Gratama JW, et al. Prevention of Epstein-Barr virus–lymphoproliferative disease by molecular monitoring and preemptive rituximab in high-risk patients after allogeneic stem cell transplantation. Blood. 2002 Jun 15;99(12):4364–9.
127. Petropoulou AD, Porcher R, de Latour RP, Xhaard A, Weisdorf D, Ribaud P, et al. Increased Infection Rate After Preemptive Rituximab Treatment for Epstein-Barr Virus Reactivation After Allogeneic Hematopoietic Stem-Cell Transplantation. Transplantation. 2012 Oct 27;94(8):879–83.
128. Stocker N, Labopin M, Boussen I, Paccoud O, Bonnin A, Malard F, et al. Pre-emptive rituximab treatment for Epstein–Barr virus reactivation after allogeneic hematopoietic stem cell transplantation is a worthwhile strategy in high-risk recipients: a comparative study for immune recovery and clinical outcomes. Bone Marrow Transplantation. 2020 Mar;55(3):586–94.

129. Heslop HE, Slobod KS, Pule MA, Hale GA, Rousseau A, Smith CA, et al. Long-term outcome of EBV-specific T-cell infusions to prevent or treat EBV-related lymphoproliferative disease in transplant recipients. Blood. 2010 Feb 4;115(5):925–35.
130. Papadopoulos EB, Ladanyi M, Emanuel D, Mackinnon S, Boulad F, Carabasi MH, et al. Infusions of Donor Leukocytes to Treat Epstein-Barr Virus-Associated Lymphoproliferative Disorders after Allogeneic Bone Marrow Transplantation. New England Journal of Medicine. 1994 Apr 28;330(17):1185–91.
131. Barker JN, Doubrovina E, Sauter C, Jaroscak JJ, Perales MA, Doubrovin M, et al. Successful treatment of EBV-associated posttransplantation lymphoma after cord blood transplantation using third-party EBV-specific cytotoxic T lymphocytes. Blood. 2010 Dec 2;116(23):5045–9.
132. Leen AM, Bollard CM, Mendizabal AM, Shpall EJ, Szabolcs P, Antin JH, et al. Multicenter study of banked third-party virus-specific T cells to treat severe viral infections after hematopoietic stem cell transplantation. Blood. 2013 Jun 27;121(26):5113–23.
133. Aalto SM, Juvonen E, Tarkkanen J, Volin L, Haario H, Ruutu T, et al. Epstein-Barr Viral Load and Disease Prediction in a Large Cohort of Allogeneic Stem Cell Transplant Recipients. Clinical Infectious Diseases. 2007 Nov 15;45(10):1305–9.
134. Münz C. Redirecting T Cells against Epstein–Barr Virus Infection and Associated Oncogenesis. Cells. 2020 Jun 4;9(6):1400.
135. Greenspan JS, Greenspan D, Lennette ET, Abrams DI, Conant MA, Petersen V, et al. Replication of Epstein–Barr Virus within the Epithelial Cells of Oral Hairy Leukoplakia, an AIDS-Associated Lesion. New England Journal of Medicine. 1985 Dec 19;313(25):1564–71.
136. Chan K, Zhang J, Chan A, Lei K, Leung S, Chan L, et al. Molecular characterization of circulating EBV DNA in the plasma of nasopharyngeal carcinoma and lymphoma patients. Cancer Research. 2003 May 1;63(9):2028–32.
137. Feng W, Israel B, Raab-Traub N, Busson P, Kenney S. Chemotherapy induces lytic EBV replication and confers ganciclovir susceptibility to EBV-positive epithelial cell tumors. Cancer Research. 2002 Mar 15;62(6):1920–6.
138. Shirley CM, Chen J, Shamay M, Li H, Zahnow CA, Hayward SD, et al. Bortezomib induction of C/EBPβ mediates Epstein-Barr virus lytic activation in Burkitt lymphoma. Blood. 2011 Jun 9;117(23):6297–303.
139. Sides MD, Block GJ, Shan B, Esteves KC, Lin Z, Flemington EK, et al. Arsenic mediated disruption of promyelocytic leukemia protein nuclear bodies induces ganciclovir susceptibility in Epstein–Barr positive epithelial cells. Virology. 2011 Jul 20;416(1-2):86–97.
140. Wildeman MA, Novalić Z, Verkuijlen SAWM, Juwana H, Huitema ADR, Tan IB, et al. Cytolytic Virus Activation Therapy for Epstein-Barr Virus–Driven Tumors. Clinical Cancer Research. 2012 Sep 15;18(18):5061–70.
141. Bollard CM, Gottschalk S, Leen AM, Weiss H, Straathof KC, Carrum G, et al. Complete responses of relapsed lymphoma following genetic modification of tumor-antigen presenting cells and T-lymphocyte transfer. Blood. 2007 Oct 15;110(8):2838–45.
142. Lucas KG, Salzman D, Garcia A, Sun Q. Adoptive immunotherapy with allogeneic Epstein-Barr virus (EBV)-specific cytotoxic T-lymphocytes for recurrent, EBV-positive Hodgkin disease. Cancer. 2004 May 1;100(9):1892–901.
143. Louis CU, Straathof K, Bollard CM, Ennamuri S, Gerken C, Lopez TT, et al. Adoptive Transfer of EBV-specific T Cells Results in Sustained Clinical Responses in Patients With Locoregional Nasopharyngeal Carcinoma. Journal of Immunotherapy. 2010 Nov;33(9):983–90.
144. Cohen JI, Fauci AS, Varmus H, Nabel GJ. Epstein-Barr Virus: An Important Vaccine Target for Cancer Prevention. Science Translational Medicine. 2011 Nov 2;3(107):107fs7.
145. Lees J, Arrand J, Pepper S, Stewart J, Mackett M, Arrand J. The Epstein-Barr virus candidate vaccine antigen gp340/220 is highly conserved between virus types A and B. Virology. 1993 Aug;195(2):578–86.
146. Cohen JI. Epstein–barr virus vaccines. Clinical & Translational Immunology. 2015 Jan 23;4(1):e32.

147. Thorley-Lawson D, Poodry C. Identification and isolation of the main component (gp350-gp220) of Epstein-Barr virus responsible for generating neutralizing antibodies in vivo. Journal of Virology. 1982 Aug;43(2):730–6.
148. Janz A, Oezel M, Kurzeder C, Mautner J, Pich D, Kost M, et al. Infectious Epstein-Barr Virus Lacking Major Glycoprotein BLLF1 (gp350/220) Demonstrates the Existence of Additional Viral Ligands. Journal of Virology. 2000 Nov 1;74(21):10142–52.
149. Gu S, Huang T, Ruan L, Miao Y, Lu H, Chu C, et al. First EBV vaccine trial in humans using recombinant vaccinia virus expressing the major membrane antigen. Developments in Biological Standardization. 1995;84:171–7.
150. Sokal EM, Hoppenbrouwers K, Vandermeulen C, Moutschen M, Léonard P, Moreels A, et al. Recombinant gp350 Vaccine for Infectious Mononucleosis: A Phase 2, Randomized, Double-Blind, Placebo-Controlled Trial to Evaluate the Safety, Immunogenicity, and Efficacy of an Epstein-Barr Virus Vaccine in Healthy Young Adults. The Journal of Infectious Diseases. 2007 Dec 15;196(12):1749–53.
151. Brooks JM, Croom-Carter DS, Leese AM, Tierney RJ, Habeshaw G, Rickinson AB. Cytotoxic T-Lymphocyte Responses to a Polymorphic Epstein-Barr Virus Epitope Identify Healthy Carriers with Coresident Viral Strains. Journal of Virology. 2000 Feb;74(4):1801–9.
152. Sitki-Green D, Covington M, Raab-Traub N. Compartmentalization and Transmission of Multiple Epstein-Barr Virus Strains in Asymptomatic Carriers. Journal of Virology. 2003 Feb;77(3):1840–7.
153. Cohen JI. Vaccine Development for Epstein-Barr Virus. In: Kawaguchi Y, Mori Y, Kimura H, editors. Human Herpesviruses. Singapore: Springer Nature; 2018. p. 477–93.
154. Bharadwaj M, Moss DJ. Epstein-Barr virus vaccine: a cytotoxic T-cell-based approach. Expert Review of Vaccines. 2002 Dec;1(4):467–76.
155. Cui X, Snapper CM. Epstein Barr Virus: Development of Vaccines and Immune Cell Therapy for EBV-Associated Diseases. Frontiers in Immunology. 2021 Oct 8;12:734471.
156. Elliott SL, Suhrbier A, Miles JJ, Lawrence G, Pye SJ, Le TT, et al. Phase I Trial of a CD8+ T-Cell Peptide Epitope-Based Vaccine for Infectious Mononucleosis. Journal of Virology. 2008 Feb;82(3):1448–57.

Human Herpesvirus 6A, 6B, 7, and 8 in Hematopoietic Stem Cell Transplant: A Clinical Review

Guy Handley

Abstract Immunosuppression related to allogeneic hematopoietic stem cell transplant increases the risk of reactivation of latent herpesviral infections. This can include the roseolaviruses, human herpesvirus-6A, human herpesvirus-6B, human herpesvirus-7, and human herpesvirus-8 or Kaposi sarcoma herpesvirus. Reactivation is typically asymptomatic, and many cases have unclear clinical significance. Defined syndromes and diseases may be associated with some, such as human herpesvirus-6B. Epidemiology, clinical syndromes, and associated transplant outcomes are reviewed for each of these viruses. Available evidence and guidelines are discussed with a narrow focus on the most pertinent clinical information, but substantial further evaluation and studies are required to characterize the significance of these viruses in stem cell transplant patients.

Keywords Human herpesvirus-6 · Human herpesvirus-7 · Human herpesvirus-8 · Stem cell transplant

1 Introduction

Herpesvirus reactivation commonly occurs following allogeneic hematopoietic stem cell transplantation (HSCT). The net state of immunosuppression following HSCT may permit reactivation of DNA viral infections from a broad spectrum of latently infected patient cell populations with variable progression to disease [1]. While several herpesviruses, including herpes simplex 1, herpes simplex 2, varicella

G. Handley (✉)
Division of Infectious Disease and International Medicine, Department of Internal Medicine, Morsani College of Medicine, University of South Florida, Tampa, FL, USA
e-mail: Guyhandley@usf.edu

C. Somboonwit et al. (eds.), *Global Virology IV: Viral Disease Diagnosis and Treatment Delivery in the 21st Century*,
https://doi.org/10.1007/978-3-031-57369-9_14

zoster virus, Epstein-Barr virus (EBV), and cytomegalovirus (CMV), are covered elsewhere in this book, this chapter will focus on epidemiology and clinical presentations of four herpesviruses in patients following HSCT. This includes three β-herpesviruses, human herpesvirus-6A (HHV-6A), human herpesvirus-6B (HHV-6B) and human herpesvirus-7 (HHV-7) also known as the three human roseolaviruses, which are most closely genetically related to CMV, and one γ-herpesvirus, human herpesvirus-8 (HHV-8) or Kaposi sarcoma herpesvirus (KSHV). Due to the constraints of this chapter, it serves as a brief review oriented toward the most pertinent clinical information, specifically in patients following HSCT, for providers directly caring for such patients. Each of these viruses is discussed extensively elsewhere, including the citations contained within this chapter [2–4]. Evidence demonstrating clinical significance or efficacy of therapeutics is limited for most, but the chapter will initially focus on HHV-6B, which has the most defined clinical syndrome and benefits of therapy in the setting of invasive disease. Notable studies or guidelines will be discussed, but further controlled studies and evaluation of these viruses are required to elucidate their role in patients following HSCT.

2 Human Herpesvirus-6

2.1 Human Herpesvirus-6B

Human herpesvirus-6B is a β-herpesvirus that frequently reactivates after allogeneic HSCT and may be associated with complications and disease. The virus infects nearly all individuals before adulthood, establishing a lifelong latency. It may reactivate after the significant immunosuppression accompanying allogeneic HCST [5, 6]. Most primary infections are asymptomatic; however, patients may present with roseola infantum [5, 6]. While two species of HHV-6, HHV-6A and HHV-6B, exist, only HHV-6B has been definitively implicated in disease following allogeneic HSCT in adult patients [7]. Asymptomatic reactivation commonly occurs in 30–80% of patients after transplant and has been associated with increased all-cause mortality, grade 3–4 acute graft-versus-host disease (GVHD), delayed monocyte engraftment, and delayed platelet engraftment [6, 8–14]. Reactivation typically resolves spontaneously without treatment [6, 10, 13, 14]. Known risk factors for reactivation include umbilical cord blood transplantation, myeloablative conditioning, allelic-mismatch donors, higher grades of acute GVHD, corticosteroids, male gender, and receipt of anti-T-cell monoclonal antibodies [6, 11–14]. Use of cyclophosphamide following HSCT for prevention of GVHD may also increase the risks of reactivation but not necessarily disease [15]. Though most remain asymptomatic, pneumonitis, rash, hepatitis, fever, and myelosuppression have been attributed to reactivation; however, in most cases, alternative etiologies exist [6, 10, 16, 17]. Consensus guidelines note that many of these associations are not clearly established and are based on weak evidence [7]. When considered, alternative etiologies should be evaluated and ruled out.

A defined complication associated with HHV-6B reactivation is encephalitis. Though not universal, this may present with a distinct syndrome of post-transplant acute limbic encephalitis (PALE) characterized by anterograde amnesia, temporal lobe seizures, and/or enhancement of the uncus, amygdala, or hippocampus on magnetic resonance imaging (MRI) [6, 18, 19]. Consensus guidelines emphasize the need to separate asymptomatic HHV-6B reactivation from true disease, but when present, such as in the case of HHV-6B encephalitis, full dosage intravenous foscarnet (90 mg/kg every 12 h or 60 mg/kg every 8 h) or ganciclovir (5 mg/kg every 12 h) is recommended as first-line therapy, with a potential role for cidofovir in salvage therapy [7, 20, 21]. Detection of HHV-6B DNA in CSF is insufficient for diagnosis and occurs in asymptomatic patients [22]. Combination therapy may have a role in severe cases and may be associated with reduced long-term sequelae, as suggested by one retrospective study [23]. No trials exist establishing an optimal duration of therapy, but most guidelines suggest at least 3 weeks with symptom resolution and absence of HHV-6B DNA on repeat testing [7, 20]. Even with treatment, outcomes are poor and long-term sequelae common [11, 14, 20]. Mortality may approach 25–50% with 20–40% of survivors having sequelae [3]. Receipt of foscarnet was independently associated with a reduction in 30-day all-cause mortality in one retrospective study [24]. In that study however, no significant difference was found when patients receiving monotherapy ganciclovir were compared to patients receiving monotherapy foscarnet. Prospective monitoring, pre-emptive and prophylactic strategies with ganciclovir or foscarnet have not demonstrated a benefit in outcomes or incidence of HHV-6B encephalitis [14, 25–28].

Poor immune reconstitution, particularly T-cell reconstitution, correlates with the detection of HHV-6 DNA following HSCT [29]. Importantly, while HHV-6 reactivation in various patient populations has been associated with delayed monocyte or platelet engraftment, increased rates of grade 3–4 acute GVHD, and overall mortality, a direct causal link has not been established. HHV-6 reactivation may potentially be a consequence of poor engraftment and immune reconstitution rather than a direct cause. As noted, pre-emptive and prophylactic strategies have not been shown to mitigate those risks, and anti-viral therapy may potentially expose patients to adverse side effects. Guidelines therefore recommend against pre-emptive or prophylactic strategies and suggest that prospective monitoring should only be considered in the highest risk populations for a limited time period without recommendations for therapy should reactivation occur [7, 20, 21].

Additional controversies exist. A small proportion of the population maintains chromosomal integration of HHV-6B DNA (ciHHV-6) through Mendelian inheritance, rendering most commercial testing platforms difficult to interpret [7, 30]. Such patients often have repeatedly detectable HHV-6B DNA at high levels without symptoms refractory to anti-viral agents. A potential for reactivation and disease attributable to HHV-6B in such patients remains controversial, and has not yet been proven [3, 7]. Treatment of such patients may lead to avoidable toxicities of anti-viral agents [31]. In HSCT recipients with ciHHV-6, grades 2–4 acute GVHD may be more common; however, no significant difference in rates of chronic GVHD, engraftment, non-relapse, or overall mortality was found [32]. Detection of ciHHV-6 may be performed using droplet digital polymerase chain reaction (PCR) or fluorescence in situ hybridization, but neither is readily available in most commercial laboratories [7]. Donor-derived ciHHV-6 or HHV-6 infection is also theoretically possible [3, 7]. Additionally, testing platforms lack standardization, and no specific threshold for HHV-6B DNAemia has been clearly proven to predict the development of disease likely in part due to viral kinetics [10, 25, 26, 33]. Discordant results have been reported among testing platforms [34]. While most literature focuses on encephalitis, myelitis may also occur with a presentation similar to calcineurin inhibitor pain syndrome [16]. Allogeneic HSCT may carry the highest risks, but cases of both encephalitis and myelitis have been described after autologous transplant or after chimeric antigen receptor T-cell therapy [34–37]. Further studies are needed both in these patients and those receiving other types of allogeneic transplants, such as haploidentical transplantation, to characterize risk factors and evaluate targeted patient populations who may benefit from monitoring or therapeutic interventions.

2.2 *Human Herpesvirus-6A*

Human herpesvirus-6A differs from HHV-6B. Similar to HHV-6B, associations have been found between HHV-6A and syndromes, including roseola infantum, in pediatric patients, as well as several neurologic disorders, including seizures, Alzheimer's disease, and multiple sclerosis [38–40]. It similarly establishes lifelong latency after primary infection with the potential for chromosomal integration [30]. No clear clinical disease related to HHV-6A reactivation following HSCT has been definitively found [7]. Nevertheless, cases exist suggesting a possible role for HHV-6A-mediated encephalitis after reactivation even in the setting of ciHHV-6; however, this remains controversial [3, 41–43]. Many commercial testing platforms do not differentiate between the two variants, which may be critical when evaluating symptomatic patients [7, 33, 39]. In at least one series, 3% of patient samples with detectable HHV-6 DNA by real time PCR assay on plasma samples in a population of HSCT recipients were found to be due to HHV-6A [44]. Critically, such patients were not evaluated for ciHHV-6. If HHV-6A DNA is detected following HSCT, it may reflect chromosomal integration rather than active infection [45]. In at least some of these case reports, viral load remained elevated despite therapy with ganciclovir and foscarnet, raising the possibility of chromosomal integration or an alternative diagnosis and in two cases the patient expired despite therapy [41, 42]. Additionally, the possibility of reactivation of chromosomally integrated HHV-6A leading to disease, though controversial, may be considered [7, 45, 46]. Ganciclovir, foscarnet, and cidofovir maintain anti-viral activity, though ganciclovir may have reduced activity in HHV-6A-infected astroglioma cells compared to HHV-6B [47]. The clinical significance is unclear given the lack of established disease states associated with HHV-6A and must be balanced with the potential adverse effects of anti-viral therapy. As available testing platforms are better able to distinguish between HHV-6A and HHV-6B, controlled studies are needed to evaluate the potential associations of HHV-6A with transplant-related outcomes and disease.

3 Human Herpesvirus-7

Along with HHV-6A and HHV-6B, human herpesvirus-7 is considered one of the roseolaviruses. The natural history consists of childhood primary infection, though later than HHV-6, with over 95% seroprevalence by adulthood [48, 49]. Most infections are asymptomatic, but symptoms are similar to those of other roseolaviruses. Seizures in pediatric patients may be more common with HHV-7 than HHV-6 [49]. Case reports suggest more severe neurologic complications with primary infection are possible [50–52]. The virus presumably establishes lifelong latency, though the process is less understood than HHV-6. Reactivation of HHV-7 from $CD4^+$ T-cells as well as peripheral blood mononuclear cells has been demonstrated through T-cell activation, and infection of peripheral blood mononuclear cells by HHV-7 has been shown to potentially facilitate reactivation of HHV-6B [53, 54]. Following HSCT, studies suggest that HHV-7 may reactivate, but incidence ranges from less than 10% of patients to greater than 50% in some populations; reactivation may be associated with CMV reactivation and delayed neutrophil engraftment [55–58]. A causative link has not been established. Ganciclovir has some reduced anti-viral activity, and higher doses are required to achieve similar degrees of inhibition compared to HHV-6 [59]. Foscarnet and cidofovir exhibit activity [59, 60]. Though not clearly the causative agent, case reports describing myelitis, meningoencephalitis, and other neurologic complications in the presence of detectable HHV-7 DNA have been reported [61–63]. While one of these patients expired despite anti-viral therapy, the two others that survived received only supportive care or high-dose glucocorticoids, with symptomatic improvement suggesting an unclear significance of detectable HHV-7 DNA. Routine monitoring for HHV-7 following HSCT is not typically performed, and a benefit of anti-viral agents for treatment or prophylaxis has not been demonstrated or evaluated.

4 Human Herpesvirus-8

Also known as Kaposi sarcoma herpesvirus, human herpesvirus-8 is separate from the previously discussed viruses. As a γ-herpesvirus, it is genetically more closely related to EBV, or human herpesvirus-4, than HHV-6 or HHV-7 [64]. Seroprevalence varies dramatically worldwide, less than 10% in the United States, and latency is poorly understood but thought to occur in $CD19^+$ B-cells, epithelial cells, or other lymphoid cell lines [65]. While typically associated with patients with acquired immunodeficiency syndrome, where it may present with Kaposi sarcoma, other distinct forms exist [64]. Kaposi sarcoma occurs in less than 1% of solid organ transplant recipients but is exceedingly rare following allogeneic HSCT, with rates as low as 0.17% [66–72]. Additionally, other lymphoid neoplastic processes, such as primary effusion lymphoma and multicentric Castleman's disease, are associated with the virus [73–75]. Those oncogenic pathways may similarly present following infection of human tonsillar B cells, likely a different but not entirely dissimilar process compared to EBV [76]. At least one study of HSCT patients found a seroprevalence of 12% and no instances of detectable HHV-8 DNA following transplant by PCR, though only the 5 seropositive patients were evaluated [77]. In one large series of 187 patients, no specific HHV-8 related disease or complications were found in patients who seroconverted following HSCT [78]. Though data is limited in HSCT patients, at least one guideline recommends against screening for HHV-8 serologies prior to transplant due to a lack of data to guide recommendations [79]. Lymphoproliferative diseases associated with HHV-8 rarely occur following HSCT; however, case reports exist [67, 80]. There are at least 2 cases of patients with poor engraftment following HSCT with detectable HHV-8 DNA or antigens, though the significance or a causal link is not clear [81, 82]. Though ganciclovir, foscarnet, and cidofovir have anti-viral activity, importantly in a majority of HHV-8-related malignancies, the vast majority of the DNA detected is from latently infected cells rather than active lytic replication, limiting the clinical utility of these agents [72, 83]. Anti-neoplastic therapies that have been trialed may include surgical excision, anthracyclines, cyclophosphamide, vincristine, rituximab, or interferon, though data demonstrating clinical efficacy is lacking, particularly in patients following HSCT [72, 84]. However, a reduction in immunosuppression alone may be sufficient [72]. Similar to HHV-7, no clear role for routine monitoring, pre-emptive or prophylactic therapy for HHV-8 following allogeneic HSCT has been established, and controlled studies are needed to identify at risk populations or a potential benefit.

5 Conclusions

Detection of herpesvirus DNA is common in patients following HSCT due to the net state of immunosuppression and variable degrees of immune reconstitution; however, only a subset of patients will develop invasive disease. For most of the viruses discussed in this chapter, clinical disease has not been established after HSCT. Reactivation of HHV-6B following allogeneic HSCT correlates with several poor transplant-related outcomes, but a direct causal link has not been established to distinguish from uncomplicated DNAemia which is common. HHV-6B-mediated encephalitis has been clearly defined and should be treated with full-dose ganciclovir or foscarnet, but associations with pneumonitis, rash, hepatitis, fever and myelosuppression are not, but may be possible. In such scenarios, alternative etiologies should be evaluated and ruled out when HHV-6B DNA is detected, as asymptomatic DNAemia is common and typically resolves spontaneously without treatment.

Detection of HHV-6A DNA typically suggests chromosomal integration, but reactivation and disease may be possible. HHV-7 and HHV-8 are not clearly clinically significant, and routine monitoring is not generally recommended. Rare case reports exist, however, suggesting possible but not definitive correlation with an invasive disease. Existing diagnostic platforms and current understanding of the clinical significance these viruses following HSCT are limited at this time. Substantial controlled studies are needed to better understand the complex patient and viral factors associated with these viruses. These may determine if they have a direct causal link to invasive disease or poor transplant-related outcomes, and evaluate for potential screening, prevention, treatment, and risk stratification strategies in this complex patient population.

References

1. Cohen JI. Herpesvirus latency. The Journal of Clinical Investigation. 2020 Jul 1;130(7):3361–9.
2. Agut H, Bonnafous P, Gautheret-Dejean A. Human Herpesviruses 6A, 6B, and 7. Microbiology Spectrum. 2016 Jun 2;4(3).
3. Hill JA, Zerr DM. Human Herpesvirus 6A, 6B, 7, and 8 Infections After Hematopoietic Stem Cell Transplantation. In: Ljungman P, Snydman D, Boeckh M, editors. Transplant Infections. Springer, Cham; 2016. p. 547–61.
4. Ljungman P, Styczynski J, Einsele H. Viral Infections. In: Carreras E, Dufour C, Mohty M, Kröger N, editors. The EBMT Handbook: Hematopoietic Stem Cell Transplantation and Cellular Therapies. Springer, Cham; 2019. p. 281–90.
5. Levy JA, Ferro F, Greenspan D, Lennette ET. Frequent isolation of HHV-6 from saliva and high seroprevalence of the virus in the population. The Lancet. 1990 May;335(8697):1047–50.
6. Zerr DM, Corey L, Kim HW, Huang M, Nguy L, Boeckh M. Clinical Outcomes of Human Herpesvirus 6 Reactivation after Hematopoietic Stem Cell Transplantation. Clinical Infectious Diseases. 2005 Apr 1;40(7):932–40.
7. Ward KN, Hill JA, Hubacek P, de la Camara R, Crocchiolo R, Einsele H, et al. Guidelines from the 2017 European Conference on Infections in Leukaemia for management of HHV-6 infection in patients with hematologic malignancies and after hematopoietic stem cell transplantation. Haematologica. 2019 Nov;104(11):2155–63.
8. Le Bourgeois A, Labopin M, Guillaume T, Delaunay J, Foucher Y, Tessoulin B, et al. Human herpesvirus 6 reactivation before engraftment is strongly predictive of graft failure after double umbilical cord blood allogeneic stem cell transplantation in adults. Experimental Hematology. 2014 Nov;42(11):945–54.
9. Lee YJ, Su Y, Cho C, Tamari R, Perales MA, Jakubowski AA, et al. Human Herpesvirus 6 DNAemia Is Associated With Worse Survival After Ex Vivo T-Cell–Depleted Hematopoietic Cell Transplant. The Journal of Infectious Diseases. 2022 Feb 1;225(3):453–64.
10. Ogata M, Kikuchi H, Satou T, Kawano R, Ikewaki J, Kohno K, et al. Human Herpesvirus 6 DNA in Plasma after Allogeneic Stem Cell Transplantation: Incidence and Clinical Significance. The Journal of Infectious Diseases. 2006 Jan 1;193(1):68–79.
11. Hill JA, Koo S, Guzman Suarez BB, Ho VT, Cutler C, Koreth J, et al. Cord-Blood Hematopoietic Stem Cell Transplant Confers an Increased Risk for Human Herpesvirus-6-Associated Acute Limbic Encephalitis: A Cohort Analysis. Biology of Blood and Marrow Transplantation. 2012 Nov 18;18(11):1638–48.
12. Yamane A, Mori T, Suzuki S, Mihara A, Yamazaki R, Aisa Y, et al. Risk Factors for Developing Human Herpesvirus 6 (HHV-6) Reactivation after Allogeneic Hematopoietic Stem Cell

Transplantation and Its Association with Central Nervous System Disorders. Biology of Blood and Marrow Transplantation. 2007 Jan 13;13(1):100–6.
13. Yoshikawa T, Asano Y, Ihira M, Suzuki K, Ohashi M, Suga S, et al. Human Herpesvirus 6 Viremia in Bone Marrow Transplant Recipients: Clinical Features and Risk Factors. The Journal of Infectious Diseases. 2002 Apr 1;185(7):847–53.
14. Ogata M, Satou T, Inoue Y, Takano K, Ikebe T, Ando T, et al. Foscarnet against human herpesvirus (HHV)-6 reactivation after allo-SCT: breakthrough HHV-6 encephalitis following antiviral prophylaxis. Bone Marrow Transplantation. 2013 Feb;48(2):257–64.
15. Singh A, Dandoy CE, Chen M, Kim S, Mulroney CM, Kharfan-Dabaja MA, et al. Post-Transplantation Cyclophosphamide Is Associated with an Increase in Non-Cytomegalovirus Herpesvirus Infections in Patients with Acute Leukemia and Myelodysplastic Syndrome. Transplantation and Cellular Therapy. 2022 Jan;28(1):48.e1–10.
16. Miyashita N, Endo T, Onozawa M, Hashimoto D, Kondo T, Fujimoto K, et al. Risk factors of human herpesvirus 6 encephalitis/myelitis after allogeneic hematopoietic stem cell transplantation. Transplant Infectious Disease. 2017 Jun;19(3).
17. Shiroshita K, Mori T, Kato J, Sakurai M, Koda Y, Abe R, et al. Clinical characteristics of human herpesvirus-6 myelitis after allogeneic hematopoietic stem cell transplantation and its favorable outcome by early intervention. Bone Marrow Transplantation. 2020 May;55(5):939–45.
18. Wainwright MS, Martin PL, Morse RP, Lacaze M, Provenzale JM, Coleman RE, et al. Human herpesvirus 6 limbic encephalitis after stem cell transplantation. Annals of Neurology. 2001 Nov;50(5):612–9.
19. Seeley WW, Marty FM, Holmes TM, Upchurch K, Soiffer RJ, Antin JH, et al. Post-transplant acute limbic encephalitis: Clinical features and relationship to HHV6. Neurology. 2007 Jul 10;69(2):156–65.
20. Ogata M, Uchida N, Fukuda T, Ikegame K, Kamimura T, Onizuka M, et al. Clinical practice recommendations for the diagnosis and management of human herpesvirus-6B encephalitis after allogeneic hematopoietic stem cell transplantation: the Japan Society for Hematopoietic Cell Transplantation. Bone Marrow Transplantation. 2020 Jun;55(6):1004–13.
21. Olson AL, Politikos I, Brunstein C, Milano F, Barker J, Hill JA. Guidelines for Infection Prophylaxis, Monitoring and Therapy in Cord Blood Transplantation. Transplantation and Cellular Therapy. 2021 May;27(5):359–62.
22. Hill JA, Boeckh MJ, Sedlak RH, Jerome KR, Zerr DM. Human herpesvirus 6 can be detected in cerebrospinal fluid without associated symptoms after allogeneic hematopoietic cell transplantation. Journal of Clinical Virology. 2014 Oct;61(2):289–92.
23. Toomey D, Phan TL, Nguyen V, Phan TT, Ogata M. Retrospective case analysis of antiviral therapies for HHV-6 encephalitis after hematopoietic stem cell transplantation. Transplant Infectious Disease. 2021 Feb;23(1):e13443.
24. Ogata M, Oshima K, Ikebe T, Takano K, Kanamori H, Kondo T, et al. Clinical characteristics and outcome of human herpesvirus-6 encephalitis after allogeneic hematopoietic stem cell transplantation. Bone Marrow Transplantation. 2017 Nov;52(11):1563–70.
25. Ogata M, Satou T, Kawano R, Goto K, Ikewaki J, Kohno K, et al. Plasma HHV-6 viral load-guided preemptive therapy against HHV-6 encephalopathy after allogeneic stem cell transplantation: a prospective evaluation. Bone Marrow Transplantation. 2008 Feb;41(3):279–85.
26. Ishiyama K, Katagiri T, Hoshino T, Yoshida T, Yamaguchi M, Nakao S. Preemptive therapy of human herpesvirus-6 encephalitis with foscarnet sodium for high-risk patients after hematopoietic SCT. Bone Marrow Transplantation. 2011 Jun;46(6):863–9.
27. Wang H, Zhu L, Xue M, Liu J, Guo Z. Low-Dose Foscarnet Preemptive Therapy for Cytomegalovirus Viremia after Haploidentical Bone Marrow Transplantation. Biology of Blood and Marrow Transplantation. 2009 Apr;15(4):519–20.
28. Olson AL, Dahi PB, Zheng J, Devlin SM, Lubin M, Gonzales AM, et al. Frequent Human Herpesvirus-6 Viremia But Low Incidence of Encephalitis in Double-Unit Cord Blood Recipients Transplanted Without Antithymocyte Globulin. Biology of Blood and Marrow Transplantation. 2014 Jun;20(6):787–93.

29. de Koning C, Admiraal R, Nierkens S, Boelens JJ. Human herpesvirus 6 viremia affects T-cell reconstitution after allogeneic hematopoietic stem cell transplantation. Blood Advances. 2018 Feb 27;2(4):428–32.
30. Aimola G, Beythien G, Aswad A, Kaufer BB. Current understanding of human herpesvirus 6 (HHV-6) chromosomal integration. Antiviral Research. 2020 Apr;176:104720.
31. Heldman MR, Job C, Maalouf J, Morris J, Xie H, Davis C, et al. Association of Inherited Chromosomally Integrated Human Herpesvirus 6 with Neurologic Symptoms and Management after Allogeneic Hematopoietic Cell Transplantation. Transplantation and Cellular Therapy. 2021 Sep;27(9):795.e1–8.
32. Hill JA, Magaret AS, Hall-Sedlak R, Mikhaylova A, Huang ML, Sandmaier BM, et al. Outcomes of hematopoietic cell transplantation using donors or recipients with inherited chromosomally integrated HHV-6. Blood. 2017 Aug 24;130(8):1062–9.
33. Handley G. Current Role of Prospective Monitoring and Preemptive and Prophylactic Therapy for Human Herpesvirus 6 After Allogeneic Stem Cell Transplantation. Open Forum Infectious Diseases. 2022 Aug 3;9(8):ofac398.
34. Handley G, Hasbun R, Okhuysen P. Human herpesvirus 6 and central nervous system disease in oncology patients: A retrospective case series and literature review. Journal of Clinical Virology. 2021 Mar;136:104740.
35. Handley G, Khawaja F, Kondapi DS, Lee HJ, Kaufman GP, Neelapu SS, et al. Human herpesvirus 6 myelitis after chimeric antigen receptor T-cell therapy. International Journal of Infectious Diseases. 2021 Nov;112:327–9.
36. Rebechi MT, Bork JT, Riedel DJ. HHV-6 Encephalitis After Chimeric Antigen Receptor T-cell Therapy (CAR-T): 2 Case Reports and a Brief Review of the Literature. Open Forum Infectious Diseases. 2021 Sep 24;8(11):ofab470.
37. Shah M, Kuhnl A, Shields G, Sudhanva M, Metaxa V, Wong S, et al. Human Herpesvirus 6 Encephalitis Following Axicabtagene Ciloleucel Treatment for Refractory Diffuse Large B Cell Lymphoma. HemaSphere. 2021 Feb 17;5(3):e535.
38. Ablashi D, Agut H, Alvarez-Lafuente R, Clark DA, Dewhurst S, DiLuca D, et al. Classification of HHV-6A and HHV-6B as distinct viruses. Archives of virology. 2014 May;159(5):863–70.
39. Agut H, Bonnafous P, Gautheret-Dejean A. Laboratory and Clinical Aspects of Human Herpesvirus 6 Infections. Clinical Microbiology Reviews. 2015 Apr;28(2):313–35.
40. Bahramian E, Furr M, Wu JT, Ceballos RM. Differential Impacts of HHV-6A versus HHV-6B Infection in Differentiated Human Neural Stem Cells. Frontiers in Immunology. 2022 Jul 13;13:847106.
41. Zerr DM, Gupta D, Huang ML., Carter R, Corey L. Effect of Antivirals on Human Herpesvirus 6 Replication in Hematopoietic Stem Cell Transplant Recipients. Clinical Infectious Diseases. 2002 Feb 1;34(3):309–17.
42. Bosi A, Zazzi M, Amantini A, Cellerini M, Vannucchi A, De Milito A, et al. Fatal herpesvirus 6 encephalitis after unrelated bone marrow transplant. Bone Marrow Transplantation. 1998 Aug;22(3):285–8.
43. de Labarthe A, Gauthert-Dejean A, Bossi P, Vernant JP, Dhedin N. HHV-6 Variant A Meningoencephalitis after Allogeneic Hematopoietic Stem Cell Transplantation Diagnosed by Quantitative Real-Time Polymerase Chain Reaction. Transplantation. 2005 Aug 27;80(4):539.
44. Reddy S, Manna P. Quantitative Detection and Differentiation of Human Herpesvirus 6 Subtypes in Bone Marrow Transplant Patients by Using a Single Real-Time Polymerase Chain Reaction Assay. Biology of Blood and Marrow Transplantation. 2005 Jul;11(7):530–41.
45. Hill JA. Human herpesvirus 6 in transplant recipients. Current Opinion in Infectious Diseases. 2019 Dec;32(6):584–90.
46. Hill JA, Sedlak RH, Zerr DM, Huang ML, Yeung C, Myerson D, et al. Prevalence of Chromosomally Integrated Human Herpesvirus 6 in Patients with Human Herpesvirus 6–Central Nervous System Dysfunction. Biology of Blood and Marrow Transplantation. 2015 Feb;21(2):371–3.
47. Akhyani N, Fotheringham J, Yao K, Rashti F, Jacobson S. Efficacy of antiviral compounds in human herpesvirus-6–infected glial cells. Journal of NeuroVirology. 2006 Aug;12(4):284–93.

48. Clark DA, Freeland JML, Mackie PLK, Jarrett RF, Onions DE. Prevalence of Antibody to Human Herpesvirus 7 by Age. The Journal of Infectious Diseases. 1993 Jul;168(1):251–2.
49. Caserta MT, Hall CB, Schnabel K, Long CE, D'Heron N. Primary human herpesvirus 7 infection: A comparison of human herpesvirus 7 and human herpesvirus 6 infections in children. The Journal of Pediatrics. 1998 Sep;133(3):386–9.
50. Mihara T, Mutoh T, Yoshikawa T, Yano S, Asano Y, Yamamoto H. Postinfectious Myeloradiculoneuropathy With Cranial Nerve Involvements Associated With Human Herpesvirus 7 Infection. Archives of Neurology. 2005 Nov;62(11):1755–7.
51. Kawada J, Kimura H, Yoshikawa T, Ihira M, Okumura A, Morishima T, et al. Hemiconvulsion–hemiplegia syndrome and primary human herpesvirus 7 infection. Brain and Development. 2004 Sep;26(6):412–4.
52. Fay AJ, Noetzel MJ, Mar SS. Pediatric Hemorrhagic Brainstem Encephalitis Associated With HHV-7 Infection. Pediatric Neurology. 2015 Dec;53(6):523–6.
53. Katsafanas GC, Schirmer EC, Wyatt LS, Frenkel N. In vitro activation of human herpesviruses 6 and 7 from latency. Proceedings of the National Academy of Sciences. 1996 Sep 3;93(18):9788–92.
54. Frenkel N, Schirmer EC, Wyatt LS, Katsafanas G, Roffman E, Danovich RM, et al. Isolation of a new herpesvirus from human CD4+ T cells. Proceedings of the National Academy of Sciences. 1990 Jan;87(2):748–52.
55. Chan PKS, Peiris JSM, Yuen KY, Liang RHS, Lau YL, Chen FE, et al. Human herpesvirus-6 and human herpesvirus-7 infections in bone marrow transplant recipients. Journal of Medical Virology. 1997 Nov;53(3):295–305.
56. Zawilinska B, Kopec J, Szostek S, Piatkowska-Jakubas B, Skotnicki AB, Kosz-Vnenchak M. Lymphotropic herpesvirus DNA detection in patients with active CMV infection – a possible role in the course of CMV infection after hematopoietic stem cell transplantation. Medical Science Monitor. 2011 Aug;17(8):CR432–41.
57. Tomaszewska A, Kryśko A, Dzieciątkowski T, Przybylski M, Basak GW, Hałaburda K, et al. Co-Infections with Cytomegalovirus and Human Herpesvirus Type 7 in Adult Polish Allogeneic Haematopoietic Stem Cell Transplant Recipients. Archivum Immunologiae et Therapiae Experimentalis. 2014 Feb;62(1):77–80.
58. Inazawa N, Hori T, Hatakeyama N, Yamamoto M, Yoto Y, Nojima M, et al. Large-scale multiplex polymerase chain reaction assay for diagnosis of viral reactivations after allogeneic hematopoietic stem cell transplantation. Journal of Medical Virology. 2015 Aug;87(8):1427–35.
59. Yoshida M, Yamada M, Tsukazaki T, Chatterjee S, Lakeman FD, Nii S, et al. Comparison of antiviral compounds against human herpesvirus 6 and 7. Antiviral Research. 1998 Dec;40(1–2):73–84.
60. De Clercq E, Naesens L, De Bolle L, Schols D, Zhang Y, Neyts J. Antiviral agents active against human herpesviruses HHV-6, HHV-7 and HHV-8. Reviews in Medical Virology. 2001;11(6):381–95.
61. Chan PKS, Chik KW, To KF, Li CK, Shing MMK, Ng KC, et al. Case report: Human herpesvirus 7 associated fatal encephalitis in a peripheral blood stem cell transplant recipient. Journal of Medical Virology. 2002 Apr;66(4):493–6.
62. Ward K, White R, Mackinnon S, Hanna M. Human herpesvirus-7 infection of the CNS with acute myelitis in an adult bone marrow recipient. Bone Marrow Transplantation. 2002 Dec;30(12):983–5.
63. Yoshikawa T, Yoshida J, Hamaguchi M, Kubota T, Akimoto S, Ihira M, et al. Human herpesvirus 7-associated meningitis and optic neuritis in a patient after allogeneic stem cell transplantation. Journal of Medical Virology. 2003 Jul;70(3):440–3.
64. Antman K, Chang Y. Kaposi's sarcoma. The New England Journal of Medicine. 2000 Apr 6;342(14):1027–38.
65. Damania BA, Cesarman E. Kaposi's Sarcoma–Associated Herpesvirus. In: Knipe DM, Howley PM, editors. Fields Virology. Philadelphia: Lippincott Williams & Wilkins; 2013. p. 2080–128.

66. Farge D. Kaposi's sarcoma in organ transplant recipients. The Collaborative Transplantation Research Group of Ile de France. The European Journal of Medicine. 1993;2(6):339–43.
67. Luppi M, Barozzi P, Rasini V, Torelli G. HHV-8 Infection in the Transplantation Setting: A Concern Only for Solid Organ Transplant Patients? Leukemia & Lymphoma. 2002 Mar;43(3):517–22.
68. Vivancos P, Sarrá J, Palou J, Valls A, Garćia J, Grañena A. Kaposi's sarcoma after autologous bone marrow transplantation for multiple myeloma. Bone Marrow Transplantation. 1996 Apr;17(4):669–71.
69. Helg C, Adatto M, Salomon D, Roux E, Miralbell R, Chapuis B, et al. Kaposi's sarcoma following allogeneic bone marrow transplantation. Bone Marrow Transplantation. 1994 Dec;14(6):999–1001.
70. Gluckman E, Parquet N, Scieux C, Deplanche M, Traineau R, Betheau P, et al. KS-associated herpesvirus-like DNA sequences after allogeneic bone-marrow transplantation. The Lancet. 1995 Dec 9;346(8989):1558–9.
71. Deauna-Limayo D, Rajabi B, Qiu W, Htut M, Sweetenham J. Kaposi sarcoma after non-myeloablative hematopoietic stem cell transplant: response to withdrawal of immunosuppressant therapy correlated with whole blood human herpesvirus-8 reverse transcriptase-polymerase chain reaction levels. Leukemia & Lymphoma. 2013 Oct;54(10):2299–302.
72. Cesaro S, Tridello G, van der Werf S, Bader P, Sociè G, Ljungman P, et al. Incidence and outcome of Kaposi sarcoma after hematopoietic stem cell transplantation: a retrospective analysis and a review of the literature, on behalf of infectious diseases working party of EBMT. Bone Marrow Transplantation. 2020 Jan;55(1):110–6.
73. Soulier J, Grollet L, Oksenhendler E, Cacoub P, Cazals-Hatem D, Babinet P, et al. Kaposi's sarcoma-associated herpesvirus-like DNA sequences in multicentric Castleman's disease. Blood. 1995 Aug 15;86(4):1276–80.
74. Cesarman E, Chang Y, Moore PS, Said JW, Knowles DM. Kaposi's Sarcoma–Associated Herpesvirus-Like DNA Sequences in AIDS-Related Body-Cavity–Based Lymphomas. New England Journal of Medicine. 1995 May 4;332(18):1186–91.
75. Rose H, Kulik M, Standish R, Prewett E, McLennan R, Szer J, et al. Constrictive pericarditis and primary effusion lymphoma following allogeneic BMT for CML. Bone Marrow Transplantation. 2012 Sep;47(9):1250–1.
76. Hassman LM, Ellison TJ, Kedes DH. KSHV infects a subset of human tonsillar B cells, driving proliferation and plasmablast differentiation. Journal of Clinical Investigation. 2011 Feb;121(2):752–68.
77. Sergerie Y, Abed Y, Roy J, Boivin G. Comparative Evaluation of Three Serological Methods for Detection of Human Herpesvirus 8-Specific Antibodies in Canadian Allogeneic Stem Cell Transplant Recipients. Journal of Clinical Microbiology. 2004 Jun;42(6):2663–7.
78. Gentile G, Capobianchi A, Volpi A, Palù G, Pica F, Calistri A, et al. Human Herpesvirus 8 DNA in Serum During Seroconversion in Allogeneic Bone Marrow Transplant Recipients. JNCI: Journal of the National Cancer Institute. 2005 Jul 6;97(13):1008–11.
79. Ljungman P, de la Camara R, Cordonnier C, Einsele H, Engelhard D, Reusser P, et al. Management of CMV, HHV-6, HHV-7 and Kaposi-sarcoma herpesvirus (HHV-8) infections in patients with hematological malignancies and after SCT. Bone Marrow Transplantation. 2008 Aug;42(4):227–40.
80. Matsushima AY, Strauchen JA, Lee G, Scigliano E, Hale EE, Weisse MT, et al. Posttransplantation Plasmacytic Proliferations Related to Kaposi's Sarcoma–Associated Herpesvirus. The American Journal of Surgical Pathology. 1999 Nov;23(11):1393–400.
81. Luppi M, Barozzi P, Schulz TF, Setti G, Staskus K, Trovato R, et al. Bone Marrow Failure Associated with Human Herpesvirus 8 Infection after Transplantation. New England Journal of Medicine. 2000 Nov 9;343(19):1378–85.
82. Cuzzola M, Irrera G, Iacopino O, Cuzzocrea A, Messina G, Console G, et al. Bone Marrow Failure Associated with Herpesvirus 8 Infection in a Patient Undergoing Autologous Peripheral Blood Stem Cell Transplantation. Clinical Infectious Diseases. 2003 Oct 1;37(7):e102–6.

83. Gantt S, Casper C. Human herpesvirus 8-associated neoplasms. Current Opinion in Infectious Diseases. 2011 Aug;24(4):295–301.
84. Riva G, Luppi M, Barozzi P, Forghieri F, Potenza L. How I treat HHV8/KSHV-related diseases in posttransplant patients. Blood. 2012 Nov 15;120(20):4150–9.

Cytomegalovirus Infections in Hematopoietic Cell Transplant and Solid Organ Transplant Recipients

Joseph Sassine and Shivan Shah

Abstract Cytomegalovirus is a common opportunistic infection in solid organ and hematopoietic cell transplant patients. The symptoms of patients with this infection can vary and target many organs. Delays in diagnosing and treating patients can increase morbidity and mortality. This chapter describes current studies and guidelines related to the diagnosis, treatment, and prevention of this disease.

Keywords Cytomegalovirus (CMV) · Virus · Opportunistic infection · Hematopoietic cell transplant (HCT) · Solid organ transplant (SOT)

1 Introduction

Cytomegalovirus (CMV) is one of the most common infections affecting hematopoietic cell transplant (HCT) and solid organ transplant (SOT) recipients [1]. A CMV is a member of the *Betaherpesvirinae* subfamily and is the largest and most complex herpesvirus; it has a double-stranded DNA genome comprising unique long (UL) and unique short segments [2]. Like other herpesviruses, CMV persists in the host after the primary infection and reactivation is triggered by defects in cell-mediated immunity. CMV primary infection is prevalent worldwide and is generally acquired during early childhood [3]. Its latency sites in cells include myeloid stem cells, including CD34+ cells, and peripheral blood mononuclear cells (PBMNCs) [4, 5]. Control of the infection relies on a marked adaptive immune

J. Sassine
Department of Internal Medicine, Oklahoma University Physicians, Oklahoma City, OK, USA

S. Shah (✉)
Department of Internal Medicine, University of South Florida, Morsani College of Medicine, Tampa, FL, USA
e-mail: shivanshah@usf.edu

C. Somboonwit et al. (eds.), *Global Virology IV: Viral Disease Diagnosis and Treatment Delivery in the 21st Century*,
https://doi.org/10.1007/978-3-031-57369-9_15

response involving CD4+ and CD8+ T cells. In contrast, the innate immune response can help trigger adaptive immunity through a cytokine response involving alpha- and beta-interferon, tumor necrosis (factor alpha (TNF-α)-, and interleukin-12 [6, 7]. In addition to its direct effects causing end-organ disease (detailed below), CMV has been associated with graft-versus-host disease (GVHD) [8] and associated with an increased risk of bacterial and fungal infections [9]. These are the reasons that CMV is referred to as an immunomodulatory virus [10].

2 Definitions

The Cytomegalovirus Drug Development Group established definitions of CMV infection and disease in transplant patients, which were updated in 2017 [11]. While these definitions are primarily intended to provide consistent definitions across clinical trials, they delineate a critical diagnostic framework for clinical use.

CMV infection is defined as the isolation of the virus or detection of viral proteins or nucleic acid in any body fluid or tissue specimen. Recurrent CMV infection is defined as a new CMV infection in a patient, who had a prior CMV infection and has not had any detected virus for at least 4 weeks between the two episodes. CMV can be detected in the blood through various techniques; the most common currently is the detection of CMV DNA in plasma or whole blood, referred to as DNAemia. Older assays relied on detecting the CMV pp65 antigen in peripheral blood lymphocytes, referred to as antigenemia. Viremia is defined as the isolation of CMV by viral culture [11].

CMV disease encompasses end-organ disease caused by CMV and CMV syndrome, an entity exclusive to solid organ transplant recipients. CMV disease entities are classified as proven, probable, and possible according to the level of diagnostic evidence. All three categories require compatible clinical signs and symptoms, which are discussed below. These definitions are summarized in Table 1.

Clinically significant CMV infection (CS-CMVi) is defined as CMV disease or CMV viremia leading to preemptive antiviral treatment. It was first used as the primary outcome of phase 3 letermovir primary prophylaxis trial [12] and remained a clinically oriented outcome of interest in CMV studies.

Refractory and resistant CMV have long had heterogeneous definitions that precluded accurately determining their accurate incidence. Recently, the Cytomegalovirus Drug Development Forum established standardized definitions for these entities.

For the patients with CMV viremia without end-organ disease, the definition relies on the comparison between the peak viral load within the first week of appropriate antiviral therapy and the peak viral load after at least 2 weeks, using the same assay in the same laboratory. Refractory CMV infection is defined as more than 1 log10 increase in CMV DNA levels in blood or serum, and probable refractory CMV infection is defined as a persistent (same or < 1 log10 increase) viral load over this 2-week time period [13].

Table 1 Definitions of proven, probable, and possible end-organ diseases caused by CMV [11]

Organ	Proven disease	Probable disease	Possible disease
Pneumonia	Clinical signs/symptoms AND CMV documented in lung tissue (culture, histopathology, immunohistochemistry, DNA hybridization)	Clinical signs/symptoms AND CMV detected in bronchoalveolar lavage (culture, DNA detection)	Clinical signs/ symptoms +/− CMV detection with quantitative NAT in tissue
Gastrointestinal disease	Clinical symptoms AND macroscopic mucosal lesions AND CMV documented in tissue (histopathology, culture, immunohistochemistry, DNA hybridization)	Clinical symptoms AND CMV documented in tissue (histopathology, culture, immunohistochemistry, DNA hybridization)	Clinical symptoms +/− CMV detection with quantitative NAT in tissue
Hepatitis	Abnormal liver function tests AND CMV documented in tissue (histopathology, culture, immunohistochemistry, DNA hybridization) AND no other cause of hepatitis	Not defined	Not defined
Retinitis	Typical ophthalmological signs judged by an experienced ophthalmologist (can be supported by CMV NAT in vitreous fluid)	Not defined	Not defined
Encephalitis/ ventriculitis	Clinical symptoms AND CMV detected in tissue (culture, immunohistochemistry, in situ hybridization, quantitative PCR (polymerase chain reaction))	Clinical symptoms AND CMV detection in CSF (nonbloody) AND abnormal imaging or encephalitis on EEG	Not defined
Nephritis	Renal dysfunction AND histologic features on biopsy AND detection of CMV in tissue	Not defined	Not defined
Cystitis	Cystitis AND histologic features on biopsy AND detection of CMV in tissue	Not defined	Not defined
Myocarditis	Myocarditis AND histologic features on biopsy AND detection of CMV in tissue	Not defined	Not defined
Pancreatitis	Pancreatitis AND histologic features on biopsy AND detection of CMV in tissue	Not defined	Not defined
Other	Clinical signs/symptoms AND CMV documentation in tissue biopsy	Not defined	Not defined

Abbreviations: *CMV* cytomegalovirus, *NAT* nucleic acid amplification test, *PCR* polymerase chain reaction, *CSF* cerebrospinal fluid, *EEG* electroencephalogram

Refractory CMV end-organ disease is defined as worsening in the signs and symptoms or progression into end-organ disease after at least 2 weeks of appropriate antivirals, and probable refractory disease refers to the lack of improvement in signs and symptoms over the same time. Finally, resistant CMV is determined by detecting a mutation in the CMV genome that decreases susceptibility to one or more antiviral drugs (Table 2) [13].

3 Risk Factors

CMV reactivation is significantly more common among recipients of an allogeneic HCT than recipients of an autologous HCT or patients with hematological malignancy who have not received an HCT. A recent study reported 84 patients with lymphoma or myeloma, 30% of whom had received an autologous HCT, who developed Clinically Significant-CMV infection (CS-CMVi) in a quaternary cancer center over almost 4 years, and multiple myeloma was associated with a lower risk of CMV end-organ disease compared to lymphoma [14]. For certain higher-risk patients with hematological malignancies, such as those receiving the anti-CD52 agent alemtuzumab, the National Comprehensive Cancer Network (NCCN) recommends CMV surveillance [15]. Regarding chimeric antigen receptor (CAR)-T cell recipients, no major study has looked in-depth into CMV. However, a recent report showed an 11% incidence of CS-CMVi, 17% of whom had end-organ disease. CMV was the most common herpesvirus to reactivate in this population that received acyclovir or valacyclovir prophylaxis. In this study, higher grade cytokine release syndrome, higher grade immune effector cell associated neurotoxicity syndrome, use of anakinra, and a higher cumulative dose of steroids within the first 30 days after CAR-T cell were associated with a significantly higher risk of herpesviruses reactivation, including CMV [16].

Among recipients of an allogeneic HCT, recipient seropositivity remains the most significant risk factor for CMV reactivation after transplant. It is estimated that about 80% of CMV-seropositive recipients will develop CMV infection without any prophylaxis [17]. In contrast, among seropositive recipients, donor CMV seropositivity is associated with a lower risk of CMV reactivation, improved 5-year survival, decreased transplant mortality [18], and a lower risk of refractory or resistant CMV [19].

Table 2 Antiviral agents used in CMV treatment or prophylaxis and genes with reported mutations associated with antiviral resistance [13]

Antiviral agent	Genes with mutations associated with resistance
Ganciclovir, valganciclovir	UL97, UL54
Foscarnet	UL54
Cidofovir	UL54
Letermovir	UL51, UL56, UL89
Maribavir	UL97, UL27

Beyond recipient and donor serostatus, additional risk factors for CMV infection after allogeneic HCT include [10, 19–21]:

- Advanced age
- Acute GVHD
- Steroid use, particularly at doses higher than 1 mg/kg/day of prednisone equivalents
- T-cell depletion, particularly with the use of alemtuzumab or anti-thymocyte globulin
- Haploidentical donor compared to matched donor
- Mismatched or unrelated donor compared to a matched related donor
- Cord blood transplant
- Lymphopenia, particularly a low CD4 count
- Post-HCT cyclophosphamide
- Undetectable CMV-specific T-cell immunity

Similar risk factors predispose SOT patients to CMV disease such as immunodeficiency from innate CD4 and CD8 dysfunction or secondary to drug immunosuppression [64]. The use of CMV-specific serologies can help us risk stratify these patients as to who is at higher and lower risk for development of CMV disease. Type of organ transplant also has a significant impact on risk, from highest to lowest, the order is as follows; small bowel, lung, pancreas, heart, and liver/kidney [66]. The disease caused by CMV occurs mainly between 30 and 90 days after transplantation and is rare after 180 days [66]. Onset may be delayed when antiviral prophylaxis is used. Table 3 shows the highest risk factors causing disease [65]:

Most of the above risk factors also increase the risk of refractory or resistant CMV. Some additional risk factors include prolonged antiviral exposure, subtherapeutic drug levels (particularly with variable clearance), higher CMV viral loads, and recurrent CMV infection [22]. Donor CMV seropositivity and primary prophylaxis with letermovir are associated with a lower risk of refractory or resistant CMV infections [19].

4 Clinical Presentation

4.1 CMV Pneumonia

A potentially fatal disease, CMV pneumonia, can occur in up to 30% of allogeneic HCT recipients without prophylaxis. While most of the cases have been described in the early period (first 30 days after HCT), they can occur late after HCT, particularly after discontinuation of CMV prophylaxis at day 100 in patients who might still be at high risk, such as those with ongoing needs for immunosuppression due to GVHD [23]. Rates of CMV Pneumonia have significantly decreased in SOT patients due to universal screening and prophylaxis.

Table 3 Highest risk factors for CMV disease in SOT [65]

RISK factor	Level of evidence	Comments
D–/R+ CMV IgG serology	I	Increased risk of CMV replication and disease exceeds 50%. In D+/R+ and D–/R+ situations, without prophylaxis, it is about 15–20%. In D–/R– combinations, the risk of CMV replication and disease is very low
Use of Muromonab-CD3 (OTK3) Or Anti-Thymocyte Globulin (ATG)	I	Use associated with higher risk of CMV replication and disease. Targeted antiviral prophylaxis decreased the risk of CMV disease in patients who received these drugs
Acute graft rejection	I	Increased risk of CMV replication and disease due to proinflammatory cytokines and subsequent anti-rejection treatment
Mofetil mycophenolate (MMF)	II	Associated with higher risk when MMF dose >2 g/daily. MMF has been particularly linked to the development of CMV gastrointestinal disease in kidney transplant recipients
Alemtuzumab	II	Higher risk only when used for rejection treatment not during induction treatment

Patients typically present with nonspecific symptoms, which can be seen in viral pneumonia, such as fever, cough, tachypnea, dyspnea, and hypoxia [24]. Similar to the clinical presentation, the radiographic findings are also nonspecific, most commonly in the form of diffuse, bilateral ground-glass opacities, which are difficult to distinguish from other viral or atypical pneumonia such as Pneumocystis pneumonia [25]. Unlike other end-organ diseases discussed below, CMV pneumonia is frequently accompanied by DNAemia [26]. Thus, diagnosing CMV pneumonia requires identifying CMV in the lower respiratory tract or the blood, in addition to a compatible clinical and radiographic picture. Since lung biopsies are not readily feasible in HCT recipients due to safety concerns, detection of CMV in the bronchoalveolar lavage (BAL) has emerged as an important diagnostic tool with a good negative predictive value [2]. While some viral shedding can occur in the lower airways and might not necessarily represent true infection, using quantitative CMV PCR in the BAL with a cutoff value of 500 IU/mL can help distinguish CMV pneumonia from pulmonary shedding [27]. Prior to the CMV antiviral era, pneumonia used to be a fatal disease; mortality has significantly decreased with the advent of anti-CMV therapy and high-dose immunoglobulins [24].

4.2 CMV Gastrointestinal Disease

While not as common as CMV pneumonia, gastrointestinal (GI) disease can carry a significant burden in allogeneic HCT and SOT recipients. In HCT, the cumulative incidence is estimated at 2% over 2 years [28]. It usually manifests as an erosive or ulcerative disease that can theoretically affect any part of the GI tract. Compared to patients with solid tumors, the upper GI tract tends to be more commonly affected than the lower GI tract [29]. The symptoms depend on the affected site, ranging from dysphagia and odynophagia with esophagitis, epigastric pain, and nausea with gastritis, to GI bleeding and diarrhea with lower tract involvement. This clinical presentation tends to be similar to that of GI GVHD. While therapeutic purposes need to distinguish the two entities (with GVHD requiring an increase in the net immunosuppressive therapy as opposed to CMV), they also tend to occur simultaneously, rendering both diagnosis and management exceptionally challenging [10]. This is best achieved with endoscopic evaluation and tissue examination. Endoscopic abnormalities, such as inflammation, ulceration, and mucosal hemorrhage, can also be seen in GVHD. The ultimate diagnosis relies on the identification of viral inclusions or CMV-specific immunohistochemistry on histopathological exams of the biopsied tissue [10]. The use of CMV PCR on tissue is evolving and counts as a possible diagnosis for now [11]. Unlike CMV pneumonia, DNAemia and antigenemia are infrequent in GI disease [30].

4.3 CMV Central Nervous System Disease

Central nervous system (CNS) involvement is more common in patients with HIV/AIDS than in allogeneic HCT and SOT recipients. It usually presents ventriculitis and/or encephalitis, with cognitive dysfunction, mutism, encephalopathy, cranial nerve palsies, and periventricular hyperintensity on magnetic resonance imaging [31]. Compared to other viral meningoencephalitides, hypoglycorrhachia occurs more frequently with CMV [32]. The diagnosis is usually confirmed with a CSF CMV PCR.

4.4 CMV Retinitis

The clinical presentation of CMV retinitis after HCT and SOT is similar to patients with AIDS, although infrequent. The diagnosis is suspected in patients with blurring or loss of central vision, floaters scotomata, and relies on fundoscopic examination by an experienced ophthalmologist. Findings on fundoscopy include perivascular exudates, hemorrhages, white macular exudates, and retinal detachment in advanced stages [33]. A positive CMV PCR in the vitreous fluid can assist in the diagnosis [11].

5 Antiviral Agents

5.1 *Ganciclovir*

Ganciclovir remains the agent of choice for the treatment of CMV infections. It is the nucleoside analog of 2′-deoxyguanosine and is administered intravenously. It requires initial phosphorylation by the viral kinase encoded by the UL97 open reading frame. This step is essential to transform ganciclovir into its active triphosphorylated form, inhibiting DNA synthesis by the CMV DNA polymerase (encoded by UL54) and incorporating it into the viral DNA [10]. Resistance to ganciclovir can develop via mutations in either UL97 or UL54 [22]. The main toxicity associated with ganciclovir is cytopenia [20], particularly neutropenia and thrombocytopenia, which can limit its use in the early days after HCT. It is important to note that dose reduction is not recommended to mitigate neutropenia that develops during therapy. The patient can be either supported by granulocyte colony-stimulating factor (G-CSF) or switched to a different antiviral agent [20].

5.2 *Valganciclovir*

Valganciclovir is the pro-drug of ganciclovir, formulated as its L-valyl ester [10]. It is administered orally, hydrolyzed by the human intestinal mucosal valine esterase, and absorbed into the systemic circulation as ganciclovir. This formulation provides a ten-fold higher systemic bioavailability than oral ganciclovir [22]. It is not usually used in the initial stage of therapy in patients with CMV end-organ disease. It shares a similar side effect and resistance profile as ganciclovir.

5.3 *Foscarnet*

Foscarnet is an organic analog of inorganic pyrophosphate. It is administered intravenously. Its activity relies on inhibiting pyrophosphate binding to the viral DNA polymerase [10]. Resistance to foscarnet develops via mutations in UL54, some of which can lead to cross-resistance with ganciclovir or cidofovir [22]. The main limiting factor in the use of foscarnet is its toxicity profile. It has been associated with acute kidney injury, electrolyte abnormalities, and mucosal or genital ulcerations [2]. Consequently, its use is usually reserved for patients who cannot tolerate ganciclovir or do not respond to ganciclovir [20].

5.4 *Cidofovir*

Cidofovir is a nucleotide analog of cytosine. It is administered intravenously; its long half-life allows for a weekly administration. It requires phosphorylation by cellular kinases to become active. Cidofovir triphosphate, the active form, is a competitive inhibitor of DNA polymerase and is incorporated into the viral DNA [10, 22]. Resistance to cidofovir is also associated with mutations in UL54, with some mutations leading to cross-resistance to foscarnet or ganciclovir [22]. The high incidence of nephrotoxicity and the reports of variable efficacy against CMV have limited the use of cidofovir to those patients that do not respond to either ganciclovir or foscarnet [20].

5.5 *Brincidofovir*

Brincidofovir is the lipid conjugate of cidofovir. This proprietary technology allowed for oral administration twice weekly. It has a potential broad-spectrum activity against multiple DNA viruses. Its main advantage over cidofovir is the absence of nephrotoxicity since it is not concentrated in the proximal tubule and is not a substrate of organic ion transporter 1 [34]. While brincidofovir showed promise in early clinical experience for refractory or resistant infections [35]. In phase 2 dose-escalation trial for primary CMV prophylaxis [36], it failed to demonstrate a significant reduction in CS-CMVi when administered as a primary prophylaxis in a phase 3 trial and was associated with significant gastrointestinal toxicity [34]. This has halted any further use of brincidofovir for CMV.

5.6 *Letermovir*

Letermovir is the first approved anti-CMV agent that does not act at the level of the DNA polymerase. It targets at the later step in the virus lifecycle, by interacting with the viral terminase complex which supports viral DNA processing and packaging [10]. It can be administered intravenously or orally and requires dose adjustment with cyclosporine [20]. It is usually well tolerated. Resistance has been reported with mutations in UL51, UL56, and UL89 [13]. It has been approved for primary prophylaxis (discussed below) but has been used occasionally in refractory or resistant cases. Of note, letermovir has no activity against other herpesviruses, and prophylaxis against herpes simplex and varicella-zoster should be provided through a separate agent [20].

5.7 *Maribavir*

Maribavir is another novel antiviral with a different mechanism of action. It is a benzimidazole riboside that affects three different steps in the lifecycle of CMV through its inhibition of the viral kinase UL97: DNA replication, encapsidation, and nuclear egress of viral capsids [37, 38]. It is administered orally, and its major toxicity is dysgeusia [38]. Resistance has been associated with mutations in UL97, some of which cross-react with ganciclovir or in UL27 [13, 38]. Since it inhibits the viral kinase required to activate ganciclovir, maribavir should not be coadministered with ganciclovir or valganciclovir. While it failed its phase 3 primary prophylaxis trial with a primary end-point of end-organ disease [39], it has recently been approved for posttransplant refractory or resistant CMV infections not responding to ganciclovir, foscarnet, or cidofovir. This approval was based on the results of the phase 3 SOLSTICE trial that showed a significantly higher CMV viremia clearance with maribavir with a much more favorable safety profile leading to fewer discontinuations of therapy when compared with ganciclovir, foscarnet, or cidofovir [38].

5.8 *Other Agents*

Leflunomide, an agent approved for treating rheumatoid arthritis, has been used in refractory or resistant CMV. It is presumed to inhibit late stages of virion assembly, and, as such, circumvent resistance mechanisms affecting viral kinase or DNA polymerase [10]. Reports of successful outcomes are mixed, might be affected by the viral load, and antedate the approval of maribavir [40, 41]. Intravenous immunoglobulins have been used in combination with antiviral therapy in patients with CMV end-organ disease, despite the lack of evidence from randomized clinical trials [10]. The high mortality associated with CMV pneumonia and the relatively tolerated safety profile of intravenous immunoglobulins have led to their use in patients with CMV pneumonia [42]. The adoptive transfer of donor-derived CMV-specific cytotoxic T lymphocytes has been evaluated [43], especially in high-risk patients such as haploidentical HCT recipients [44] or in patients with refractory CMV infection [45], and remains a promising option under investigation in ongoing clinical trials.

More details for each antiviral are in each chapter dedicated to antivirals.

6 Management Approach

6.1 Primary Prophylaxis, Preemptive Therapy, or Both?

Primary prophylaxis is a strategy that consists of administering an antiviral agent with anti-CMV activity to all patients who are at risk for CMV infection or disease in the absence of active infection. Primary prophylaxis offers several advantages, such as efficacy in preventing CMV's direct and indirect effects, a positive impact on mortality (detailed below), and ease of administration. Nevertheless, it might result in overtreatment when exposing some patients who might not develop CMV disease to an antiviral agent with potential side effects and a significant financial cost [46].

In contrast, preemptive therapy refers to the routine surveillance for active CMV infection (currently with serial PCR tests in whole blood or plasma) and initiation of antiviral therapy when the viral load exceeds a certain threshold, which is typically determined by the clinician [20]. This strategy permits targeting the antiviral therapy to those patients at the highest risk for CMV disease, thus minimizing the risks of overtreatment and unnecessary toxicity. Nevertheless, such strategy that relies on DNAemia (or antigenemia in the past) might miss some cases of CMV disease that are not preceded by DNAemia (as discussed previously) and relies on the availability of a sensitive CMV testing with a quick turnaround time [20, 46]. The viral load threshold that should trigger preemptive therapy depends on the individual patient's risk. Patients at highest risk include cord blood or haploidentical or HLA-mismatched HCT recipients, those with T-cell depleted transplants, or patients receiving higher dose corticosteroids (at least 1 mg/kg/day of prednisone equivalents) [47]. The risk for these patients is highest during the first 100 days after HCT, particularly if they are not receiving letermovir prophylaxis [47].

A recent paper outlined expert-recommended thresholds [47], although, in our experience, different centers have different practices that also depend on the testing algorithm and the overall institutional practices. CMV monitoring is typically started at the time of transplant and continued weekly until at least Day+100, and can be extended in patients with lymphopenia, CMV infection before D+100, GVHD requiring high-dose steroids or absence of CMV-specific T-cell immunity if measured [20]. Once the defined viral load threshold is reached, preemptive therapy is usually divided into two phases:

- Induction: A therapeutic dose antiviral is administered and continued until clearance of DNAemia or a substantial decrease in viral load and for a minimum of two weeks.
- Maintenance: A lower dose antiviral is administered after induction therapy to complete the treatment course with a lower risk of toxicities [20].

Within the scope of SOT, universal or preemptive prophylaxis has been used to prevent CMV disease. Universal prophylaxis relies on giving antiviral therapy to all at-risk recipients (except D−/R−) for 3–12 months, according to the type of transplantation and serological status [66]. A preemptive strategy relies on the weekly monitoring of CMV DNAemia and the initiation of antiviral therapy when the viral load is above a predetermined threshold [66]. A survey-based study done by Grossi et al. [68] showed there was a mixed approach to prophylaxis in clinical practice. Findings revealed a heterogenous approach to CMV diagnosis and management and, sometimes, significant divergence from international guidelines. Valganciclovir prophylaxis (of variable duration) was administered by 201/224 (90%) respondents in D+/R− SOT and by 40% in R+ cases, with preemptive strategies generally reserved for R+ cases: DNA thresholds to initiate treatment ranged across 10–10,000 copies/ml [68].

The use of universal vs. preemptive prophylaxis studies has been mixed. Recently a meta-analysis from Yadav et al. showed a pooled incidence of CMV infection and CMV disease in the overall Liver Transplant recipients using antiviral prophylaxis and preemptive therapy were 24.7% vs. 40.4% and 6.4% vs. 9.4%, respectively [67]. In contrast, there was a significant reduction in the incidence of late onset of CMV disease in preemptive therapy compared to antiviral prophylaxis in the high-risk group. However, the incidence of CMV disease, allograft rejection, graft loss, drug-related adverse effects, opportunistic infections, and mortality did not differ significantly between the interventions [67].

Duration of prophylaxis differs between the type of transplant, type of immunosuppression, and donor/recipient serologies. Table 4 below shows the recommended duration of prophylaxis [66].

Following preemptive therapy, patients remain at a high risk of recurrence. They, thus, can be either maintained on a secondary prophylaxis regimen with either valganciclovir or letermovir or monitored with a plan for preemptive therapy in case of recurrence [20]. The agents and dosing regimens used for preemptive therapy and the treatment of end-organ disease are outlined in Table 5.

Table 4 Duration of CMV prophylaxis (ppx) in solid organ transplant patients

Type of transplant	Serologies	Duration of universal ppx	Comment on preemptive ppx
Kidney	D+/R- R+	6 months 3 months	If preemptive would get weekly CMV QNAT (or pp65 antigenemia) for 12 wk. after transplantation and if a positive CMV threshold is reached start treatment
Kidney/ pancreas or pancreas alone	D+/R- R+	3–6 months 3 months	As above
Liver	D+/R- R+	6 months 3 months	As above
Heart	D+/R- R+	3–6 months 3 months	As above
Heart/lung or lung alone	D+/R- R+	6–12 months 6–12 months	As above
Small intestine	D+/R- R+	3–6 months 3 months	As above

Table 5 Antiviral agents and dosing regimens [20, 38]

Antiviral agent	Dosing by indication (normal renal function)		
	Prophylaxis	Induction	Maintenance
Ganciclovir	–	5 mg/kg every 12 h	5 mg/kg/day
Valganciclovir	–	900 mg every 12 h	900 mg/day
Foscarnet	–	90 mg/kg every 12 h Or 60 mg/kg every 8 h	90 mg/kg/day
Cidofovir	–	5 mg/kg/week	5 mg/kg every other week
Letermovir	480 mg/day	–	–
Maribavir	–	400 mg every 12 h	400 mg every 12 h

The 2021 guidelines by the American Society of Transplantation and Cellular Therapy (ASTCT) recommend that all centers performing allogeneic HCT should have a prophylactic strategy and a preemptive therapy strategy [20]. These guidelines offer some nonpharmacological measures to prevent CMV infection in HCT recipients, including:

- Preferential selection of a CMV-seronegative donor for a CMV-seronegative recipient to lower the risk of CMV transmission in the allograft
- Blood products from seronegative donors or leukoreduction of blood products for seronegative patients
- Good hand hygiene, especially when in contact with children, to prevent primary infection acquired from the community
- Preferential selection of a CMV-seropositive donor for a CMV-seropositive recipient as long as this satisfies the HLA matching and age criteria [20]

In terms of pharmacological prophylaxis, the first agent to be considered was intravenous acyclovir. In a case-control study published in 1988, the administration of intravenous acyclovir from D-5 to D+30 was associated with a significant reduction in CMV disease (22% versus 38%) and lower all-cause mortality at D+100 (29% versus 54%) [48]. Another study published in 1997 demonstrated a 19% survival advantage at 1 year for the group who received intravenous acyclovir from D-5 to D+30, followed by oral acyclovir for an additional six months, as compared to acyclovir from D-5 to D+30, followed by placebo [49]. Following the advent of ganciclovir, an antiviral with established anti-CMV activity, a randomized clinical trial published in 1993 evaluating ganciclovir prophylaxis from D-7 to D0 then resumed after resolution of neutropenia, compared to placebo, demonstrated a significant reduction in CMV infection (20% versus 56%), CMV disease (10% versus 24%), including within the first 120 days, with ganciclovir; however, the rates of neutropenia were higher in the ganciclovir arm and overall survival was similar in both arms [50]. Another placebo-controlled study published in the same year evaluated ganciclovir through D+100. It demonstrated a significant reduction in CMV infection (3% versus 45%) and CMV disease (0% versus 29%) within the first 100 days, at the expense of higher neutropenia (30% versus 0%) and a higher risk for bacterial infection in the patients who were on ganciclovir and developed neutropenia; however, overall survival at D+100 and D+180 were similar [51]. Subsequently, the use of ganciclovir for primary prophylaxis was restricted to very high-risk patients due to the high risks for toxicity, particularly cytopenias, leading to increased invasive bacterial or fungal infections, neutralizing any improvements in survival rates [47]. A few years later, some centers started considering the oral prodrug of ganciclovir and valganciclovir due to ease of administration and decreased risk of neutropenia. A randomized multicenter trial published in 2015 compared valganciclovir primary prophylaxis to PCR-guided preemptive therapy and showed a similar composite primary outcome of death, CMV disease, or other invasive infection by D+270 between both arms; however, the valganciclovir primary prophylaxis arm had fewer CMV DNAemia and a higher administration of hematopoietic growth factors [52]. These results have discouraged centers from using prophylaxis and have shifted the balance towards preemptive therapy, until the development of letermovir.

Letermovir was first evaluated in a phase 2 randomized clinical trial published in 2014, which randomized patients into escalating doses of letermovir versus placebo as primary prophylaxis for 12 weeks postengraftment, with the primary endpoint of all-cause prophylaxis failure [53]. The trial showed a lower rate of all-cause prophylaxis failure and virologic failure in the letermovir arms compared to placebo, which attained statistical significance for the 120 mg and 240 mg daily doses [53]. The 240 mg daily dose was determined to have the greatest antiviral activity while maintaining an acceptable safety profile [53]. This has led to a subsequent phase 3 trial published in 2017 that paved the way for FDA approval. This trial randomized CMV-seropositive allogeneic HCT recipients in a 2:1 fashion to either letermovir 480 mg daily (240 mg daily when co-administered with cyclosporine) versus placebo through week 14 post-HCT, with a primary endpoint of CS-CMVi through week 24 [12]. The study demonstrated a significant decrease in CS-CMVi in the letermovir arm. Very few CMV end-organ disease events occurred to detect a difference [12]. This difference was maintained in subgroup analysis, stratifying patients into low-risk and high-risk groups. Nevertheless, there was a notable increase in CS-CMVi in the letermovir arm after discontinuing prophylaxis, which merits additional attention [12]. A post hoc mortality analysis of this trial, including patients with no detectable CMV DNA at randomization and who had vital status data through week 48, showed a significant reduction in all-cause mortality in the letermovir arm in the total population, mostly driven by improved survival among the patients who had CS-CMVi [54]. A subsequent real-world retrospective study of letermovir primary prophylaxis demonstrated a significant reduction in CS-CMVi, CMV end-organ disease, refractory or resistant CMV infection, and nonrelapse mortality at week 48 among the patients who received primary prophylaxis with letermovir, compared to no prophylaxis [19].

As such, according to the 2021 ASTCT guidelines, the recommended agent for primary prophylaxis is letermovir, to be continued through D+100 [20]. Extension of prophylaxis beyond 100 days, and use of letermovir in children are currently under evaluation in ongoing clinical trials.

6.2 Treatment of End-Organ Disease

The treatment of end-organ disease follows the same principles as preemptive therapy; however, the length of induction therapy might vary according to the organ involved. Pneumonia and gastrointestinal or central nervous system disease are treated for 2–3 weeks with therapeutic dose antivirals, whereas retinitis requires 6–8 weeks of systemic therapy [10]. This is usually followed by maintenance therapy and/or secondary prophylaxis. The agents used and the dosing regiments are outlined in Table 3. As previously discussed, intravenous immunoglobulins have been used as an adjuvant agent in patients with CMV pneumonia, usually at 0.2–0.5 mg/kg [10]. Response to therapy is usually evaluated by serial blood viral loads in those patients with DNAemia, resolution of symptoms, and serial imaging or fundoscopy whenever abnormal findings are detected at baseline [55].

6.3 Postprophylaxis Delayed-Onset CMV Disease

Despite prophylaxis in SOT patients, delayed onset CMV disease is a known phenomenon that occurs. A study by Räihä et al. on kidney transplants showed 71/136 (52%) patients developed postprophylaxis primary CMV infection and 52/136 (38%) patients' probable postprophylaxis CMV disease. The median interval was 260 days after transplantation and 72 days after the end of prophylaxis [69]. Based on this, it would be reasonable to perform CMV PCR screening 3–6 months after stopping prophylaxis, especially in high-risk SOT patients. It should also be noted that patients should have testing done if they develop symptoms of Acute CMV disease. Another way clinical physicians decide when to stop prophylaxis is by using T-cell assays. In a discussion article by Hall et al., they show several observational studies, including high-risk (D+R−) and intermediate-risk (R+) patients, have shown that a positive CMV-CMI assay predicts protection from subsequent CMV reactivation and clinical disease development, thereby potentially useful for personalizing the duration of CMV antiviral prophylaxis [70].

6.4 Refractory or Resistant CMV

Until the recent FDA approval of maribavir for the treatment of posttransplant refractory or resistant CMV, the treatment of these complex infections or end-organ diseases has relied on switching agents according to the detected mutation in resistant cases, and according to which agent was used as first-line in refractory cases [22]. The 2021 ASTCT guidelines recommend the following strategies in Table 6 [20]:

Table 6 Treatment for CMV resistance

Type of resistance	Recommended medication
Refractory CMV with no known mutations	Optimize dosing of the first line agent, switch to foscarnet if on ganciclovir, or to ganciclovir if on foscarnet
UL97 mutation with high-level resistance to ganciclovir	Switch to foscarnet or cidofovir
UL97 mutation with low-level resistance to ganciclovir	High dose ganciclovir 7.5–10 mg/kg every 12 h if tolerated, otherwise switch to foscarnet or cidofovir
UL54 mutation with resistance to foscarnet and ganciclovir	Switch to cidofovir, consider leflunomide or artesunate
UL54 mutation with resistance to ganciclovir and cidofovir	Switch to foscarnet, consider leflunomide or artesunate
UL54 mutation with resistance to foscarnet only	Switch to ganciclovir
Switch to ganciclovir UL54 mutation with resistance to ganciclovir, foscarnet, and cidofovir	Foscarnet with high-dose ganciclovir, G-CSF support
UL56, UL51, or UL89 mutations with resistance to letermovir	Switch to ganciclovir or foscarnet

Patients should also be considered for referral for trials of investigational agents, including CMV-specific cytotoxic T lymphocytes. Nonrandomized clinical studies have evaluated donor-derived or third-party adoptive T-cell immunotherapy with good response rates ranging between 74% and 93% [56, 57]. Of note, concurrent administration of high-dose steroids has been an exclusion criterion in these studies, and the use of adoptive T-cell immunotherapy in these patients is unclear.

Maribavir was initially studied as an agent for preemptive therapy. In phase 2 randomized trial published in 2019, patients with randomized to escalating doses of maribavir compared to standard dose valganciclovir for a maximum of 12 weeks, CMV viremia clearance was assessed at week 6 of therapy and found to be similar across all doses of maribavir to standard dose valganciclovir, with a higher incidence of dysgeusia and a lower rate of neutropenia in the maribavir arms [58]. Another phase 2 trial published in the same year evaluated maribavir for treating refractory or resistance CMV infections in both HCT and solid organ transplant (SOT) recipients [59]. In this trial, patients were randomized to three arms of escalating doses of maribavir, with the primary endpoint being CMV viremia clearance within 6 weeks of therapy. This study showed a similar proportion of patients meeting the primary endpoint across all three dosing regimens, ranging between 62.5% and 70%, a significantly positive outcome for refractory or resistant infections that have been difficult to treat [59]. Finally, the phase 3 clinical trial was published in 2021. In this multicenter randomized trial, HCT or SOT recipients with refractory or resistant CMV were randomized to either maribavir or investigator-assigned treatment for 8 weeks and assessed for CMV viremia clearance at 8 weeks as the primary endpoint [38]. Notably, 78% of the patients on maribavir completed the 8-week course versus only 32% on the control arm, mostly due to adverse events. At week 8, 56% of the patients on maribavir achieved viremia clearance, compared to

24% on the control arm. Mortality was similar between the two groups. Dysgeusia occurred more commonly in the maribavir arm, whereas neutropenia, other cytopenia, and acute kidney injury occurred more commonly in the control group [38]. This trial has led to the FDA approval of maribavir for posttransplant refractory or resistant CMV in November 2021. Given the results of this trial, in terms of efficacy and safety profile, it is reasonable to consider maribavir as the first-line agent for refractory or resistant infections. The exact duration of therapy with maribavir remains to be determined with more real-world experience since the trial allowed for 8 weeks of therapy and no subsequent maintenance or secondary prophylaxis [38]. Of concern is the development of de novo maribavir resistance in 27% of the patients who received maribavir in the phase 3 trial, particularly in patients with a high baseline viral load [38]. The therapeutic approach to these patients is unclear. It would likely involve one of the strategies recommended by the ASTCT guidelines, particularly with the UL97 mutations, some of which might confer cross-resistance to ganciclovir.

7 Future Directions

Our understanding of the immunobiology of CMV reactivations after allogeneic HCT and SOT remains an actively evolving field. Multiple assays have been developed to evaluate CMV-specific cell-mediated immunity with the goal of predicting which transplant recipients are at the highest risk of CS-CMVi [60, 61]. These assays have been evaluated in single-center studies; however, their incorporation into daily practice remains a matter of expert opinion and institutional practices, until a randomized trial can better inform this strategy [47]. Developing an effective CMV vaccine also remains a matter of ongoing research. A DNA vaccine candidate failed to reduce overall mortality or CMV end-organ disease in a recent phase 3 trial [62]. Other candidate vaccines are under evaluation. The role of adoptive immunotherapy in prophylaxis and preemptive therapy should also be assessed in randomized trials [63, 64].

Conclusions: CMV is a vast and diverse disease, and using Risk factors and diagnostic techniques can help identify the most susceptible patients. Between SOT and HCT patients, risk factors and disease symptoms can be similar; however, duration of treatment can vary depending on the type of transplant. It is also important to note that there is developing resistance in this immunocompromised population. Mediations such as foscarnet and ganciclovir, considered the staple of treatment for CMV disease, are now useless against UL 54 and UL 97 mutations. Newer agents such as Letermovir have allowed us to expand our anti-viral prophylaxis; however, much data on treatment still needs to be provided. Marabavir, a newer agent, has shown promise in helping patients with refractory/resistant CMV, but has also shown de novo resistance. Further methods and clinical trials will be needed to help diagnose and treat these patients and to reduce morbidity and mortality in this population.

References

1. Parody, R., et al., Severe infections after unrelated donor allogeneic hematopoietic stem cell transplantation in adults: comparison of cord blood transplantation with peripheral blood and bone marrow transplantation. Biol Blood Marrow Transplant, 2006. **12**(7): p. 734–48.
2. Pande, A. and E.R. Dubberke, Cytomegalovirus Infections of the Stem Cell Transplant Recipient and Hematologic Malignancy Patient. Infect Dis Clin North Am, 2019. **33**(2): p. 485–500.
3. Britt, W.J., Cytomegalovirus, in Mandell, Douglas, and Bennett's Principles and Practice of Infectious Diseases, J.E. Bennett, R. Dolin, and M.J. Blaser, Editors. 2020, Elsevier: Philadelphia. p. 1857–1871.
4. Khaiboullina, S.F., et al., Human cytomegalovirus persists in myeloid progenitors and is passed to the myeloid progeny in a latent form. Br J Haematol, 2004. **126**(3): p. 410–7.
5. Larsson, S., et al., Cytomegalovirus DNA can be detected in peripheral blood mononuclear cells from all seropositive and most seronegative healthy blood donors over time. Transfusion, 1998. **38**(3): p. 271–8.
6. Sylwester, A.W., et al., Broadly targeted human cytomegalovirus-specific CD4+ and CD8+ T cells dominate the memory compartments of exposed subjects. J Exp Med, 2005. **202**(5): p. 673–85.
7. Smith, C. and R. Khanna, Immune regulation of human herpesviruses and its implications for human transplantation. Am J Transplant, 2013. **13 Suppl 3**: p. 9–23; quiz 23.
8. Miller, W., et al., Cytomegalovirus infection after bone marrow transplantation: an association with acute graft-v-host disease. Blood, 1986. **67**(4): p. 1162–7.
9. Nichols, W.G., et al., High risk of death due to bacterial and fungal infection among cytomegalovirus (CMV)-seronegative recipients of stem cell transplants from seropositive donors: evidence for indirect effects of primary CMV infection. J Infect Dis, 2002. **185**(3): p. 273–82.
10. Ariza-Heredia, E.J., L. Nesher, and R.F. Chemaly, Cytomegalovirus diseases after hematopoietic stem cell transplantation: a mini-review. Cancer Lett, 2014. **342**(1): p. 1–8.
11. Ljungman, P., et al., Definitions of Cytomegalovirus Infection and Disease in Transplant Patients for Use in Clinical Trials. Clin Infect Dis, 2017. **64**(1): p. 87–91.
12. Marty, F.M., et al., Letermovir Prophylaxis for Cytomegalovirus in Hematopoietic-Cell Transplantation. N Engl J Med, 2017. **377**(25): p. 2433–2444.
13. Chemaly, R.F., et al., Definitions of Resistant and Refractory Cytomegalovirus Infection and Disease in Transplant Recipients for Use in Clinical Trials. Clin Infect Dis, 2019. **68**(8): p. 1420–1426.
14. Sassine, J., et al., 930. Clinically Significant CMV Infections in Patients with Lymphoma or Multiple Myeloma Open Forum Infectious Diseases, 2021. **8**(Supplement 1): p. S557–S558.
15. Baden, L.R., et al., Prevention and Treatment of Cancer-Related Infections, Version 2.2016, NCCN Clinical Practice Guidelines in Oncology. J Natl Compr Canc Netw, 2016. **14**(7): p. 882–913.
16. Khawaja, F., et al., Herpesviruses Infections in CAR T Cell Recipients. Transplant Cell Ther, 2022. **28**(3): p. S381–S382.
17. Ljungman, P., M. Hakki, and M. Boeckh, Cytomegalovirus in hematopoietic stem cell transplant recipients. Infect Dis Clin North Am, 2010. **24**(2): p. 319–37.
18. Ljungman, P., et al., Donor CMV serologic status and outcome of CMV-seropositive recipients after unrelated donor stem cell transplantation: an EBMT megafile analysis. Blood, 2003. **102**(13): p. 4255–60.
19. Sassine, J., et al., Refractory and Resistant Cytomegalovirus After Hematopoietic Cell Transplant in the Letermovir Primary Prophylaxis Era. Clin Infect Dis, 2021. **73**(8): p. 1346–1354.
20. Hakki, M., et al., American Society for Transplantation and Cellular Therapy Series: #3-Prevention of Cytomegalovirus Infection and Disease After Hematopoietic Cell Transplantation. Transplant Cell Ther, 2021. **27**(9): p. 707–719.

21. Goldsmith, S.R., et al., Posttransplant cyclophosphamide is associated with increased cytomegalovirus infection: a CIBMTR analysis. Blood, 2021. **137**(23): p. 3291–3305.
22. El Chaer, F., D.P. Shah, and R.F. Chemaly, How I treat resistant cytomegalovirus infection in hematopoietic cell transplantation recipients. Blood, 2016. **128**(23): p. 2624–2636.
23. Kotloff, R.M., V.N. Ahya, and S.W. Crawford, Pulmonary complications of solid organ and hematopoietic stem cell transplantation. Am J Respir Crit Care Med, 2004. **170**(1): p. 22–48.
24. Vigil, K.J., J.A. Adachi, and R.F. Chemaly, Viral pneumonias in immunocompromised adult hosts. J Intensive Care Med, 2010. **25**(6): p. 307–26.
25. Horger, M.S., et al., Cytomegalovirus pneumonia after stem cell transplantation: correlation of CT findings with clinical outcome in 30 patients. AJR Am J Roentgenol, 2006. **187**(6): p. W636–43.
26. Green, M.L., et al., Efficacy of a viral load-based, risk-adapted, preemptive treatment strategy for prevention of cytomegalovirus disease after hematopoietic cell transplantation. Biol Blood Marrow Transplant, 2012. **18**(11): p. 1687–99.
27. Boeckh, M., et al., Cytomegalovirus (CMV) DNA Quantitation in Bronchoalveolar Lavage Fluid From Hematopoietic Stem Cell Transplant Recipients With CMV Pneumonia. J Infect Dis, 2017. **215**(10): p. 1514–1522.
28. van Burik, J.A., et al., Cytomegalovirus enteritis among hematopoietic stem cell transplant recipients. Biol Blood Marrow Transplant, 2001. **7**(12): p. 674–9.
29. Torres, H.A., et al., Gastrointestinal cytomegalovirus disease in patients with cancer: a two decade experience in a tertiary care cancer center. Eur J Cancer, 2005. **41**(15): p. 2268–79.
30. Ruell, J., et al., Active CMV disease does not always correlate with viral load detection. Bone Marrow Transplant, 2007. **40**(1): p. 55–61.
31. Ando, T., et al., Cytomegalovirus ventriculoencephalitis in a reduced- intensity conditioning cord blood transplant recipient. Transpl Infect Dis, 2010. **12**(5): p. 441–5.
32. Handley, G., et al., Distinguishing cytomegalovirus meningoencephalitis from other viral central nervous system infections. J Clin Virol, 2021. **142**: p. 104936.
33. Xhaard, A., et al., Increased incidence of cytomegalovirus retinitis after allogeneic hematopoietic stem cell transplantation. Transplantation, 2007. **83**(1): p. 80–3.
34. Marty, F.M., et al., A Randomized, Double-Blind, Placebo-Controlled Phase 3 Trial of Oral Brincidofovir for Cytomegalovirus Prophylaxis in Allogeneic Hematopoietic Cell Transplantation. Biol Blood Marrow Transplant, 2019. **25**(2): p. 369–381.
35. El-Haddad, D., et al., Brincidofovir (CMX-001) for refractory and resistant CMV and HSV infections in immunocompromised cancer patients: A single-center experience. Antiviral Res, 2016. **134**: p. 58–62.
36. Marty, F.M., et al., CMX001 to prevent cytomegalovirus disease in hematopoietic-cell transplantation. N Engl J Med, 2013. **369**(13): p. 1227–36.
37. Prichard, M.N., Function of human cytomegalovirus UL97 kinase in viral infection and its inhibition by maribavir. Rev Med Virol, 2009. **19**(4): p. 215–29.
38. Avery, R.K., et al., Maribavir for Refractory Cytomegalovirus Infections With or Without Resistance Post-Transplant: Results from a Phase 3 Randomized Clinical Trial. Clin Infect Dis, 2021.
39. Marty, F.M., et al., Maribavir prophylaxis for prevention of cytomegalovirus disease in recipients of allogeneic stem-cell transplants: a phase 3, double-blind, placebo-controlled, randomised trial. Lancet Infect Dis, 2011. **11**(4): p. 284–92.
40. Avery, R.K., et al., Utility of leflunomide in the treatment of complex cytomegalovirus syndromes. Transplantation, 2010. **90**(4): p. 419–26.
41. Gokarn, A., et al., Use of Leflunomide for Treatment of Cytomegalovirus Infection in Recipients of Allogeneic Stem Cell Transplant. Biol Blood Marrow Transplant, 2019. **25**(9): p. 1832–1836.
42. Alexander, B.T., et al., Use of cytomegalovirus intravenous immune globulin for the adjunctive treatment of cytomegalovirus in hematopoietic stem cell transplant recipients. Pharmacotherapy, 2010. **30**(6): p. 554–61.

43. Blyth, E., et al., Donor-derived CMV-specific T cells reduce the requirement for CMV-directed pharmacotherapy after allogeneic stem cell transplantation. Blood, 2013. **121**(18): p. 3745–58.
44. Pei, X.Y., et al., Adoptive therapy with cytomegalovirus-specific T cells for cytomegalovirus infection after haploidentical stem cell transplantation and factors affecting efficacy. Am J Hematol, 2022. **97**(6): p. 762–769.
45. Bao, L., et al., Adoptive immunotherapy with CMV-specific cytotoxic T lymphocytes for stem cell transplant patients with refractory CMV infections. J Immunother, 2012. **35**(3): p. 293–8.
46. Razonable, R.R. and A. Humar, Cytomegalovirus in solid organ transplant recipients-Guidelines of the American Society of Transplantation Infectious Diseases Community of Practice. Clin Transplant, 2019. **33**(9): p. e13512.
47. Einsele, H., P. Ljungman, and M. Boeckh, How I treat CMV reactivation after allogeneic hematopoietic stem cell transplantation. Blood, 2020. **135**(19): p. 1619–1629.
48. Meyers, J.D., et al., Acyclovir for prevention of cytomegalovirus infection and disease after allogeneic marrow transplantation. N Engl J Med, 1988. **318**(2): p. 70–5.
49. Prentice, H.G., et al., Long-term survival in allogeneic bone marrow transplant recipients following acyclovir prophylaxis for CMV infection. The European Acyclovir for CMV Prophylaxis Study Group. Bone Marrow Transplant, 1997. **19**(2): p. 129–33.
50. Winston, D.J., et al., Ganciclovir prophylaxis of cytomegalovirus infection and disease in allogeneic bone marrow transplant recipients. Results of a placebo-controlled, double-blind trial. Ann Intern Med, 1993. **118**(3): p. 179–84.
51. Goodrich, J.M., et al., Ganciclovir prophylaxis to prevent cytomegalovirus disease after allogeneic marrow transplant. Ann Intern Med, 1993. **118**(3): p. 173–8.
52. Boeckh, M., et al., Valganciclovir for the prevention of complications of late cytomegalovirus infection after allogeneic hematopoietic cell transplantation: a randomized trial. Ann Intern Med, 2015. **162**(1): p. 1–10.
53. Chemaly, R.F., et al., Letermovir for cytomegalovirus prophylaxis in hematopoietic-cell transplantation. N Engl J Med, 2014. **370**(19): p. 1781–9.
54. Ljungman, P., et al., A Mortality Analysis of Letermovir Prophylaxis for Cytomegalovirus (CMV) in CMV-seropositive Recipients of Allogeneic Hematopoietic Cell Transplantation. Clin Infect Dis, 2020. **70**(8): p. 1525–1533.
55. Yong, M.K., et al., American Society for Transplantation and Cellular Therapy Series: #4 - Cytomegalovirus treatment and management of resistant or refractory infections after hematopoietic cell transplantation. Transplant Cell Ther, 2021. **27**(12): p. 957–967.
56. Withers, B., et al., Long-term control of recurrent or refractory viral infections after allogeneic HSCT with third-party virus-specific T cells. Blood Adv, 2017. **1**(24): p. 2193–2205.
57. Leen, A.M., et al., Multicenter study of banked third-party virus-specific T cells to treat severe viral infections after hematopoietic stem cell transplantation. Blood, 2013. **121**(26): p. 5113–23.
58. Maertens, J., et al., Maribavir for Preemptive Treatment of Cytomegalovirus Reactivation. N Engl J Med, 2019. **381**(12): p. 1136–1147.
59. Papanicolaou, G.A., et al., Maribavir for Refractory or Resistant Cytomegalovirus Infections in Hematopoietic-cell or Solid-organ Transplant Recipients: A Randomized, Dose-ranging, Double-blind, Phase 2 Study. Clin Infect Dis, 2019. **68**(8): p. 1255–1264.
60. El Haddad, L., et al., The Ability of a Cytomegalovirus ELISPOT Assay to Predict Outcome of Low-Level CMV Reactivation in Hematopoietic Cell Transplant Recipients. J Infect Dis, 2019. **219**(6): p. 898–907.
61. Chemaly, R.F., et al., Cytomegalovirus (CMV) Cell-Mediated Immunity and CMV Infection After Allogeneic Hematopoietic Cell Transplantation: The REACT Study. Clin Infect Dis, 2020. **71**(9): p. 2365–2374.
62. Ljungman, P., et al., A randomised, placebo-controlled phase 3 study to evaluate the efficacy and safety of ASP0113, a DNA-based CMV vaccine, in seropositive allogeneic haematopoietic cell transplant recipients. EClinicalMedicine, 2021. **33**: p. 100787.

63. Boeckh, M. and L. Corey, Adoptive Immunotherapy of Viral Infections: Should Infectious Disease Embrace Cellular Immunotherapy? J Infect Dis, 2017. **216**(8): p. 926–928.
64. M. I. Morris et al. (eds.), Emerging Transplant Infections, https://doi.org/10.1007/978-3-030-01751-4_24-1
65. C. Lumbreras, O. Manuel, O. Len, I.J.M. ten Berge, D. Sgarabotto, H.H Hirsch, Cytomegalovirus infection in solid organ transplant recipients, Clinical Microbiology and Infection, Volume 20, Supplement 7, 2014, Pages 19–26, ISSN 1198-743X, https://doi.org/10.1111/1469-0691.12594. (https://www.sciencedirect.com/science/article/pii/S1198743X14604997)
66. Azevedo LS, Pierrotti LC, Abdala E, Costa SF, Strabelli TM, Campos SV, Ramos JF, Latif AZ, Litvinov N, Maluf NZ, Caiaffa Filho HH, Pannuti CS, Lopes MH, Santos VA, Linardi Cda C, Yasuda MA, Marques HH. Cytomegalovirus infection in transplant recipients. Clinics (Sao Paulo). 2015 Jul;70(7):515–23. https://doi.org/10.6061/clinics/2015(07)09. Epub 2015 Jul 1. PMID: 26222822; PMCID: PMC4496754.
67. Yadav DK, Adhikari VP, Yadav RK, Singh A, Huang X, Zhang Q, Pandit P, Ling Q, Liang T. Antiviral prophylaxis or preemptive therapy for cytomegalovirus after liver transplantation?: A systematic review and meta-analysis. Front Immunol. 2022 Nov 10;13:953210. https://doi.org/10.3389/fimmu.2022.953210. PMID: 36439159; PMCID: PMC9685424.
68. Kotton CN, Kamar N. New Insights on CMV Management in Solid Organ Transplant Patients: Prevention, Treatment, and Management of Resistant/Refractory Disease. Infect Dis Ther. 2023 Feb;12(2):333–342. https://doi.org/10.1007/s40121-022-00746-1. Epub 2022 Dec 30. PMID: 36583845; PMCID: PMC9925645.
69. Räihä, J, Ortiz, F, Mannonen, L, et al. The burden of cytomegalovirus infection remains high in high-risk kidney transplant recipients despite six-month valganciclovir prophylaxis. *Transpl Infect Dis*. 2021; 23:e13577. https://doi.org/10.1111/tid.13577
70. Hall VG, Humar A, Kumar D. Utility of Cytomegalovirus Cell-Mediated Immunity Assays in Solid Organ Transplantation. J Clin Microbiol. 2022 Aug 17;60(8):e0171621. https://doi.org/10.1128/jcm.01716-21. Epub 2022 May 11. PMID: 35543099; PMCID: PMC9383112.

Herpes Simplex Virus and Varicella Zoster Virus in Hematopoietic Stem Cell Transplant Recipients

Olga Klinkova and Aliyah Baluch

Abstract Patients who undergo hematopoietic stem cell transplant (HCT) have a high incidence of herpes simplex virus (HSV) and varicella zoster virus (VZV) infections, despite universal prophylaxis post-transplant. Unique risk factors, such as the utilization of aggressive conditioning chemotherapy regimens, acute or chronic graft-versus-host disease (GVHD), and decline in T-cell-mediated immunity, contribute to the high incidence of these infections post-HCT. Epidemiology, specific risk factors, and unique clinical features of HSV and VZV infections post-HCT are described in this chapter. In addition, acyclovir-resistant (ACVr) HSV infections have the highest incidence in HCT recipients. In this chapter, we describe specific risk factors, diagnostic challenges, and treatment strategies for ACVr HSV infection management.

Keywords Herpes simplex virus · Varicella-zoster virus · Hematopoietic stem cell transplant · Mucositis · Immunocompromised · Resistance · Graft-versus-host disease · Acyclovir · Foscarnet · Cidofovir

1 Key Concepts

- HSV and VZV infections have a high incidence in HCT recipients, despite universal antiviral prophylaxis post-HCT.
- Unique risk factors such as utilization of aggressive conditioning chemotherapy regimens, acute or chronic graft-versus-host disease (GVHD), and decline in T-cell- mediated immunity contribute to the high incidence of these infections post-HCT.

O. Klinkova (✉) · A. Baluch
Division of Infectious Diseases, Department of IHM/Internal Medicine, Moffitt Cancer Center, Tampa, FL, USA
e-mail: olga.klinkova@moffitt.org

C. Somboonwit et al. (eds.), *Global Virology IV: Viral Disease Diagnosis and Treatment Delivery in the 21st Century*,
https://doi.org/10.1007/978-3-031-57369-9_16

- The incidence of HSV infections resistant to acyclovir (ACV) treatment is rising among the recipients of HCT. Treatment options such as use of intravenous foscarnet, cidofovir, high dose intravenous acyclovir, and topical agents are described.

2 Herpes Simplex Viruses 1 and 2

2.1 Epidemiology and Risk Factors

Herpes simplex viruses (HSV) 1 and 2 are prevalent worldwide. Based on the study by Looker et al., it was estimated that the prevalence of HSV-1 virus worldwide was averaged at 67% across all ages between 0 and 49 years old population [1]. After primary infection, HSV becomes latent within the neuronal cells and persists for life. In HSV seropositive recipients of allogeneic hematopoietic stem cell transplant (HSCT), HSV reactivation occurs in 60–80% of the patients in the absence of viral prophylaxis [2, 3]. Therefore, universal antiviral prophylaxis with acyclovir (ACV) is currently accepted as a standard of care in this patient population.

Despite the antiviral prophylaxis of HSV seropositive HCT recipients, HSV infection/reactivation is very common. Based on the retrospective analysis of 382 seropositive recipients of allogeneic stem cell transplants, conducted by Uhlin et al., the cumulative incidence of early (within the first 90 days post-transplant) HSV reactivation was 18%, with the majority of infections occurring within the first month post-transplant (67%) [3].

2.2 Clinical Presentation and Diagnosis

In patients undergoing HCT, HSV causes a spectrum of conditions, from asymptomatic shedding to severe disease, with the involvement of the gastrointestinal, pulmonary, and central nervous systems. While mucocutaneous disease remains the most common presentation of HSV-1, especially in settings of pre-engraftment or early engraftment period due to the high incidence of chemotherapy-induced mucositis early post-transplant, painful genital ulcerations are the most common manifestation of HSV-2.

Real-time HSV polymerase chain reaction (PCR) assays are most commonly used for the diagnosis of acute HSV infection. HSV PCR testing of the scrapings of the vesicle base or a tissue sample of the atypical skin lesion is performed to provide a rapid diagnosis. Even though viral culture is very specific for the diagnosis of HSV infection, its sensitivity is suboptimal compared to PCR testing, especially when low amounts of HSV are present [4]. Therefore, the majority of hospital-based microbiology labs no longer perform viral culture methods and only utilize PCR-based techniques for the diagnosis of HSV.

3 Varicella Zoster Virus (VZV) Infections During HCT

3.1 *Epidemiology and Risk Factors*

The cumulative incidence of VZV infections after HCT has been reported to be as high as 40% [5, 6]. However, the most recent data from a multicenter, retrospective cohort study conducted by Schuster et al. on the incidence of infections among 444 HCT recipients followed for up to 30 months after HCT demonstrated only 7% incidence of VZV infections in these patients. This might be reflective of the recent advances in prophylactic strategies and vaccines compared to the earlier studies, in addition to the shorter follow-up period for the patients included in this analysis [7].

Even though primary VZV infection can occur in patients not previously exposed to VZV, the majority of VZV infections in HCT patients are related to reactivation of the latent VZV. Reactivation can occur at any point post-transplant; however, it is more common in the late post-transplantation period, as demonstrated in prior studies [5]. As demonstrated in the retrospective analysis of Koc et al., the median onset of VZV infection was 227 days from the day of HCT, with the range of 45 days to 3.7 years; only 12% of infections occurred within the first 3 months of HCT [5]. Use of antiviral agents to prevent other viral infections such as HSV and cytomegalovirus delayed the onset of VZV reactivation [5].

Post-HCT, specific risk factors that have been shown to be associated with a high risk of VZV reactivation include being the recipient of allogeneic HCT, utilization of pre-transplant radiation as a part of a conditioning regimen, having a diagnosis other than chronic myelogenous leukemia, administration of post-transplant antithymocyte globulin, and acute or chronic GVHD [5, 6, 8].

3.2 *Clinical Presentation and Diagnosis*

Painful vesicular rash in a dermatomal distribution remains the most common manifestation of VZV infection in HCT patients [5, 8]. Based on earlier studies, complicated infections have been reported in up to 40% of the HCT recipients; the most common complications include multidermatomal involvement in more than two dermatomes, disseminated varicella-like skin rash, visceral involvement, and meningitis/encephalitis [5, 8]. It was determined in the study by Baumrin et al. that complicated VZV infections occurred in about 1% of HCT patients and occurred in bimodal distribution, with earlier cases occurring within the first 90 days after HCT and a later peak occurring between 12 and 24 months after HCT [9].

Postherpetic neuralgia is another common long-term complication of shingles that occurs in 35–40% of the patients [5, 10]. In the study conducted by Onozawa et al. of 78 patients who developed VZV infection after HCT, advanced age, male gender, and graft-versus-host disease prophylaxis with a tacrolimus-based regimen were associated with an increased risk of post-herpetic neuralgia in patients who underwent allogeneic HCT [10, 11].

In most cases, diagnosis of VZV infections is made based on the history and clinical findings of a typical vesicular rash in a dermatomal distribution. The real-time polymerase chain reaction (PCR) assay is the most used and can be performed on the scraping of the vesicle base, vesicle fluid, or other body fluid such as cerebrospinal fluid [12]. Compared to the shell vial culture, real-time PCR assays have been shown to be more sensitive for the diagnosis of VZV infection [12]. Serum or whole blood VZV PCR may be helpful in the diagnosis of visceral VZV infection when cutaneous lesions might be absent or appear later in the course of the disease [13]. In the cases of atypical cutaneous lesions, immunohistochemical identification of VZV on a pathology sample as well as tissue PCR testing can assist in diagnosis.

4 Prophylaxis and Treatment of HSV and VZV Infections Post-HCT

Universal HSV prophylaxis is recommended for all HSV-seropositive HCT recipients [14, 15]. Prophylaxis is started at the initiation of conditioning chemotherapy and continues for at least 30 days post-HCT [14, 15]. Longer prophylaxis beyond 30 days and at least for 1 year is recommended for VZV-seropositive HCT recipients, patients with prior frequent HSV infections, and patients that continue immunosuppression for GVHD prophylaxis/treatment [14, 15]. Specific options for prophylaxis and treatment are outlined in the chapter "Antiviral Agents" of this book.

5 Acyclovir-Resistant HSV Infections in Patients Undergoing HSCT

5.1 Epidemiology and Risk Factors

Patients undergoing HCT have the highest incidence of infections with ACVr-HSV compared to other immunosuppressed patient populations [16, 17]. Isolation of ACVr-HSV has been linked to poorer outcomes and a trend toward increased mortality in this patient population [18]. Kakiuchi et al. conducted a 2-year prospective surveillance study on the reactivation of HSV-1 in the patients 1 week before and 100 days after HSCT [19]. Acyclovir-resistant (ACVr) HSV-1 was isolated in 28% of the patients included in this study.

Long-term prophylaxis with antiviral agents predisposes hematopoietic transplant patients to the development of resistance [20]. As demonstrated in a study by Ariza-Heredia et al., the median duration of previous acyclovir exposure was 12 months, with the range of 2–28 months [21]. In addition, prolonged viral replication in transplant recipients due to impaired host defenses may allow less pathogenic but resistant viral units to survive and cause clinical infection [20, 22]. In addition, patients undergoing HSCT for relapsed underlying malignancies were at statistically higher risk for the emergence of ACVr-HSV compared to other subtypes of patients. Other risk factors for harboring antiviral drug- resistant HSV include GVHD requiring systemic steroids [18, 21, 23], and the pre-engraftment period [21].

5.2 Clinical Presentation and Diagnosis

Infection with ACVr-HSV can occur early in the post-transplant period with the median onset of the first episode reported being 10.5 days after transplant, with more late presentation at a median onset of 89.5 days post-transplant [21]. Painful oral hemorrhagic mucositis and hemorrhagic labial lesions are most commonly described; facial (Fig. 1), intranasal, as well as subungual lesions on the hands have been reported as well in patients with ACVr-HSV [24]. Even though clinical manifestations are quite similar to the presentation of a non-resistant type as describes above, the clinicians should have high index of suspicion for ACVr-HSV infection if no appreciable improvement in the size the lesions occurred, lesions develop atypical appearance or new satellite lesions observed after 1 week of appropriate anti-viral therapy [20].

When ACVr-HSV infection is suspected, analysis of HSV resistance to antiviral agents should be performed on a specimen. Both phenotypic and genotypic resistance assays are commercially available. Phenotypic resistance assay determines

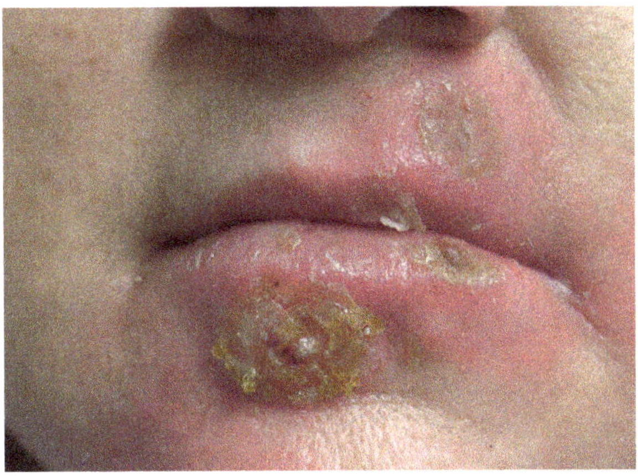

Fig. 1 Title: Patient with refractory acute myelogenous leukemia with perioral herpetic lesions that were confirmed to be ACVr-HSV based on the resistance testing

the in-vitro susceptibility of the cultured HSV isolate to ACV and other antiviral agents. The plaque reduction assay is the most commonly used method that measures the inhibitory concentration (IC50) of the drugs sufficient to inhibit plaque formation by 50% [25]. This assay is time-consuming and typically takes 7–14 days to receive results, as it requires the isolation of the virus in culture to perform in vitro drug susceptibility testing [26].

Genotypic resistance assay relies on the identification of specific mutations in the viral genome encoding thymidine kinase and DNA polymerase and accounting for resistance. However, due to the naturally high polymorphism of HSV-1 gene [26], the distinction between random variations due to polymorphism and true drug resistance mutations is challenging, therefore, making the use of these assays limited in clinical practice [20]. The most common mechanism of resistance to acyclovir is related to thymidine kinase (TK) mutation; mutations associated with DNA polymerase are less common due to the essential role of DNA polymerase in viral replication [5, 6].

5.3 Treatment

For the patients with suspected ACVr-HSV infection, it is recommended to change antiviral therapy to an alternative agent while awaiting the results of the resistance testing due to the anticipated delay in resistance testing results as described above.

Intravenous foscarnet is FDA-approved for the treatment of suspected ACVr-HSV infection [27, 28]. It directly inhibits DNA polymerase by blocking the pyrophosphate binding site and does not require phosphorylation for activation [27]. It is administered by dosing 40 mg/kg/dose IV every 8–12 days for 14–21 days or until resolution of symptoms. Hydration with either normal saline or D5W should be given prior to the initial foscarnet infusion. Daily monitoring of renal function and electrolytes is recommended while on foscarnet t therapy. Foscarnet is highly nephrotoxic drug that requires appropriate dose adjustments based on the patient's clearance of creatinine. It is documented that more than half of the patients develop elevated creatinine while on foscarnet therapy, with creatinine returning to baseline in 2–4 weeks after discontinuation of the drug [29]. Foscarnet-induced genital ulcerations have been reported to occur in up to 10% of the patients during high-dose foscarnet use [30]. The exact etiology of this side effect is unclear but could be related to foscarnet-induced irritant dermatitis secondary to cutaneous contact with foscarnet, as the drug is mostly excreted unchanged in the urine [31].

Intravenous cidofovir has an off-label indication for the treatment of ACVr HSV infection. It is a thymidine kinase-independent viral DNA polymerase inhibitor. In the induction phase, it is administered at 5 mg/kg/dose once weekly for 2 weeks, followed by the maintenance phase at 5 mg/kg/dose once every 2 weeks for three doses [27, 28]. Probenecid and intravenous hydration are given prior to and after intravenous cidofovir administration to prevent kidney injury. Cidofovir is also considered highly nephrotoxic, with a reported rate of acute kidney injury up to 25% [32].

Continuous infusion of intravenous acyclovir has been reported to be effective in small case series for the treatment of ACVr HSV infection [33, 34]. In the study by Kim et al., IV acyclovir with the dose range of 30–45 mg/kg daily was administered as a continuous infusion until resolution of symptoms in six patients [34]. One of the patients in this case series was treated with the combination of continuous infusions of IV acyclovir as well as IV foscarnet [34]. The exact mechanism by which a high-dose continuous infusion acyclovir can overcome drug resistance is unclear. It is proposed that this method "may result in favorable pharmacokinetics leading to higher intracellular concentrations of the drug, which may be able to overcome altered substrate specificity or bypass the need for viral kinase phosphorylation when mutations result in missing or truncated viral thymidine kinase" [34]. High dose IV acyclovir can be considered as an alternative in patients that are not candidates for IV foscarnet or IV cidofovir use due to the latter toxicity profiles or intolerance.

Topical therapy options include topical formulations of cidofovir, foscarnet, and topical imiquimod. Both topical cidofovir and foscarnet are not commercially available but can be compounded from the injectable formulations at the specialty pharmacies [28].

6 Conclusions

In summary, the incidence of HSV and VZV infections persists, notably among HCT recipients, despite prophylactic antiviral strategies. The complex interplay of high-dose chemotherapeutic conditioning, manifestations of GVHD, and T-cell immunity attenuation are key factors underpinning this prevalence. With the rise of acyclovir-resistant HSV strains, clinicians need to adapt through prompt diagnosis, incorporating alternative antivirals such as foscarnet and cidofovir, and considering escalated doses of intravenous acyclovir. Ongoing research and clinical vigilance must evolve in response to these challenges, ensuring that prophylactic and therapeutic regimens are optimized for efficacy in this immunocompromised cohort.

Conflict of Interest The authors report no conflicts of interest.

References

1. Looker KJ, Magaret AS, May MT, Turner KM, Vickerman P, Gottlieb SL, et al. Global and Regional Estimates of Prevalent and Incident Herpes Simplex Virus Type 1 Infections in 2012. PLoS One. 2015;10(10):e0140765.

2. Meyers JD, Flournoy N, Thomas ED. Infection with herpes simplex virus and cell-mediated immunity after marrow transplant. J Infect Dis. 1980;142(3):338–46.
3. Michael Uhlin AS, Mikael Sundin, Jonas Mattsson, Mats Remberger. Risk Factors and Clinical Outcome for Herpes Simplex Virus Reactivation in Patients after Allogeneic Hematopoietic Stem Cell Transplantation Biol Blood Marrow Transplant. 2015(21):S147.
4. Wald A, Huang ML, Carrell D, Selke S, Corey L. Polymerase chain reaction for detection of herpes simplex virus (HSV) DNA on mucosal surfaces: comparison with HSV isolation in cell culture. J Infect Dis. 2003;188(9):1345–51.
5. Koc Y, Miller KB, Schenkein DP, Griffith J, Akhtar M, DesJardin J, et al. Varicella zoster virus infections following allogeneic bone marrow transplantation: frequency, risk factors, and clinical outcome. Biol Blood Marrow Transplant. 2000;6(1):44–9.
6. Locksley RM, Flournoy N, Sullivan KM, Meyers JD. Infection with varicella-zoster virus after marrow transplantation. J Infect Dis. 1985;152(6):1172–81.
7. Schuster MG, Cleveland AA, Dubberke ER, Kauffman CA, Avery RK, Husain S, et al. Infections in Hematopoietic Cell Transplant Recipients: Results From the Organ Transplant Infection Project, a Multicenter, Prospective, Cohort Study. Open Forum Infect Dis. 2017;4(2):ofx050.
8. Han CS, Miller W, Haake R, Weisdorf D. Varicella zoster infection after bone marrow transplantation: incidence, risk factors and complications. Bone Marrow Transplant. 1994;13(3):277–83.
9. Baumrin E, Cheng MP, Kanjilal S, Ho VT, Issa NC, Baden LR. Severe Herpes Zoster Requiring Intravenous Antiviral Treatment in Allogeneic Hematopoietic Cell Transplantation Recipients on Standard Acyclovir Prophylaxis. Biol Blood Marrow Transplant. 2019;25(8):1642–7.
10. Onozawa M, Hashino S, Haseyama Y, Hirayama Y, Iizuka S, Ishida T, et al. Incidence and risk of postherpetic neuralgia after varicella zoster virus infection in hematopoietic cell transplantation recipients: Hokkaido Hematology Study Group. Biol Blood Marrow Transplant. 2009;15(6):724–9.
11. Kanbayashi Y, Matsumoto Y, Kuroda J, Kobayashi T, Horiike S, Hosokawa T, et al. Predicting risk factors for varicella zoster virus infection and postherpetic neuralgia after hematopoietic cell transplantation using ordered logistic regression analysis. Ann Hematol. 2017;96(2):311–5.
12. Stranska R, Schuurman R, de Vos M, van Loon AM. Routine use of a highly automated and internally controlled real-time PCR assay for the diagnosis of herpes simplex and varicella-zoster virus infections. J Clin Virol. 2004;30(1):39–44.
13. de Jong MD, Weel JF, van Oers MH, Boom R, Wertheim-van Dillen PM. Molecular diagnosis of visceral herpes zoster. Lancet. 2001;357(9274):2101–2.
14. Styczynski J, Reusser P, Einsele H, de la Camara R, Cordonnier C, Ward KN, et al. Management of HSV, VZV and EBV infections in patients with hematological malignancies and after SCT: guidelines from the Second European Conference on Infections in Leukemia. Bone Marrow Transplant. 2009;43(10):757–70.
15. Tomblyn M, Chiller T, Einsele H, Gress R, Sepkowitz K, Storek J, et al. Guidelines for preventing infectious complications among hematopoietic cell transplantation recipients: a global perspective. Biol Blood Marrow Transplant. 2009;15(10):1143–238.
16. Stranska R, Schuurman R, Nienhuis E, Goedegebuure IW, Polman M, Weel JF, et al. Survey of acyclovir-resistant herpes simplex virus in The Netherlands: prevalence and characterization. J Clin Virol. 2005;32(1):7–18.
17. Danve-Szatanek C, Aymard M, Thouvenot D, Morfin F, Agius G, Bertin I, et al. Surveillance network for herpes simplex virus resistance to antiviral drugs: 3-year follow-up. J Clin Microbiol. 2004;42(1):242–9.
18. Patel D, Marchaim D, Marcus G, Gayathri R, Lephart PR, Lazarovitch T, et al. Predictors and outcomes of acyclovir-resistant herpes simplex virus infection among hematopoietic cell transplant recipients: case-case-control investigation. Clin Transplant. 2014;28(1):1–5.
19. Kakiuchi S, Tsuji M, Nishimura H, Yoshikawa T, Wang L, Takayama-Ito M, et al. Association of the Emergence of Acyclovir-Resistant Herpes Simplex Virus Type 1 With Prognosis in Hematopoietic Stem Cell Transplantation Patients. J Infect Dis. 2017;215(6):865–73.

20. Piret J, Boivin G. Resistance of herpes simplex viruses to nucleoside analogues: mechanisms, prevalence, and management. Antimicrob Agents Chemother. 2011;55(2):459–72.
21. Ariza-Heredia EJ, Chemaly RF, Shahani LR, Jang Y, Champlin RE, Mulanovich VE. Delay of alternative antiviral therapy and poor outcomes of acyclovir-resistant herpes simplex virus infections in recipients of allogeneic stem cell transplant – a retrospective study. Transpl Int. 2018;31(6):639–48.
22. Field HJ. Persistent herpes simplex virus infection and mechanisms of virus drug resistance. Eur J Clin Microbiol Infect Dis. 1989;8(8):671–80.
23. Chakrabarti S, Pillay D, Ratcliffe D, Cane PA, Collingham KE, Milligan DW. Resistance to antiviral drugs in herpes simplex virus infections among allogeneic stem cell transplant recipients: risk factors and prognostic significance. J Infect Dis. 2000;181(6):2055–8.
24. Anton-Vazquez V, Mehra V, Mbisa JL, Bradshaw D, Basu TN, Daly ML, et al. Challenges of aciclovir-resistant HSV infection in allogeneic bone marrow transplant recipients. J Clin Virol. 2020;128:104421.
25. Antiviral Susceptibility Testing: Herpes Simplex Virus by Plaque Reduction Assay; Approved Standard. NCCLS M33-A 2004;24(7).
26. Sauerbrei A, Bohn-Wippert K. Phenotypic and Genotypic Testing of HSV-1 and HSV-2 Resistance to Antivirals. Methods Mol Biol. 2020;2060:241–61.
27. Poole CL, James SH. Antiviral Therapies for Herpesviruses: Current Agents and New Directions. Clin Ther. 2018;40(8):1282–98.
28. Jiyeon Joy Park P, BCOP. Management of Acyclovir-Resistant Herpes Simplex Virus Infection in Patients Undergoing Hematopoietic Stem-Cell Transplantation. Journal of Hematology Oncology Pharmacy. 2021;11(4).
29. Deray G, Martinez F, Katlama C, Levaltier B, Beaufils H, Danis M, et al. Foscarnet nephrotoxicity: mechanism, incidence and prevention. Am J Nephrol. 1989;9(4):316–21.
30. Moyle G, Barton S, Gazzard BG. Penile ulceration with foscarnet therapy. AIDS. 1993;7(1):140.
31. Torres T, Fernandes I, Sanches M, Selores M. Foscarnet-induced penile ulceration. Acta Dermatovenerol Alp Pannonica Adriat. 2011;20(1):39–40.
32. Lalezari JP, Holland GN, Kramer F, McKinley GF, Kemper CA, Ives DV, et al. Randomized, controlled study of the safety and efficacy of intravenous cidofovir for the treatment of relapsing cytomegalovirus retinitis in patients with AIDS. J Acquir Immune Defic Syndr Hum Retrovirol. 1998;17(4):339–44.
33. Ikawa Y, Fujiki T, Nishimura R, Noguchi K, Koshino E, Fujiki A, et al. Improvement of refractory acyclovir-resistant herpes simplex virus type 1 infection by continuous acyclovir admintistration. J Infect Chemother. 2019;25(1):65–7.
34. Kim JH, Schaenman JM, Ho DY, Brown JM. Treatment of acyclovir-resistant herpes simplex virus with continuous infusion of high-dose acyclovir in hematopoietic cell transplant patients. Biol Blood Marrow Transplant. 2011;17(2):259–64.

The Virology of Hemophagocytic Lymphohistiocytosis

David Thomas

Abstract The hemophagocytic lymphohistiocytosis (HLH) syndrome encompasses a group of disorders with a similar clinical and laboratory presentation, most frequently with the pathologic finding of hemophagocytosis within the bone marrow. The disease exists in genetic and spontaneous forms, resulting in dysregulation of the immune system, specifically the inflammatory response through various mechanisms. The syndrome can be incited by many different events and underlying conditions such as hematologic malignancies, rheumatic and immunologic disorders, medications, and infections. This includes several viral pathogens. Given the tropism of certain viruses toward the bone marrow and immune cells, the herpesviruses, particularly Epstein-Barr, are frequently associated with this disease along with many others. This chapter will review the current understanding of the HLH syndrome and review the role of viruses in HLH as well as proposed virologic mechanisms and current therapies.

Keywords Hemophagocytic · Lymphohistiocytosis · Herpesviruses · Cytomegalovirus · Epstein-barr · Adenovirus · COVID-19 · Cytokine · COVID-19 · Cancer · Viruses associated with hemophagocytic lymphohistiocytosis · Current treatment paradigms · Pathophysiology of hemophagocytic syndrome · HLH diagnostic criteria

---

D. Thomas (✉)
Division of Infectious Disease, University of Louisville School of Medicine, Louisville, KY, USA
e-mail: david.thomas.9@louisville.edu

C. Somboonwit et al. (eds.), *Global Virology IV: Viral Disease Diagnosis and Treatment Delivery in the 21st Century*,
https://doi.org/10.1007/978-3-031-57369-9_17

1 The Hemophagocytic Syndrome

1.1 Definition and Background

Hemophagocytic lymphohistiocytosis (HLH), sometimes abbreviated as hemophagocytic syndrome, is a disease of the immune system principally characterized as a dysregulation of the activation cascade leading to a highly exaggerated pro-inflammatory state affecting multiple organ systems and ultimately culminating with significant morbidity and mortality in those affected. Originally termed histiocytic medullary reticulosis when it was first described in the literature by Scott & Robb-Smith in their seminal article published in the Lancet, this disease has been identified in patients of any and all conceivable backgrounds [1]. While in this initial report the syndrome was noted in an adult patient without specific clues to etiology, there was a case reported by Farquhar and Claireaux in 1952 in two infants who were siblings [2]. The incidence of this disease is largely unknown. Still, while it is considered a rare disorder, cases do appear to be increasing in recent years and in populations previously underrepresented in the literature. While most of the conventional literature on this subject has been in pediatrics, there has also been a marked increase in adult patients [3, 4]. It is unclear if this is truly an increased incidence or if knowledge of this disorder has improved detection and expanded the number of cases.

The literature admits a limited understanding of this disorder or spectrum of disorders. It has taken many forms with many different investigated triggers. There is a significant body of evidence that shows that genetic factors play a vital role in determining which individuals are more susceptible. These factors include clusters within the same family similar to the 1952 case. This has led to further classification of the disease into primary familial disease and secondary sporadic cases due to a defined trigger. Secondary disease is further subclassified as viral, autoimmune, or neoplasia-related; many cases are considered multifactorial as well [4]. Particular to adult patients, external (infections and drugs) factors play the largest role in antagonizing the immune system into activation; these factors can either initiate disease or are underlying diseases or disorders that increase the risk of HLH [4]. This can include bacterial, viral, parasitic, and fungal infections as well as many different medications and other drugs. This contrasts with many cases reported in children; primary HLH is typically considered due to an underlying genetic defect or defects, playing a predominant role. Although as we will discuss later, this is not always considered to be the case, and potential secondary triggers have been described elsewhere.

This chapter aims to focus on the exploration of secondary HLH, specifically with some mentioning of genetic HLH in relation to viral infections as the proposed trigger. Here, we will discuss the current understanding of the pathogenesis of secondary HLH, diagnosis and clinical concerns, as well as treatment advances. We will discuss the most standard recognized viral triggers of HLH and explore within the literature the reports of more novel viral agents suspected of triggering the condition. We will also attempt to discuss how viruses associated with this condition contribute to the pathology of the HLH syndrome itself.

1.2 Pathophysiology of HLH

The activation of the immune system culminating in HLH is understood to be an overreactive process resulting from prolonged and excessive activation of antigen-presenting cells (macrophages, histiocytes) as well as CD8+ T cells. Hemophagocytosis occurs via the CD163 heme-scavenging receptor. This is considered a hallmark in the physiology of activated macrophages/histiocytes and is considered the most specific and characteristic finding of the syndrome [3]. The genetic traits that have been identified thus far in familial disease and otherwise appear to be targeted in the functioning of NK and cytotoxic T lymphocytes, leading to excessive immune activation with ineffective internal control mechanisms [2, 3, 4]. It is theorized that perforin/granzyme and Fas maintain dendritic cells and T lymphocytes, leading to enhanced antigen presentation and release of interferons as well as the stimulation of Toll-like receptors crucial in mediating cytotoxicity [4, 10], specifically, a defect in granule-mediated cytotoxicity. This has been observed in both genetic and secondary forms of the disease. The result is a massive release of proinflammatory cytokines, including TNF-alpha, IL-2, IL-6, and others, leading to a massive cytokine storm.

While the etiology and pathogenesis of an individual case of HLH can be confusing, many cases share common fluctuating elements of systemic disease, such as fever, commonly observed laboratory-based abnormalities, and clinical presentation. As mentioned above, internal organ dysfunction is common as well [3, 4]. This can be due to different factors: underlying disease, infectious triggers, or complications of the HLH. In our experience, and noted within the literature, pulmonary involvement including acute respiratory distress syndrome (ARDS) is quite commonly seen in nearly half of the reported cases in one analysis [4]. Some of the abnormalities seen in laboratory testing include an extremely high elevation of ferritin and other acute phase reactants including triglycerides, lactate dehydrogenase (LDH), and frequent suppression of multiple cell lines within the bone marrow. This process will frequently culminate in multisystem organ failure and death.

It is understood that patients with cancer, particularly those with hematologic malignancies, can be prone to develop HLH [4]. The most implicated diseases are lymphomas by far. However, this has also been noted to occur in patients with other forms of cancer, autoimmune diseases, and numerous different infections, including viral disease, which is the primary focus of this chapter. It has also been observed to trigger on commencing different medications, including chemotherapy, antiretrovirals, and numerous others. Both solid organ and bone marrow transplantation is a risk for HLH (although significantly higher in cord blood transplants). Of the solid organs, kidneys are the most common for unclear reasons. These data have largely been gleaned from retrospective analyses [3, 4].

1.3 Clinical Presentation and Diagnostic Factors

HLH appears to have a male predominance of around 60%. Age at onset appears to be in the fourth through sixth decades of life. The most common symptom is fever, approaching close to 100% of cases [4]. Splenomegaly and hepatomegaly are more commonly noted than pulmonary disease. Some patients can develop lymphadenopathy as well as neurological symptoms, GI symptoms, renal dysfunction, encephalopathy, and even skin lesions. This can make the disease difficult to discern from other hyperinflammatory conditions, such as adult-onset Still's disease, when based on clinical presentation alone.

Laboratory abnormalities can include many nonspecific findings including hyperferritinemia, most cases exceeding 1000 pmol/L. High triglycerides and LDH are also seen. More specifically, sILR2 (anti-CD25) is elevated with diminished NK cell activity [3, 4, 5]. Hematological derangements, as mentioned, can involve low counts in multiple cell lines, leading to significant neutropenia and acute immune dysfunction. Acute phase reactants are frequently elevated as well, which can manifest as hemolysis and disseminated intravascular coagulation [4]. In many cases, if a bone marrow biopsy and aspiration are attempted, one can visualize the hemophagocytosis firsthand, although this is not always observed in all suspected cases. Making an accurate diagnosis is neither simple nor straightforward in many cases. Many of the commonly observed signs and symptoms are summarized in Table 1.

As alluded to above, numerous conditions can lead to a similar clinical presentation to that of HLH, and some diseases even include the development of a true HLH within their clinical course; Chediak-Higashi and Wiskott-Aldrich syndrome are well known for this [5, 6]. Viral infections can trigger HLH in primary and secondary forms of the disease. Probably the most recognizable form of secondary HLH is the macrophage activation syndrome described in rheumatoid arthritis [5].

In 1991, the first attempts to develop a set of diagnostic guidelines for HLH were devised by the Histiocyte Society utilizing commonly observed clinical and laboratory findings as well as pathologic findings. These were later revised in 2004, resulting in the HLH-2004 guidelines, which are largely still in prominent use today. The initial five criteria developed in the 1991 guidelines were retained, including fever, splenomegaly, hypertriglyceridemia, cytopenia of at least two lineages in peripheral blood samples, and hemophagocytosis itself. At this point, more specifically

Table 1 Commonly reported clinical symptoms in HLH [10]

Fever (>90%)
Splenomegaly
Hepatomegaly
Neurologic derangements
Lymphadenopathy
Skin rash
Respiratory failure

observed findings such as anti-CD25 and diminished/absent NK activity were also included. In cases analyzed in HLH-94, ferritin elevation was also observed to be above 500 micrograms per liter in most cases. At this time, most diseases continued to be observed more in the pediatric population [5, 6]. The 2004 criteria are summarized in Table 2.

Viral disease has been known to be linked with HLH for over 40 years. A study published in 1979 by Risdall et al. described 19 patients whose marrow aspirates demonstrated histiocytic hyperplasia with prominent hemophagocytosis. All patients were found to have a similar syndrome associated with viral infection. High fever and other symptoms previously noted in Part I of this chapter were described along with liver abnormalities [7]. Viral studies noted active viral infection in fifteen of these patients. The preponderance of viral pathogens isolated were the human herpesviruses, including cytomegalovirus (CMV) and EBV, which were the most common isolates in this study. Varicella-zoster encompassed a small minority of cases. A list of common infectious triggers of secondary HLH can be seen in Table 3. Several of these will be highlighted throughout the chapter.

2 Viral Infections Associated with HLH

2.1 *Epstein-Barr Virus*

In 2007, a nationwide survey of HLH in Japan was spearheaded by faculty from numerous organizations. They estimated the incidence of HLH to be 1 in 800,000 per year. A total of 799 patients were diagnosed in 292 institutions that particular year, and most commonly, the disease was found to be associated with Epstein-Barr virus (EBV) [8]. The study was conducted by questionnaires to the institutions initially for numbers and then a second questionnaire looking at clinical features, treatment, and outcomes. Half of the patients with EBV subtype were under 15 years of

Table 2 The HLH-2004 diagnostic criteria [5]

1. A molecular diagnosis that is consistent with HLH
2. Five of the eight diagnostic criteria as set below:
Fever
Splenomegaly
Cytopenia affecting at least two cell lineages in the peripheral blood
(Hb <90 g/L, platelets <100 × 10^9/L, neutrophils <1.0 × 10^9/L)
Hypertriglyceridemia, hypoferritinemia (either/or)
(fasting triglycerides at or above 3.0 mmol/L or 265 mg/dL, fibrinogen at or below 1.5 g/L)
Hemophagocytosis in the bone marrow or spleen or lymph nodes
Low or absent NK cell activity levels per local laboratory reference values
Ferritin at or above 500 ng/mL
Soluble CD25 (i.e., soluble IL-2 receptor) at or above 2400 U/mL

Table 3 Select viral pathogens associated with HLH [10]

Epstein-Barr virus
Cytomegalovirus
Varicella-Zoster virus
Herpes Simplex virus
Influenza A & B
Paramyxoviridae Measles Mumps Rubella Parainfluenza
Human immunodeficiency virus
Adenovirus
Flaviviruses (dengue)
Parvovirus B19

age [8]. Five-year survival exceeded 80% for this subtype, which was found to be similar to other etiologies.

Many cases are described in the literature involving a patient with EBV infection. It is unclear if reactivation versus primary infection is more common. One such case described in 2014 involved a 16-year-old male admitted to a local hospital for symptoms of sore throat and lymphadenopathy accompanied by fevers ongoing for the past week. During the admission, his course, initially thought to be self-limiting, paradoxically worsened. HLH was subsequently diagnosed [9]. The exact pathophysiology of this infection leading to HLH has been discussed extensively. It is believed to be due to a polymorphism acquired postinfection in the EBV-infected cytotoxic T lymphocytes involving their cytokine-related genes that may be associated with triggering HLH [6, 9].

As previously mentioned, cases in the adult population have been increasingly diagnosed as well [4, 10]. One case of a 20-year-old woman with 2 weeks history of fever, fatigue, and failure to thrive presented in the inpatient with headaches and shortness of breath. She was noted to have Holt-Oram syndrome, an autosomal-dominant condition characterized by congenital cardiac and forelimb anomalies [10]. She had been infected with primary EBV and diagnosed with positive serologies and polymerase-chain reaction (PCR) testing a year prior. On hospital day 15, she exhibited worsening respiratory distress, altered mentation, and multisystem organ failure. Four days later into the hospitalization, the patient succumbed to her disease. Of note, EBV viremia was captured throughout her stay and notably increased up to 974,000 copies per mL before her death. While this example and the preceding suggest that primary infection is implicated, it is also described elsewhere in chronic active EBV infections as well [10].

2.2 Cytomegalovirus and Other Human Herpesviruses

The Risdall group's study reported that while EBV accounted for nearly 75% of all cases, CMV was seen in several of their reported cases of viral infection associated with HLH [7]. CMV inclusions were detected in those patients at multiple sites, including lungs, esophagus, lymph nodes, bone marrow, adrenals, and liver. Some cases in the literature are less clear and have been noted to occur in CMV-positive individuals with other potential triggers, such as the case of an 80-year-old Japanese woman with reactive HLH who had an additional history of adult-onset Still's disease [11]. This patient was diagnosed initially with Still's disease during a hospitalization for high fevers, arthralgia, and skin rash with pleuritis. She was initially started on steroids. During this, her repeat serum ferritin levels and transaminases markedly increased. Aspiration of the bone marrow noted high histiocytes with hemophagocytosis and CMV-positive leukocytes. The authors noted that while HLH occurs during active Still's disease, immune suppression from the treatment of an autoimmune disorder carries the risk of secondary infections such as that of CMV. Nevertheless, in this case, CMV was not considered the primary trigger. Still's disease is sometimes confused with HLH but has been reported as such a trigger elsewhere [12]. However, evidence in the literature exists that CMV is associated with HLH in otherwise healthy patients as a true disease trigger. It activates TNF alpha expression in infected cells, resulting in the same defects in the regulation of cytotoxicity and carries a high mortality rate. This can be even more devastating in patients with impaired immunity [8].

The next most associated herpesvirus after CMV and EBV is human herpesvirus 8 (HHV-8) and there are reports in both pediatric and adult patients [7]. Most such cases are associated with clinical disease caused by the virus, such as Castleman disease or Kaposi sarcoma. This risk is even more pronounced in persons living with HIV. For the other remaining herpesviruses, as mentioned in the Risdall study, VZV was seen as well, albeit in very few patients [7]. In addition, Herpes simplex virus (HSV)-associated HLH has also been reported, accounting for 30% of neonatal cases reported in Japan in 2009. Along with these, a few isolated cases have been described in adults. The mortality rate is very high [7].

There has been at least one reported case of HLH potentially arising from human herpesvirus 6 (HHV-6). The patient was from the Dominican Republic, initially presenting with left-sided flank pain. She was 24 years old reporting 3 weeks of fever, cough, and rash. She was initially diagnosed with adult-onset Still's disease and she was too unstable for a bone marrow biopsy. Ultimately, she underwent a biopsy which confirmed secondary HLH. It is unclear if Still's disease was a true diagnosis and trigger or whether HHV-6 was principally responsible for her presentation [12]. As HHV-6 causes limited disease in immunocompetent adults, it is difficult to resolve this with certainty, although CMV is generally asymptomatic in most patients.

2.3 Human Immunodeficiency Virus (HIV)

The incidence of HLH in persons who live with HIV has likely been underreported and underestimated as a risk factor. A retrospective analysis at three tertiary care centers in France was reported in 2010 of 58 patients infected with HIV-1 and underlying secondary reactive HLH. Of the patients studied, 23 were thought to have infection-associated HLH. Of the 58 patients, 42 were male, among whom 31 were of African origin. The median duration of HIV infection was 4 years, and more than half of these patients were on antiretroviral therapy (ART) [13]. Their median CD4 count was 91, indicating more advanced disease may increase risk. This is likely the case, as advanced HIV disease also increases the risk for opportunistic lymphomas. In this analysis, only ten patients reportedly had an underlying Hodgkin's lymphoma. However, even in patients on ART with undetectable viral loads as measured by PCR, chronically low CD4 cytopenia appears to increase the risk of HLH associated with hematologic malignancy, an underlying risk from HIV infection. Most cases in that report had no malignancy and no other known triggers. Except for nine patients with tuberculosis and five patients with CMV disease, no other trigger was readily identified outside of the HIV infection itself.

Generally, approximately 10–20% of biopsied bone marrow specimens from HIV-positive patients show signs of hemophagocytosis before starting any antiretroviral treatment [6, 13]. Consequently, patients who start ART during a low CD4 count period would have some degree of overlap with risk for contracting opportunistic infections and, thus, more active triggers of HLH. Possibly, some antiretroviral medications themselves may also induce a reactive HLH in an otherwise susceptible individual. However, other common viral triggers include EBV, CMV, and HHV-8, which would naturally cause disease in larger numbers within the HIV population. Opportunistic infections, as well as cancers, contribute to an elevated incidence of HLH in these cases [6, 13].

2.4 Adenovirus

Adenovirus infection with multiple serotypes has been associated with HLH [14, 15, 16]. Many cases have specifically noted infection with serotype 7; disease provokes the development of reactive HLH. This has been reported in several pediatric cases [14, 15]. A case series involving a cluster of five pediatric cases of serotype 7 adenovirus-associated HLH was reported in 2021 in Philadelphia [14]. Ages ranged from 15 months to 16 years of age. Four of the five were male, except for the 15-month-old infant. Fever was present in all cases, as was respiratory failure. At Lovelace Respiratory Research Institute, all positive specimens were inoculated into a cell line for virus isolation and further characterization, including molecular subtyping. Whole genome sequencing for the respiratory isolates was also obtained, confirming the results. A phylogenic analysis noted that four of the five patients specifically were infected with serotype 7D. These strains are closely related to strains circulating in Pennsylvania in 2016 and 2017 [14]. Whether this observation indicates the subtype has a higher proclivity to induce reactive HLH or if this is coincidence is unknown.

Of note, a fatal case of adenovirus serotype 21 infection-associated HLH was reported in Indianapolis, Indiana, in 2010 [15]. This report also documented numerous other reports caused by this serotype. This case involved a 22-year-old adult male patient with a history of juvenile rheumatoid arthritis who had not been on immunosuppression for more than 2 years prior to this presentation. CT chest revealed diffuse dense bilateral infiltrates. The patient was treated with stress-dose corticosteroids but subsequently developed multisystem organ failure, ultimately not surviving the syndrome. Postmortem revealed macrophages with hemophagocytosis and lymphophagocytosis in the hilar lymph nodes. This presentation appears to have been consistent with the classical macrophage activation syndrome etiology. It is one hypothesis that the patient was predisposed to the development of this form of HLH, given his rheumatic disease history. Thus, the viral infection with adenovirus 21 ultimately triggered his syndrome. It is possible that the virus was a bystander, with the activation being from his juvenile rheumatoid arthritis itself. However, this would seem unlikely given that he was off treatment for a prolonged period.

Overall, it is very concerning that multiple subtypes of adenovirus can be associated with HLH, given that it has a proclivity to infect healthy children. While this typically results in a benign self-limiting illness, often upper respiratory in presentation, in patients who are immunocompromised it has also been associated with disseminated and/or fulminant disease involving numerous organs [14, 15, 16], particularly among the transplant population. One case of virus-associated HLH was documented in an 11-month-old infant who had no history of immune defects, familial HLH, or malignant disease and suffered from both viral pneumonia, which then progressed into HLH [16]. The virus itself was not subtyped in this case.

2.5 *Influenza*

Both influenza A and B have been reported to trigger secondary HLH [17, 18]. This is particularly noteworthy as a complication of the 2009 influenza A subtype HIN1 "Swine Flu" outbreak [17]. One prospective observational study of 25 critically ill patients with the 2009 H1N1 virus at a unit in Germany found that the primary outcome was the development of secondary HLH with a mortality rate of 89% [17]. Interestingly, the patients who did not develop HLH had a lower mortality rate of only 25% ($P = 0.004$). This flu epidemic was frequently noted to cause severe disease, including fulminant ARDS. At the time, in adult patients, HLH was still relatively unknown, and as such this complication may have gone underdiagnosed far more than what has been reported.

A case of HLH triggered by influenza B was reported as well in 2019 by Gao et al. [18] The patient was a 54-year-old male presenting to the emergency department in March with acute abdominal pain, fever, and diarrhea for the prior 72 h who had refused vaccination the previous year. A nasopharyngeal swab was positive for influenza B. He was treated with oseltamivir, but symptoms persisted, and he remained febrile despite an otherwise negative septic workup. He was noted to have hemophagocytosis on bone marrow biopsy after laboratory results demonstrated low NK cell activity and high CD25. This patient was successfully treated with etoposide and dexamethasone and was later discharged home [18].

2.6 Other Known Viruses Associated with HLH

Many other cases of viral-induced HLH have been reported in the literature [5, 9]. These have shown up much less frequently than others, including hepatitis, enteroviruses, the Paramyxoviridae, including parainfluenza, measles, mumps, and rubella, as well as hantavirus along with influenza and adenovirus as previously discussed [9]. In addition, Parvovirus B19, dengue virus, and many others have been reported to trigger the syndrome in lesser reported numbers [12, 19].

One such case reported in 2015 involved a 33-year-old male patient from the Dominican Republic who had no past medical history and was otherwise considered healthy, presenting with headaches, nausea, vomiting, diarrhea, and altered mentation with word-finding difficulties [12]. Dengue serology returned positive. Of note, the patient had an episode of epistaxis and hematemesis, remaining hemodynamically unstable and transferred to intensive care, where he underwent bone marrow aspiration and biopsy, which showed normocellular marrow with an increased number of histiocytes demonstrating hemophagocytosis, confirming the diagnosis of HLH. He was treated with etoposide and dexamethasone, but ultimately did not improve. He experienced a seizure, was subsequently made DNR (do not resuscitate) by the family, and passed away from his illness [12].

Parvovirus B19 has also been seen in cases and can cause persistent marrow infection leading to HLH despite negative serologies, presenting a major diagnostic challenge [19]. One such reported case was a 51-year-old female patient from Indonesia with a history of papillary thyroid cancer post-surgery and presented with fever, cough, and weight loss. Symptoms had begun over the previous month, and while laboratory abnormalities on admission demonstrated anemia, his platelets and leukocyte counts were within normal limits on admission. He was ruled out for gastrointestinal bleeding, and imaging revealed no organomegaly or other evidence of cancer. He was not in hemolysis, and rheumatologic workup was also negative. Parvovirus serology for IgG only was positive. DNA for B19 by PCR was curiously negative. Ferritin was initially over 1000 and subsequently over 3000 on recheck, prompting bone marrow biopsy and hemophagocytosis with nuclear inclusions suggesting parvovirus-associated HLH [19]. Despite lower numbers, this viral infection seems primed to induce HLH in susceptible individuals. Acute infection activates immune cells, leading to cytokine dysregulation and ultimately upregulating transcription of IL-6 via other proinflammatory cytokines. Elevated cytokine levels, when combined with persistent marrow infection, are felt to contribute to triggering HLH in these patients. Interestingly, a TNF-α polymorphism in specific Asian populations has been associated with increased susceptibility to secondary HLH, which may have been a factor in this case and suggests further genetic predisposition in patients who develop secondary HLH [19].

2.7 *HLH in the COVID-19 Era*

One of the most pro-inflammatory viruses to manifest in our lifetimes is likely SARS-CoV-2. The disease has been associated with severe and deadly outcomes in many patient populations without clear predominance. Systemic inflammatory syndromes such as the multisystem inflammatory syndrome in children and the similar cytokine release syndrome seen in adult patients, usually in combination with ARDS, are well documented [20]. Viral pneumonia is highly associated with COVID-19 disease, even in mild patients, as multiple studies have noted. As we have already mentioned, viral infections are closely linked to the development of HLH. As such, with the observation of how pro-inflammatory the infection is, it is logical to look for cases of HLH within the COVID-19 population.

Clinical overlap between the COVID-19 presentation and HLH is noted within the HLH-2004 diagnostic criteria [5], particularly fever. One such recent review cites hyperferritinemia over 1000 in 60% of studies looking at this HLH feature in COVID patients [20]. Meeting five or more clinical criteria to solidify a diagnosis of HLH was only looked at in one such study, and no patients included attained this level of overlap. As such, while many features of HLH appear to overlap with severe COVID-19 infection, true HLH appears to be quite low, estimated at less than 5% of cases by Retamozo et al. [20] The number of reported cases of HLH in COVID-19 patients notably increased throughout 2020. In another study cited by Retamozo, only 8% met five criteria and in only 5% of cases physicians measured NK cell function and anti-CD25 levels. Of those 60 cases reported, over half died during the admission.

While many suggest the link is greater between COVID-19 and HLH, ARDS appears to be much more commonly associated as well as a nondescript proinflammatory cytokine storm, which is not specifically HLH. The truth is likely somewhere in between. Severe COVID-19 involves multiple inflammatory conditions. Thus, it may be possible to cite more specific criteria to the COVID-19 patient that could reduce mortality rather than attempting to define all such cases as simply HLH. More research is needed on this topic.

3 HLH Pathogenesis Through Viral Infection

3.1 A Many-Faced Enemy

There has been significant debate on the role of viral infection in the pathogenesis of HLH. This has mainly been the topic of heated discussions within the pediatric literature. Given the different forms of the disease are associated with immune deficiencies, as previously mentioned [6], acquired HLH has also been associated with a variety of viral infections as well as multiple other infectious agents. This suggests significant confounding when attempting to tease out the specific etiology of the syndrome in different patients. EBV, as previously cited, appears to be the most common reported trigger of the syndrome and seems to have a higher prevalence in Asian patients [6, 7, 8, 21]. There also appears to be two different forms of EBV HLH [21]. The first occurs during primary infection when EBV infects and replicates in the B cells and rarely in T and NK cells, leading to potential chronic active infection. The second occurs during the reactivation of the dormant virus. Serologies and viral PCRs can sometimes assist in determining the effects of viral infections in HLH development. Clonal expansion from EBV has been seen in various EBV lymphomas, specifically T-cell lymphoma. The clonality of these cells and the distinction between the monoclonal proliferation of T cells seen in HLH as well as these T cell lymphomas may give a glimpse of the extremely disordered functioning of T cells in EBV HLH. This is the case not only in EBV but also in other herpesviruses and likely other viruses that can evade the immune system and persist within the body for years to come.

3.2 How Viruses Contribute to HLH

All viruses, especially the larger DNA viruses such as herpesviruses, are potent modulators of the cellular immune response, as alluded to above. Viruses are host to a variety of evasion mechanisms to immune recognition, including interference with cytokine balances and arresting apoptosis in infected cells; all features which may increase a given host patient's proclivity to developing reactive HLH [22]. The majority of our knowledge on the pathogenesis of HLH speaks to the commonalities of viral infection in many cases. All manner of the herpesviruses, not just EBV, have been implicated, as we have discussed at length previously in this chapter. CMV and other viruses that can retreat to sanctuary sites have also been shown to trigger HLH. Critically, those with active viral infections during HLH had a worse prognosis than those who were non-infected [21, 22].

One such theory of incitement of viral HLH involves the continuous pathogen receptor triggering hypothesis, which leads to eventual bone marrow exhaustion. Given lifelong latency, EBV, CMV, and other herpesviruses chronically burden the immune process with small amounts of antigens requiring constant immune surveillance to contain the infection. As was alluded to in this chapter's first section, examining HLH pathophysiology, pattern recognition receptors such as the Toll-like receptors stimulate innate immune response continuously and are predominantly TLR9 [22]. The excessive activation of this receptor system has already been linked to HLH in murine models from two independent study populations. This suggests that persistent triggering of the immune response would eventually lead to HLH activation.

One of the more impressive aspects of viral evasion is through molecular mimicry, which can imitate components of the host immune system response to modulate inflammation for the benefit of the invasive viral pathogen. These so-called virokines and viroceptors, are critical in arresting the inflammatory effects of the virus. However, some virokines and viroceptors counterintuitively enhance the activation of proinflammatory signal cascades, which additively interferes with the balancing of the immune response and likely also contributes to cytokine storms in HLH cases derived from these pathogens. HHV-8, for example, can stimulate secretion of a virolog of IL-6, capable of activating the JAK/STAT pathway [22]. Multiple viruses, including herpesviruses and HIV, can employ chemokine mimicry to mediate the attraction of neutrophils and monocytes, promoting viral latency. Viruses can also induce some chemokine responses, such as EBV and HHV-8 triggering nuclear factor kappa B (NFkB) and nuclear factor of activated T cells (NFAT) signaling, which can also stimulate host cytokine release. Viral-encoded FcR can also directly suppress NK cell cytotoxicity. These various mechanisms are listed in Table 3.

3.3 Is Viral Infection Enhanced During HLH?

The question of opportunistic infections during HLH is largely unanswered. Very little was gleaned during the literature review for this chapter in relation to this question. Given the extreme activation of the immune system, it is unlikely that any pathogen would be able to take a foothold of the host in such a state and, even so, would only serve to increase mortality in the short term, which would not benefit the virus at all [21, 22]. Hypothetically, given the reduction in cell lines, particularly with respect to neutropenia and lymphopenia, it is possible that such a state of immune compromise could occur if the patient were able to tolerate the syndrome at that degree for over a month. We have some idea of infection risk and understand that such risk increases with time spent in an immunocompromised state, particularly for viral pathogens, including CMV disease and other viruses like polyomaviruses. However, the 30-day mortality is high in such a compromised condition. If not controlled, those patients would likely never reach that point clinically. When patients respond favorably to the treatment of the disease, their numbers tend to return mostly to normal. However, in the discussion of treatment advances for this condition, some agents used in regaining control over the immune system have various effects on T-cell function; these include the mTOR inhibitors such as sirolimus (rapamycin), JAK inhibitors such as ruxolitinib, and others. There also does not appear to be any literature describing an enhancement of a concurrent viral infection simply by the induction of HLH.

4 An Overview of Treatment Options for HLH

4.1 Corticosteroids and Etoposide

Thus far, difficulties with performance of proper clinical trials, given the lower numbers of active cases and the high early mortality, has limited clear treatment guidelines for HLH to develop, particularly in adult patients and with secondary disease in mind [23]. The initial use of corticosteroids as a treatment option was largely anecdotal. However, one trial in 2003 was done with younger adult patients, and EBV-associated HLH who received etoposide yielded higher survival at 4 weeks from treatment. As such, etoposide has since become a first-line treatment in most physicians' armamentariums. One of the major concerns with utilizing a topoisomerase inhibitor in patients who do not have an underlying malignancy (as well as those who do) is the benefit of control of the HLH versus the very high risk of myelodysplasia within the next 5 years. However, given the degree of mortality from the HLH itself, the benefits seem to outweigh the risks, particularly in fulminant cases (Table 4).

4.2 Biologics

More targeted approaches have been attempted, with a case report in 2008 attempting to utilize daclizumab, a specific antibody to the soluble IL-2 receptor (anti-CD25) which has so far been utilized very little, if at all, outside of this case in the reported literature despite the patient having a good response. Tocilizumab has also been tried as it is often useful as an IL-6 blocker in the termination of cytokine release syndrome with PD-1 and PD-L1 inhibitors and/or within chimeric antigen receptor T-cell therapies. It has yet to be largely or formally studied in the literature based on the review for this chapter. A necdotally, there is concern therapeutic effect of this antibody would lack the necessary potency to fully arrest HLH.

Table 4 Mechanisms of viral pathogenesis for HLH [22]

Sustained triggering of pattern recognition receptors
Infection of marrow hematopoietic stem cells and bone marrow exhaustion
Infection of important immune cell types (NK and T lymphocytes, dendritic cells, B cells)
Inhibition of apoptosis (various methods such as TNFr decoys, Fas internalization)
Evasion of NK and T-lymphocyte recognition (downregulation of MHC)
Suppression of T lymphocyte cytotoxicity (downregulation of perforin)
Suppression of NK cell cytotoxicity (viral-encoded Fc receptors)
Viral interference with cytokine and chemokine balances (antagonists, scavenging)

In 2017, a major impact on refractory disease was noted with the use of ruxolitinib, a Janus kinase (JAK) inhibitor which has shown promise in murine models, and on review, there have been several reports noted in the literature, which was mentioned in a recent review article on the subject [23]. In most cases reviewed, conventional therapy with steroids and/or combination with cyclosporine was given with limited response, followed by the use of ruxolitinib, which noted dramatic response in some cases but ineffective in others.

Additionally, there have been some studies of anakinra, canakinumab, and rilonacept which have been utilized in blocking IL-1. Unfortunately, blockade of this is not sufficient to prevent HLH development within these studies [24]. This is an interesting observation as looking back at the COVID-19 dilemma, anakinra was effective, as well as tocilizumab, in arresting the cytokine storm in those patients. Rituximab would be another potential treatment option in combination with other agents. This is particularly noted to have seen some success in EBV HLH [24].

4.3 Antiviral Treatment of Viral HLH

One lingering question that has not been answered sufficiently in this chapter is: Can the process be halted by stopping viral propagation? There needed to be more sources to comb through in attempting to answer this question. Largely, there exists no concrete data that patients can be treated with direct antiviral therapy to cure viral-associated HLH [25]. It is a logical step if the virus could be eliminated. It has been observed that viremia tends to rise at least in EBV-associated HLH during the syndrome [24, 25].

One such patient treated with a combination of valacyclovir and rituximab achieved a favorable response. The case involved a 16-year-old male patient with an unremarkable medical history and mononucleosis syndrome over the preceding 6-week period, which subsequently rapidly deteriorated with significant leukocytosis, bone marrow failure with thrombocytopenia, and anemia with extreme hyperferritinemia. The patient was noted to have an EBV DNA viral load of 100,000 copies per mL at the time of admission. Axillary lymphadenopathy was noted and biopsied, but no lymphoma. The patient was initially started on etoposide and dexamethasone plus rituximab. The team decided to add oral valacyclovir to the regimen at 750 mg daily in two divided doses to target EBV. The patient improved, and the initial syndrome was controlled. The patient survived an additional 5 months from admission due to complications unrelated to his initial good response [25]. Control of the underlying viral trigger seems to be an important factor in treating EBV HLH, and most likely, this could be generalized to other viral etiologies. Unfortunately, we have very few effective antivirals outside of HAART and anti-herpesvirus treatments to combat the remainder of causative viral agents. However, with the degree of prominence herpesviruses play in inciting HLH, this could be useful in further research.

5 Conclusions

Viral infections appear to play a crucial role in developing a large proportion of secondary HLH cases. While data on the incidence and prevalence of this infection remains limited, this set of pathophysiologic syndromes is believed to have been underdiagnosed particularly in adults. However, as this syndrome is becoming more recognized, correct diagnosis and more prompt treatment have become much more likely in recent years. Secondary reactive HLH, unlike primary genetic HLH, is often treated without consideration of more invasive therapies such as hematopoietic stem cell transplantation. While such measures are sometimes utilized and can even be essential in the treatment of some HLH cases, a more detailed discussion of these is beyond the scope of this chapter. Early recognition is key to successfully managing this condition, although current data are still limited. Utilizing diagnostic criteria such as HLH-2004 provides a vital means of confirming the diagnosis even in cases where hemophagocytosis is not demonstrated in bone marrow biopsy specimens. Further studies looking at the performance of this and other criteria would likely be necessary to make further progress in this devastating disease. We have seen that viral infection causes significant morbidity, and predisposition toward HLH is reasonably postulated based on these understood mechanisms of viral propagation, particularly concerning EBV and the other herpesviruses.

Conflicts of Interest The author reports no conflicts of interest.

References

1. Scott RB, Robb-Smith AHT. Histiocytic medullary reticulosis. *Lancet*. 1939;234(6047): 194–198.
2. Farquhar JW, Claireaux AE. Familial haemophagocytic reticulosis. *Arch Dis Child*. 1952;27(136):519–525.
3. Filipovich, Aexandra H. Hemophagocytic lymphohistiocytosis (HLH) and related disorders. *Hematology Am Soc Hematol Educ Program*. 2009;1:127–131.
4. Ramos-Casals M, Brito-Zerón P, López-Guillermo A, Khamashta MA, Bosch X. Adult haemophagocytic syndrome. *Lancet*. 2014;383(9927):1503–1516.
5. Henter JI, Horne A, Arico M, et al. HLH-2004: diagnostic and therapeutic guidelines for hemophagocytic lymphohistiocytosis. *Pediatr Blood Cancer*. 2007;48:124–131.
6. Morimoto A, Nakazawa Y, Ishii E. Hemophagocytic lymphohistiocytosis: pathogenesis, diagnosis, and management. *Pediatr Int*. 2016;58:817–825.
7. Ishii E., Ohga S., Imashuku S., Yasukawa M., Tsuda H., Miura I. Nationwide survey of hemophagocytic lymphohistiocytosis in Japan. *Int J Hematol*. 2007;86:58–65.
8. Risdall RJ, McKenna RW, Nesbit ME, Krivit W, Balfour HH, Simmons RL, et al. Virus-associated hemophagocytic syndrome: a benign histiocytic proliferation distinct from malignant histiocytosis. *Cancer*. 1979;44:993–1002.
9. Mostaza-Fernandez JL, Laso, JG, Ule DC, Riz de Morales, JMG. Hemophagocytic lymphohistiocytosis associated with viral infections: Diagnostic challenge and therapeutic dilemma. *Rev Clin Esp (Barcelona, Spain)*. 2014;214(6):320–327.

10. Rouphael-Maakaroun N., Moanna A., Jacob J.T., Albrecht H. Infections associated with haemophagocytic syndrome. *Rev Med Virol*. 2010;20:93–105.
11. Amenomori M., Migita K., Miyashita T., Yoshida S., Ito M., Eguchi K. Cytomegalovirus-associated hemophagocytic syndrome in a patient with adult onset Still's disease. *Clin Exp Rheumatol*. 2005;23:100–102.
12. Thomas D, Shah N, Patel H, et al. Hemophagocytic lymphohistiocytosis in Adult Patients with Variable Etiology: A series of five cases with review of the literature. *J Nat Sci Biol Med*. 2015;7:415–20.
13. Fardet L., Lambotte O., Meynard J-L., Kamouh W., Galicier L., Marzac C. Reactive haemophagocytic syndrome in 58 HIV-1-infected patients: clinical features, underlying diseases and prognosis. *AIDS*. 2010;24:1299–1306.
14. Otto, WR, Behrens EM, Teachey DT, et al. Human Adenovirus 7-Associated Hemophagocytic Lymphohistiocytosis-like Illness: Clinical and Virological Characteristics in a Cluster of Five Pediatric Cases. *Clin Infect Dis*. 2021;73(7):1532–1538.
15. Lahm T, Spatz HN, Hawley DA, Leland DS, Wolf KM, Twigg HL, Ober MD. Fatal adenovirus serotype 21 infection associated with hemophagocytic lymphohistiocytosis and multiorgan failure. *Resp Med CME*. 2010;3(4):223–225.
16. Hosnut, FO, Ozcay F, Malbora B, Hizli S, Ozbek N. Severe Adenovirus Infection associated with hemophagocytic Lymphohistiocytosis. *Turk J Haematol*. 2014;31(1):103–105.
17. Beutel G, Wiesner O, Eder M, et al. Virus-associated hemophagocytic syndrome as a major contributor to death in patients with 2009 influenza A (H1N1) infection. *Critical Care*. 2011;15:R80.
18. Gao J, Pasion J, Sanchez-Petitto G, Law J. Influenza virus B-associated hemophagocytic lymphohistiocytosis. *Hematol Rep*. 2019;11(3):8268.
19. Kalmuk J, Matar S, Feng G, Kilb E, Lim MY. Parvovirus B19-induced hemophagocytic lymphohistiocytosis: Case report and review of the literature. *Clin Case Rep*. 2019;7(11):2076–2081.
20. Retamozo S, Brito-Zeron P, Siso-Almirall A, Florez-Chavez A, Soto-Cardenas MJ, Ramos-Casals M. Haemophagocytic syndrome and COVID-19. *Clinical Rheumatology*. 2021;40:1233–1244.
21. Ansuini V., Rigante D., Esposito S. Debate around infection-dependent hemophagocytic syndrome in paediatrics. *BMC Infect Dis*. 2013;13:15.
22. Brisse, Ellen, Wouters CH, Andrei G, Matthys P. How Viruses Contribute to the Pathogenesis of Hemophagocytic Lymphohistiocytosis. *Front Immunol*. 2017;8:1102.
23. Thomas D, Pacioles T. Adult Secondary Hemophagocytic Lymphohistiocytosis: A Mini-Review on Recent Developments. *Ann Hematol Onco*. 2023;10(5):1434.
24. Soy, Mehmet, Atagunduz, P, Atagundus I, Sucak, GT. Hemophagocytic lymphohistiocytosis: a review inspired by the COVID-19 pandemic. *Rheumatology International*. 2021;41:7–18.
25. Stefanou C, Tzortzi C, Georgiou F, Timiliotou C. Combining an antiviral with rituximab in EBVV-related haemophagocytic lymphohistiocytosis led to rapid viral clearance; and a comprehensive review. *BMJ Case Rep*. 2016;2016:bcr2016216488.

Antiviral Agents: Cidofovir and Brincidofovir

Yanina Pasikhova

Abstract Cidofovir (CDV) was the first nucleotide analog approved for clinical use, for the systemic treatment of human cytomegalovirus (HCMV) retinitis in patients with human immunodeficiency virus infection and acquired immunodeficiency syndrome. To overcome side effects, particularly nephrotoxicity of CDV, brincidofovir (BrinCDV) an oral prodrug of CDV has been developed. In vitro studies demonstrate that CDV and BrinCDV are active against a comprehensive range of DNA viruses, particularly important is CDV's activity against acyclovir-resistant herpes simplex and varicella zoster viruses and human cytomegalovirus mutants with alterations in the viral UL97 gene. Despite broad in vitro activity, clinical benefit has not been demonstrated for most viruses. In-depth discussion of CDV and BrinCDV pharmacokinetics, clinical utility, and toxicities is explored.

Keywords Cidofovir · Brincidofovir · Nucleotide analog · Antivirals · Acyclovir-resistant herpes · Cytomegalovirus UL97 · Cytomegalovirus retinitis

1 Background

Cidofovir (CDV) was the first nucleotide analog approved for clinical use (Fig. 1). The antiviral properties of CDV [(S)-1-(3-hydroxy-2-phosphonylmethoxypropyl) cytosine] were first reported in 1986 and intravenous formulation was approved a decade later for the systemic treatment of human cytomegalovirus (HCMV) retinitis in patients with human immunodeficiency virus (HIV) infection and acquired

---

Y. Pasikhova (✉)
Infectious Diseases/Antimicrobial Stewardship Pharmacist, Moffitt Cancer Center, Tampa, FL, USA
e-mail: Yanina.Pasikhova@moffitt.org

C. Somboonwit et al. (eds.), *Global Virology IV: Viral Disease Diagnosis and Treatment Delivery in the 21st Century*,
https://doi.org/10.1007/978-3-031-57369-9_18

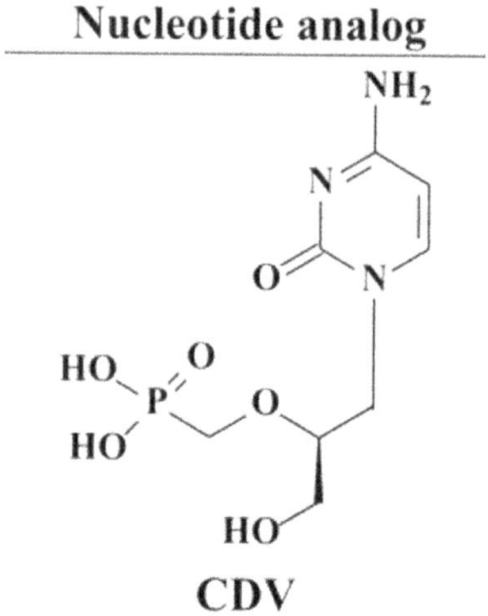

Fig. 1 Cidofovir chemical structure [1]

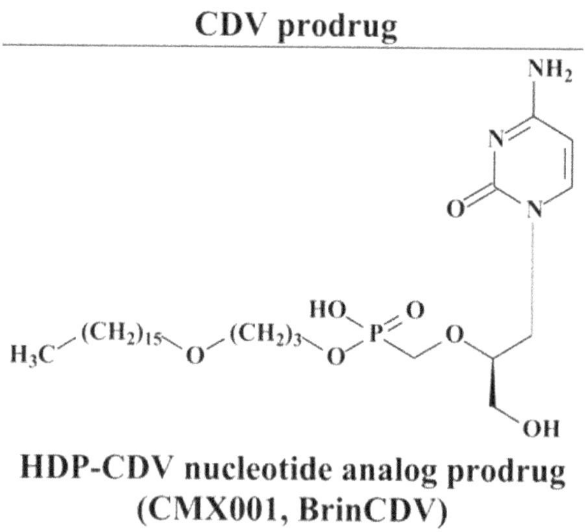

Fig. 2 Brincidofovir chemical structure [1]

immunodeficiency syndrome (AIDS) [1]. In addition to antiviral properties, CDV demonstrates antitumor and antiproliferative properties [1]. Nephrotoxicity is a well-recognized dose-limiting clinical adverse effect that has prompted research for alternative formulations, dosing, and supportive care strategies.

To overcome side effects, particularly nephrotoxicity of CDV, an alkoxy-alkyl esters of CDV and its cyclic form, cyclic-CDV (c-CDV), was developed [1]. In vitro, the alkoxy-alkyl esters of CDV and c-CDV are significantly more active than the parent compounds against a number of DNA viruses [1, 2]. Brincidofovir is an oral prodrug of CDV in which a natural fatty acid (lysophosphatidylcholine) molecule is used as carrier to facilitate drug adsorption in the gastrointestinal tract [1, 2]. Brincidofovir (BrinCDV; CMX001; hexadecyloxypropyl-cidofovir; cidofovir-HDP; HDP-CDV) (Fig. 2) was originally designed in response to the need for oral prophylaxis or treatment of smallpox, in case variola virus was used as a weapon in bioterrorism [1]. Since then BrinCDV is being marketed by Chimerix®, under the name of Tembexa®, for the treatment of a range of viral infections, including smallpox, and investigated for other DNA viruses [1]. Currently in the United States, BrinCDV availability is limited to strategic national stockpile distribution for treatment of human smallpox disease in adult and pediatric patients, including neonates [3].

2 Mechanism of Action

As mentioned earlier, CDV requires phosphorylation before becoming active, like other nucleoside and nucleotide analogs; however, contrary to those compounds, the negative charge of CDV phosphonate moiety significantly impairs cellular uptake. Subsequently CDV enters the cells via fluid-phase endocytosis. Once internalized, and enters the cytoplasm, only two phosphorylation steps are needed to convert CDV to its active diphosphoryl derivative (CDVpp) [1]. The two phosphorylation steps are carried out by cellular enzymes, and as a result, CDV is independent of the first phosphorylation step, which, for nucleoside analogs such as acyclovir, penciclovir, and ganciclovir, is catalyzed by the HSV- or VZV-encoded thymidine kinase or the HCMV-encoded protein kinase UL97 [1]. Ultimately, this mechanism accounts for CDV's activity against thymidine kinase–deficient HSV and VZV mutants and UL97 HCMV mutants. The specificity of CDV's antiviral activity stems in part from CDVpp's higher affinity for viral DNA polymerases than for host cell polymerases, ultimately hindering viral replication by inhibiting DNA synthesis by two distinct mechanisms [1, 4]. CDVpp is a competitive inhibitor of viral DNA polymerases and competes with the natural substrate, deoxycytidine triphosphate (dCTP). In addition, CDVpp functions as an alternative substrate for viral DNA polymerases and is then incorporated into the growing DNA chain which then slows or arrests DNA synthesis [1, 4] (Fig. 3).

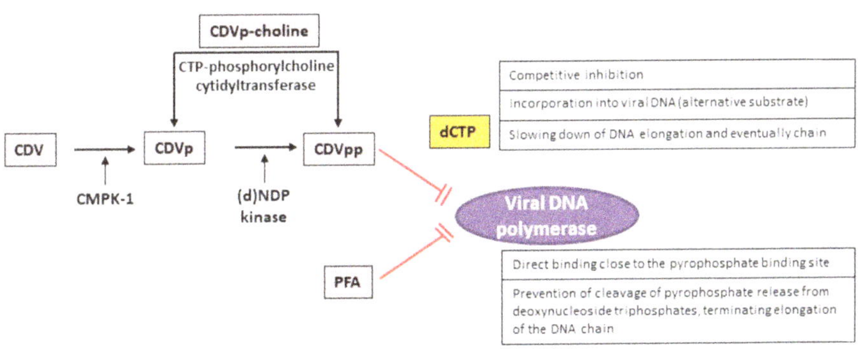

Fig. 3 Cidofovir mechanism of action [4]

BrinCDV uptake is very rapid due to the drugs ability to spontaneously insert into cell membranes, translocate to the inner membrane leaflet, and desorb promptly [1]. This allows the prodrug to then be metabolized by phospholipase C to produce CDV, which then undergoes phosphorylation to CDVpp [1, 5].

2.1 *Antiviral Activity*

2.1.1 Human Herpesviruses

Understanding the mechanism of CDV's antiviral effects explains why HSV strains exhibit increased susceptibility to CDV with altered or deficient thymidine kinase activity resistant to acyclovir and related compounds [6]. This difference appears to be due to these viruses inducing smaller increases in the intracellular concentration of dCTP than wild-type viruses. Because CDV is a competitive inhibitor of dCTP incorporation into HSV DNA, the lower concentrations of dCTP in cells infected with thymidine kinase–deficient HSV strains than in those infected with wild-type strains enhance the inhibitory effects of CDV diphosphate against HSV DNA polymerase [1, 6].

The effect of CDV incorporation into HCMV DNA varies, but it generally inhibits its polymerase activity [1, 7, 8]. The HCMV polymerase incorporates CDV into DNA where HCMV DNA polymerase-associated 3′–5′ exonuclease cannot erase CDV from the 3′ end due to the phosphonate group incorporated CDV. This then drastically slows down DNA synthesis, whereas incorporation of two consecutive molecules of CDV causes DNA chain termination. CDV can interfere with HCMV DNA synthesis in a number of ways, the most efficient being DNA chain termination after incorporation of two consecutive CDV molecules at the 3′ end. The inhibition of HCMV DNA polymerase by CDVpp and the inability of HCMV DNA polymerase to excise incorporated CDV from DNA may account for the potent and long-lasting anti-HCMV activity of CDV [1, 7].

2.1.2 Non-human Herpesviruses

Polyomaviruses and papillomaviruses viruses do not encode DNA polymerases and the mechanism of inhibition by CDV in these viruses is quite different, and less clearly understood, from that of herpesviruses and poxviruses and has recently been reviewed by Andrei and colleagues [9].

Regarding PyV, studies have demonstrated that CDV does not affect early steps of replication such as receptor binding and entry [9]. However, the drug reduced intracellular BKPyV DNA replication by over 90%. In contrast, in uninfected human renal tubular cells at equivalent concentrations, cellular DNA replication and metabolic activity only had a reduction of 7% and 11%, respectively. Furthermore, BKPyV infection increased cellular DNA replication to 142% and metabolic activity to 116%, respectively, which were reduced by CDV to levels of uninfected untreated cells [9]. Using African green kidney epithelial cells, investigators created in vitro CDV-resistant SV40 mutants that bear mutations in the ORI and helicase domains of the LT-ag, indicating that the helicase activity required for viral DNA unwinding during replication may be affected by CDV. Further research is required to prove that the helicase/ATPase activity of the LT-ag is affected by CDV and/or its metabolites [9]. Interference with the helicase/ATPase activity of the LT-ag may explain the activity of CDV during PyV productive infection but not against PyV-induced tumors. CDV anti-proliferative effects against PyVs can be explained by the cells infected by PyV being more sensitive to the drug because of the effects of the LT-ag on G1/S cell checkpoint deregulation. This is sustained by the higher antiproliferative effects of CDV against LT-ag transformed cells compared to the corresponding non-transformed cells [9].

The anti-proliferative effects of CDV on HPV have intensively been studied with the first studies showing the cytostatic activity of the drug against cervical carcinoma cells dating back to 1998, where CDV displayed time-dependent anti-proliferative effects, a phenomenon not typically seen with chemotherapeutic drugs [32]. HPV- and PyV-transformed cells appeared to be more sensitive to the effects of CDV since the viral oncoproteins induce cellular proliferation making the cells more sensitive to the anti-proliferative drug effects. Thus, the activity of CDV against HPV- and PyV-transformed cells may be explained, at least in part, by an inhibitory effect of the compound on rapidly dividing cells. In addition, the presence of the HPV or PyV genome might enhance the sensitivity of the cells to CDV, but when various cell lines not containing HPV such as melanomas, lung, colon, and breast carcinomas were tested, CDV still demonstrated an anti-proliferative effect [9].

3 Spectrum of Activity: In Vitro

In vitro studies demonstrate that CDV is active against a comprehensive range of DNA viruses, including adenoviruses, hepadnaviruses, herpesviruses, iridoviruses, papillomaviruses, polyomaviruses, and poxviruses [1]. Particularly important is CDV's activity against adenoviruses, polyomaviruses, and papillomaviruses for which there is currently no other active antiviral agent [1] (Table 1). For Herpes Simplex Virus (HSV) and Varicella Zoster Virus (VZV) resistant to acyclovir due to mutations in the viral thymidine kinase; against human cytomegalovirus (HCMV) mutants with alterations in the viral UL97 gene, and against poxviruses, including smallpox (variola) virus, CDV retains activity [1].

As BrinCDV is the oral prodrug of CDV, it is expected to possess an activity spectrum similar to that of CDV [1, 2]. In vitro studies have demonstrated BrinCDV's broad spectrum of activity and that it is roughly 25–150 times more potent than CDV [2]. Given its oral bioavailability and safer toxicity profile, BrinCDV may replace CDV in future therapeutic regimens for numerous indications. However, at

Table 1 Antiviral spectrum of cidofovir (CDV)

Viral family	Species susceptible
Adenoviridae	Human adenoviruses (ADV), mouse adenovirus
Herpesviridae	Herpes simplex type 1 (HSV-1) (thymidine kinase (TK) + and TK− strains); herpes simplex type 2 (HSV-2) (TK+ and TK− strains); varicella-zoster virus (VZV) (TK+ and TK− strains); human cytomegalovirus (HCMV); Epstein-Barr virus (EBV); human herpesvirus 6 (HHV-6); human herpesvirus 7 (HHV-7); human herpesvirus 8 (HHV-8); simian varicella virus (SVV); equine herpesvirus type 1 (EHV-1); bovine herpesvirus type 1 (BHV-1) and type 2; feline herpesvirus (FHV); caprine herpesvirus (CHV); murine cytomegalovirus (MCMV); rat cytomegalovirus (RCMV); Gguinea pig cytomegalovirus (GPCMV); rhesus rhadinovirus (RRV); herpesvirus saimiri (HVS); murine γ-herpesvirus 68 (MHV-68)
Iridoviridae	African swine fever virus
Poxviridae	Variola virus (VARV); molluscum contagiosum virus (MCV); vaccinia virus (VACV); cowpox virus (CPV); monkeypox virus (MPV); camelpox virus (CMLV); Orf virus
Papillomavirus	Human papillomavirus (HPV): different types; cottontail rabbit papillomavirus (CRPV)
Polyomavirus	Human polyomavirus: BK virus and JC virus

Adapted from Grayson [1]

the time of this writing, BrinCDV has not been approved for any of the indications outside of smallpox and lacks clinical data for routine use [1, 10].

It is important to note that despite broad in vitro activity, clinical benefit of CDV and BrinCDV has not been demonstrated for a number of infections. Clinical utility and data are described in more detail in Sect. 7 and summarized in Table 2 of this chapter.

3.1 Human Herpes Viruses

3.1.1 Herpes Simplex Viruses (HSV 1 [HHV1] and HSV 2 [HHV2])

CDV inhibits replication of both HSV 1 and 2 cell culture with half maximal effective concentration (EC_{50}) values in the range of 0.1–0.5 μg/ml in human embryonic lung cells, and thymidine kinase–deficient strains that are resistant to acyclovir have CDV EC_{50} values that are 3- to 20-fold lower than those of wild-type strains. [1] The mechanism of this effect is indeterminate but may be due to these strains altering intracellular deoxycytidine triphophosphate concentrations [1].

In 2010, Quenelle et al. evaluated the efficacy of BrinCDV compared to acyclovir in BALB/c mice inoculated intranasally with HSV-1 or HSV-2 [11]. BrinCDV was effective in reducing mortality using doses of 5 to 1.25 mg/kg administered orally once daily, even when treatments were delayed 48–72 h after viral inoculation. In addition, organ samples obtained from mice treated with BrinCDV had

Table 2 Clinical uses of cidofovir (CDV)

Route of administration	Clinical indication
Intravenous	HCMV retinitis in patient with HIV (approved) ACY and /or FOS-resistant HSV and VZV HCMV infections in non-HIV patients, particularly those that are resistant to ganciclovir due to mutations in the UL97 gene PyV infections due to JC (PML) and BK virus (HC and nephropathy) ADV infections Recalcitrant virus warts HHV6 encephalitis[a]
Topical	ACY and /or FOS-resistant mucocutaneous HSV 1/2 *(gel/cream)* HPV-associated papillomatous lesions *(gel/cream/injection)* Recurrent laryngeal papillomatosis Anogenital warts Common warts Cervical, vulvar, anal, penile intraepithelial neoplasia Virus warts *(intralesional)* HSV or ADV keratoconjunctivitis *(eye drops)* HCMV retinitis *(intravitreal)*
Intravenous or topical (gel/cream)	Molluscum contagiosum, orf and other poxvirus infections such as monkeypox and smallpox Complications of smallpox vaccine (vaccinia)
Intravesical	Hemorrhagic cystitis due to ADV and / or BK

Adapted from Grayson [1]
Abbreviation: *ACY* acyclovir, *FOS* foscarnet, *HC* hemorrhagic cystitis
[a]Limited clinical data; effectiveness is therefore unknown [56]

titers 3–5 log10 plaque-forming units per gram of tissue lower than samples obtained from mice treated with acyclovir, including five different regions of the brain [11].

3.1.2 Varicella-Zoster Virus (VZV [HHV3])

Like with HSV, CDV is active in vitro against VZV, and both wild-type and thymidine kinase-deficient mutant viruses have CDV EC_{50} values of ~0.2 μg/ml in human embryonic lung cells [1]. For BrinCDV, the anti-VZV activity was 4 log10 greater than that of CDV in human foreskin fibroblasts (BrinCDV EC_{50} of 0.0004 μM versus CDV EC_{50} of 0.5 μM) [1, 12].

3.1.3 Epstein–Barr Virus (EBV [HHV4])

CDV has been reported to be active against EBV with an EC_{50} value of ~0.01 μg/ml and appeared more active in inhibiting EBV DNA synthesis than acyclovir or ganciclovir. [1, 13] In a lethal immunosuppressive model in mice, CDV provided significant protection against murine gamma herpesvirus 68 (MHV-68), a rodent virus related to EBV. The activity of CDV against EBV and three different animal gamma herpesviruses was again demonstrated more recently [1, 14]. CDV also has marked in vitro activity against the EBV-associated nasopharyngeal carcinoma, decreased EBV oncoproteins, and enhanced radiosensitivity in EBV-related malignancies. [1]

BrinCDV has demonstrated in vitro activity as well, perhaps even more potent than that of CDV. The reported concentrations required to reduce EBV DNA synthesis by 50% and 90% in the EBV latently infected P3HR-1 cell line stimulated to lytic replication with phorbol esters were 1.9 and 20 μM for CDV compared to 0.3 and 5.2 μM, respectively, for BrinCDV [1].

3.1.4 Cytomegalovirus (HCMV [HHV5])

The first reported study to demonstrate in vitro activity of CDV against HCMV was in 1988, compared CDV to antivirals with know activity, including ganciclovir and foscarnet. [1] All compounds were active; however, CDV demonstrated having the best selectivity index at 1000, compared to 100 for ganciclovir and 10 for foscarnet. In clinical HCMV isolates, CDV was also shown to be more potent than ganciclovir in vitro [12, 15]. On a viral DNA synthesis level, CDV inhibits HCMV DNA at an EC_{50} of 0.1 μg/ml, a value 1000-fold lower than the 50% cytotoxicity concentration (CC_{50}) for cellular DNA synthesis [16]. Compared to ganciclovir and foscarnet against HCMV replication, the potency of CDV was confirmed to be superior when assessed as EC_{50} values via DNA hybridization and flow [17].

Like CDV, BrinCDV is highly active in vitro against HCMV, with an EC_{50} of 0.3 nM and a CC_{50} of 320 nM. However, BrinCDV exhibits approximately a 3-log increase in activity against HCMV in vitro compared to CDV [1]. Several studies have demonstrated the improved antiviral activity of BrinCDV compared to CDV in clinical and laboratory strains of HCMV, including ganciclovir resistant isolated with mutations in the UL97 and UL54 genes [18]. In human foreskin, fibroblasts BrinCDV was associated with a 2- to 3.3-log increase activity versus HCMV and a 75- to approximately 350-fold activity increase against murine, rat, and guinea pig CMV compared to CDV [1, 18].

3.1.5 Other Herpesviruses [HHV6–HHV8]

The comparative efficacy of numerous antiviral agents against HHV types 6 through 8 has been reviewed, and CDV emerged as one of the most potent and selective drugs against HHV-6 as measured by inhibition of virus-induced cytopathic effect or viral DNA synthesis, in both human T-lymphoblast cells and cord blood lymphocytes [19]. In addition, CDV demonstrated in vitro activity against HHV-7 and HHV-8 in several cell types [1, 19].

BrinCDV demonstrates enhanced activity against HHV-6 compared to CDV. The extent of increased activity varies based on the HHV-6 variant with a 900-fold for HHV-6A and 771-fold for HHV-6B [1].

3.2 *Non-herpes Human Viruses*

3.2.1 Adenovirus

In general, the reported EC_{50} values of CDV against adenovirus fall in the range of 4.6–17 μg/ml, and unlike ribavirin, that demonstrates activity only against C-group adenoviruses CDV is a relatively potent inhibitor of adenovirus serotypes from all species [1]. Similar to parent drug, BrinCDV is very potent against a broad range of adenovirus serotypes [1]. In a study using human type 5 adenovirus (Ad5) in immunosuppressed Syrian hamsters, CDV and BrinCDV demonstrated exceptional activity against Ad5 pathology and replication in the liver, even when administered therapeutically starting at 3 days post-challenge [20].

3.2.2 Papillomaviruses

The activity of CDV against human papillomavirus (HPV)-positive cells was tested in oncogenic HPV-16 or HPV-18 [1]. CDV demonstrated inhibition of HPV proliferation via a dose- (0.5–100 μg/ml) and time-dependent (3–10 days) fashion, [21] thus indicating that CDV has an antiproliferative effect on HPV-transformed cells, rather than an antiviral effect per se [1]. In rabbits infected with cottontail rabbit papillomavirus, topical CDV was effective at inhibiting papillomavirus-induced wart growth, especially with low titers of inoculum, if it was initiated early in infection [22].

Activity of BrinCDV (HDP-CDV) and two other related compounds, octadecyloxyethyl-CDV(ODE-CDV) and oleyloxyethyl-CDV (OLE-CDV), was evaluated with CDV against both HPV− and HPV+ cervical cancer cells [23]. The antiproliferation activity in HPV-16, HPV-18, HPV-68 cells C33A cells, a cervical cancer cell line lacking HPV, and nonmalignant primary human foreskin fibroblast cells was investigated. In cervical cancer cells, HDP-CDV was the least active of the three compounds, but still 18 to 1230 times more active than CDV itself, with OLE-CDV and ODE-CDV being 500 to 17,000 times more potent than CDV [23].

3.2.3 Polyomaviruses

The first evidence of CDV activity against polyomavirus (PyV) was described by Andrei et al. in 1997—for its ability to inhibit viral replication of murine and simian polyomavirus [24]. None of the other drugs tested by the authors, acyclovir, ganciclovir, brivudine, ribavirin, foscarnet, and cytarabine, showed significant activity. CDV has been reported to inhibit human polyomavirus, BK virus, with an EC_{50} of 64 μg/ml, likely above the plasma levels achievable in humans, but when the effects on BK virus replication were tested in primary human renal proximal tubule epithelial cells (RPTECs), the drug inhibited the generation of viral progeny in a dose-dependent manner, yielding a 90% reduction at 40 μg/ml [1]. In addition to inhibiting viral replication in RPTECs, CDV has been reported to inhibit BK replication in human renal cortical epithelial (HRCE) and salivary gland cells [1, 25].

BrinCDV effectively inhibits BK replication in primary human renal tubular epithelial cells, where it had a more rapid and lasting effect than CDV, and in primary human urothelial cells [1].

4 Mechanism of Resistance

For CDV and BrinCDV resistance would also expect to map to UL54 gene, given that CDVpp is a substrate inhibitor of UL54 [2]. Despite occurrences of clinical failure, currently there is no convincing evidence demonstrating selection of resistant viruses to CDV in patients receiving therapy with CDV, and in clinical studies of BrinCDV, for HCMV infections, treatment-emergent resistance was not reported [2]. Subsequently, lack of in vivo drug-resistant strains results in all information about CDV resistance mutations be synthesized from in vitro data [1].

4.1 Human Herpesviruses

4.1.1 Herpes Simplex Viruses (HSV 1 [HHV1] and HSV 2 [HHV2])

HSV-1 strains resistant to CDV had V573M, R700M, or K960R mutations in the viral DNA polymerase, resulting in an eightfold to tenfold EC_{50} increase compared to the wild-type virus [1, 26, 27]. There appears to be a markedly reduced pathogenicity in CDV-resistant strains; if this finding translates into human subjects, then even if CDV-resistant mutant viruses arise in vivo, they may not be associated with disease progression [1]. Additionally, CDV-resistant HSV strains remained sensitive to foscarnet and acyclovir. A CDV-resistant HSV-1 with an R700M polymerase mutation was hypersensitive to foscarnet (it had a fourfold decrease in EC_{50} values compared to the wild-type reference strain). These data support the possible use of CDV in the treatment of foscarnet-resistant HSV infections. Indeed, CDV has proven efficacious in the treatment of acyclovir and/or foscarnet-resistant HSV-associated infections [1].

4.1.2 Cytomegalovirus (HCMV [HHV5])

CDV-resistant HCMV variants, selected by in vitro passage, exhibited cross-resistance to ganciclovir and cluster in specific regions of the HCMV DNA polymerase [1]. However, clinical significance of this phenomenon is likely of limited consequences as CDV is not a first-line therapy for HCMV infections; thus, selection of strains resistant to the compound is relatively rare and occurs primarily during long-term administration of ganciclovir. [28] On the flip side, most ganciclovir-resistant clinical isolates of HCMV have mutations in the UL97 gene, because CDV does not require the UL97 gene product for initial phosphorylation such isolates remain susceptible to CDV [1]. The majority of the HCMV polymerase mutants resistant to CDV and ganciclovir remain susceptible to foscarnet, whereas foscarnet-resistant strains are not generally resistant to CDV but may exhibit cross-resistance to ganciclovir. [1] Mutations to HCMV that convey resistance to CDV also result in resistance to BrinCDV [29]. Under selective pressure over a 10-month period a novel in vitro selected BrinCDV-resistant HCMV demonstrated one mutation, D542E in UL54 DNA polymerase. This viral strain with this novel D542E mutation showed reductions in susceptibility of greater than tenfold to BrinCDV and CDV, but no resistance to foscarnet or ganciclovir. [29] However, multidrug-resistant HCMV strains have been isolated in immunocompromised patients, and previously unrecognized mutations in the HCMV polymerase gene continue to be isolated highlighting the need for new treatment options such as maribavir (see chapter "Antiviral Agents: Maribavir") and letermovir (see chapter "Antiviral Agents: Letermovir") [30–32].

4.2 Non-herpes Human Viruses

4.2.1 Adenovirus

CDV-resistant adenovirus strains have been generated by in vitro passage in CDV, which showed no significant drop in pathogenicity, demonstrated preferential replication in CDV-treated infected organ, and had mutations in conserved regions of the adenovirus DNA polymerase predicted to be involved in nucleotide binding [1].

5 Pharmacokinetics and Pharmacodynamics

Parenterally administered CDV exhibits dose-independent pharmacokinetic features [33]. However, CDV plasma pharmacokinetic variables do not reflect the in vivo duration of action of CDV because the antiviral effect depends on the intracellular concentrations of the active phosphorylated metabolites in cells [1, 33]. Specific pharmacokinetic parameters of CDV are detailed in sections below. No pharmacodynamic information is available for CDV [1].

BrinCDV pharmacokinetics and safety against double-stranded DNA viruses was first described by Painter et al. [34] BrinCDV is orally bioavailable and well tolerated in healthy volunteers and demonstrated plasma drug exposures that are expected to result in activity against multiple double-stranded DNA viruses. Similar to CDV, pharmacodynamic data for BrinCDV is unknown [35].

5.1 Absorption and Bioavailability

5.1.1 Intravenous or Subcutaneous Administration

Pharmacokinetic studies of CDV have demonstrated limited plasma protein binding and poor oral absorption; thus, when administered systemically, CDV is always given parenterally, mostly intravenously, even though it has been observed that subcutaneous administration produces equivalent exposure [1, 36]. Intravenous administration of CDV results in dose-independent pharmacokinetic parameters [33], with single intravenous infusion of CDV, at the recommended dosage of 5 mg/kg, with concomitant oral probenecid, yielding the following median pharmacokinetic values: peak serum concentration (C_{max}) of 19.6 ± 7.18 mg/l, area-under-the-concentration-time curve from zero to infinity ($AUC_{0-\infty}$) of 40.8 ± 8.97 mg/l/h, creatinine clearance (based on serum creatinine concentration) of 0.138 ± 0.036 l/h/kg, renal clearance of 0.096 ± 0.031 l/h/kg, steady-state volume of distribution of 0.39 ± 0.13 l/kg, and plasma elimination half-life ($t_{½}$ beta) of 2.2 ± 0.5 h [1, 37].

5.1.2 Topical, Intralesional, Intravesical, and Oral Administration

To minimize toxicity, CDV has been administered locally as intralesional, topical, and intravesical instillation for treatment of DNA virus infections, despite being approved only for intravenous treatment of HCMV retinitis in patients with AIDS [1]. Limited data are available regarding the systemic distribution of CDV after intralesional or topical application [1]. However, available studies suggest that the resulting serum concentrations are very low compared with intravenous administration but may be enhanced in abraded skin. Topical bioavailability and penetration of CDV may be enhanced in skin that is not intact, thus it is recommend considering a possibility of systemic exposure to CDV in such scenarios [1]. Intralesional administration of CDV in children and adults demonstrates that CDV plasma levels and the AUC were dose dependent in children, but not in adults [1, 38]. A linear relationship was found between plasma concentration and dose in children but not in adults. Furthermore, this study led to a conclusion that the CDV plasma levels were below those leading to toxicity, and the levels and the AUC were dose dependent in children but not in adults; thus, diffusion from the injected site was the greatest and most unpredictable among adults [38]. Thus, given such variability, it is recommended that topical CDV be used at a lower dose than the recommended intravenous dose to minimize the risk of systemic toxicity [1].

Intravesicular administration of CDV, for safety and pharmacokinetic parameters, has been evaluated in a small number of patients with hemorrhagic cystitis after allogeneic hematopoietic stem cell transplantation [1, 39]. Six patients with baseline serum creatinine <1.4 mg/dl, with obvious hematuria and confirmed BK or adenovirus viruria, were enrolled. Three hours after an oral probenecid dose (2 g), CDV (2.5–5 mg/kg in 50–100 ml of normal saline) was given via a transurethral catheter for up to 2 h of dwell time, with serial plasma samples obtained over 24 h. Compared with the reported AUC_{0-24} for an equivalent intravenous dose of CDV, intravesicular instillation of the drug resulted in 1–74% of the corresponding systemic exposure. In addition to the highly variable systemic exposure, this approach was further limited due to tolerability, as the authors report that only two patients were able to endure a 2-h dwell time, and one patient developed a >50% increase in serum creatinine within 7 days of administration [39].

Acyclic nucleoside phosphonates like CDV possess a double negative charge at physiologic pH, resulting in poor oral bioavailability; they enter the cells slowly and have the propensity to concentrate in the renal proximal tubules, resulting in nephrotoxicity [1]. To improve the pharmacokinetics and antiviral properties, lipid analogs of this class of compounds with increased oral adsorption and improved cellular uptake and metabolism were developed. BrinCDV was designed to look like a natural phospholipid metabolite, which is adsorbed without hydrolysis [1]. Oral administration of equimolar doses of CDV and BrinCDV to mice provided markedly greater plasma levels with BrinCDV than with CDV [40]. In a randomized double-blind, placebo-controlled, parallel group, dose-escalating study in healthy volunteers BrinCDV was given orally as single and multiple dose ranges [34]. After a single dose, BrinCDV was readily absorbed, with plasma Tmax drug concentration ranging from 2 to 3 h. Maximum plasma drug concentration and systemic exposure ($AUC_{0-\infty}$) increased in proportion to doses. The rate of appearance of CDV in plasma after a single dose of BrinCDV depended on the absorption rate of BrinCDV and the rate of conversion of BrinCDV to CDV. The Tmax of CDV ranged from 7 to 15 h and the Cmax of CDV increased somewhat less than in proportion to the dose, but the AUC increased approximately in proportion to the dose [34].

5.2 *Distribution*

Limited data on CDV distribution is available. After a single intravenous CDV dose of 5 mg/kg with probenecid, the steady-state volume of distribution (Vd) was 0.53 l/kg [1]. To date no information is available about specific tissue distribution or CDV distribution following topical administration.

BrinCDV has a an approximate Vd of 1230 l [41].

5.3 *Metabolism and Excretion*

Over 80% of the drug is excreted unchanged in the urine within 24 h, with a half-life of 2.4 to 3.2 h [41]. However, cidofovir diphosphate, an active metabolite, is eliminated more slowly with first- and second-phase intracellular half-lives of 24 and 65 h, respectively [42]. This property permits the drug to be dosed every 2 weeks. Tubular secretion of CDV is mediated primarily by the human renal anion transporter hOAT1 (SLC22A6), with a minor contribution of hOAT3 [43]. Concomitant oral probenecid results in a decrease in renal clearance of CDV, indicating that CDV is eliminated from the systemic circulation by both glomerular filtration and active tubular secretion [33, 36].

For BrinCDV, the terminal elimination half-life increased with increasing dose, and CDV exposure ($AUC_{0\text{-last}}$) was dose proportional. BrinCDV is undetected in urine, whereas CDV is detected as an elimination product of BrinCDV [34].

5.4 Drug Interactions

Probenecid decreases the renal absorption of CDV, thus decreasing the drugs' nephrotoxic effects [33, 36]. It is possible that other drugs excreted by glomerular filtration and/or tubular secretion might also alter the renal excretion of CDV, but no data support that supposition. Other nephrotoxic drugs cause synergistic nephrotoxicity with CDV, and when used together, the benefits of co-administration would outweigh the possible risk [1].

6 Adverse Reactions and Toxicity

Intravenous administration of CDV is implicated in most toxicities associated with the drug [1]. For BrinCDV, clinical trials are conducted under widely varying conditions, and adverse reaction rates observed in one trial may not be comparable to rates in another clinical trial and may not reflect the rates observed in practice. Of note, BrinCDV safety has not been studied in patients with smallpox [35].

6.1 Nephrotoxicity

CDV nephrotoxicity has been observed in most species in which it has been studied and is manifested by proteinuria; glycosuria; decreased serum phosphate, uric acid, and bicarbonate; increased serum creatinine; and degeneration and necrosis of the proximal renal tubule cells [1]. Proteinuria and creatinine elevations have been shown to be dose related and at least in part reversible in most patients upon drug discontinuation [1]. CDV nephrotoxicity is attributed to active uptake of the compound, at the basolateral membrane of renal proximal tubular cells, being faster than the efflux of it into the urine, thus causing accumulation in renal tubular epithelial cells [44]. It has been demonstrated that CDV is a substrate of human organic anion transporter 3 (hOAT3) and hOAT1, but quantitatively hOAT1 appeared to be the major renal transporter and plays a critical role in CDV-related nephrotoxicity [1, 43, 44].

Once CDV-associated nephrotoxicity and prolonged antiviral activity were recognized in the phase I/II clinical trials, focus was directed on optimizing dosing and administration of CDV, including simultaneous administration of probenecid, hydration, and extended dosing intervals [45]. In addition to intravenous hydration, accumulation of CDV in the renal tubular cells can be prevented by an infrequent administration. In animal studies, dosing of CDV no more than once weekly appeared to reduce nephrotoxicity without affecting efficacy [1]. Nephrotoxicity was also noted during the initial clinical trials of CDV in HCMV-infected patients [45–47]. Similar to pre-clinical animal studies, patients receiving weekly intravenous CDV at doses of 0.5 or 1.0 mg/kg did not have evidence of drug-induced nephrotoxicity, which was exclusively seen in patients receiving 3.0 or 10.0 mg/kg doses [45]. Based on pre-clinical and clinical data, intravenous CDV should not be given in doses higher than 5 mg/kg once weekly for 2 weeks, followed by 5 mg/kg once every 2 weeks [1]. Additional effects on proximal tubular cells might be amended by concomitant administration of probenecid without affecting CDV tissue levels [36]. Co-administration of probenecid, an inhibitor of organic anion transport that interferes with the transporter-mediated tubular uptake of CDV, results in decreased concentration of CDV and its metabolites in renal tubules thus reducing nephrotoxicity [1, 42]. Treatment recommendations have been established to lessen the risk of nephrotoxicity, especially in patients with underline renal disease, but nephrotoxicity continues to be a dose-limiting side effect in patients receiving CDV. In clinical practice, reported rates of nephrotoxicity vary, with one retrospective study of allogeneic stem cell transplant recipients treated with CDV for HCMV about a quarter of patients developed renal toxicity. Furthermore, over half of the patients with nephrotoxicity in the study continued to display signs of toxicity after drug discontinuation [1].

Masking of one of the negative charges on the phosphonate of CDV substantially reduced kidney uptake. The study by Painter et al. demonstrated that BrinCDV was well tolerated at all doses, with no dose-limiting toxicity, no evidence of nephrotoxicity or myelotoxicity. In addition, only CDV was detected in urine, confirming that unchanged BrinCDV is not eliminated by urinary excretion [34].

6.2 Ocular Toxicity

Uveitis has been observed after intravenous CDV for treatment of HCMV retinitis in AIDS patients. In one study, 89% of CDV-treated patients developed anterior uveitis [48]. However, significantly lower rates of 11% are reported in prescribing information [49].

Among the subset of patients monitored for intraocular pressure changes, a ≥50% decrease from baseline was reported in 24% patients at the 5 mg/kg maintenance dose. Severe hypotony (intraocular pressure of 0–1 mm Hg) has been reported in three patients. Risk of ocular hypotony may be increased in patients with preexisting diabetes mellitus.

6.3 Hematological Toxicity

At the doses typically used, CDV probably has no hematologic toxicities [1]. However, in clinical trials, at the 5 mg/kg maintenance dose, a decrease in absolute neutrophil count to ≤500 cells/mm^3 occurred in 24% of patients. Granulocyte colony-stimulating factor (GCSF) was used in 39% of patients [49]. It is important to note that the observed neutropenia was not related to the dose of CDV and occurred at similar rates with either 5 or 3 mg/kg dose and similar to the rates observed in patients with advanced AIDS not on drugs with obvious bone marrow toxicity [50, 51]. No significant hematological toxicities have been reported in BrinCDV studies [2].

6.4 Other Adverse Reactions

Based on toxicology studies, CDV is considered a carcinogen in rats and a potential carcinogen in humans [49]. In a study of intravenous CDV administration to rats at a dose of 0.6, 3, or 15 mg/kg once weekly for 26-weeks, a significant increase in mammary adenocarcinomas in female and significant incidence of Zymbal gland carcinomas in male and female rats were seen at the highest dose but not at the lower two doses [49].

Studies showed that CDV caused inhibition of spermatogenesis in rats and monkeys. However, no adverse effects on fertility or reproduction were seen following once weekly intravenous injections in male rats for 13 consecutive weeks at doses equivalent to 1.1 times the recommended human dose based on AUC comparisons [49]. Female rats dosed intravenously once weekly at equivalent to 0.09 times the recommended human dose based on AUC or higher, for up to 6 weeks prior to mating and for 2 weeks post–mating, had decreased litter sizes and live births per litter and increased early resorptions per litter [49]. In rabbits, an increased incidence of fetal external, soft tissue, and skeletal abnormalities occurred at the high dose (1.0 mg/kg per day) that was also toxic to the mothers [1]. Patients should be advised that CDV caused reduced testes weight and hypospermia in animals. Such changes may occur in humans and cause infertility. Women of childbearing potential should be advised that CDV is embryotoxic in animals and should not be used during pregnancy. Women of childbearing potential should be advised to use effective contraception during and for 1 month following treatment with CDV. Men should be advised to practice barrier contraceptive methods during and for 3 months after treatment with CDV [51].

Systemic side effects from topical or intralesional CDV are extremely rare, though a local inflammatory response with reactions at application site may occur [1]. Acute renal failure has been reported with high concentration (4%) of topical CDV [52]. In addition, a case of bilateral uveitis and hypotony associated with topical CDV has recently been reported [53]. To minimize risk of systemic toxicity, topical CDV should not be applied to large areas of abraded skin or exceed a concentration of 1% [1].

Palpable masses occurred in rats after as few as 26 oral doses of BrinCDV at systemic exposures less than the expected human exposure [41]. The masses diagnosed as mammary adenocarcinomas; carcinoma in squamous cell, Zymbal's gland, uterus, and small intestine; and hemangiosarcoma in mesenteric and mediastinal lymph node, liver, and abdominal cavity were observed in rats following long-term (13-weeks and 26-weeks) dosing studies. Based on these data, BrinCDV is considered a potential human carcinogen [41]. Similar to CDV, based on animal data, BrinCDV may cause fetal harm, and pregnancy testing should be performed in individuals of childbearing age prior to drug initiation [10]. Females should be advised to use effective contraception during treatment and for at least 2 months after the last dose of BrinCDV, and sexually active males with partners of childbearing potential should use condoms during treatment and for at least 4 months after last dose [10, 41]. Like CDV, BrinCDV may impact fertility. Based on testicular toxicity in animal studies, BrinCDV may irreversibly impair fertility in males of reproductive potential [10].

BrinCDV appears to lack nephrotoxicity but patients may develop hepatotoxicity. Elevations in alanine aminotransferase (ALT), aspartate aminotransferase (AST), and total bilirubin have been observed, including cases of concurrent increases in ALT and bilirubin. During the first 2 weeks of BrinCDV therapy in 392 subjects, ALT elevations greater than 3 times the upper limit of normal was reported in 7% of subjects [41]. These elevations in hepatic laboratory tests were generally reversible and did not require discontinuation. Severe hepatobiliary adverse events including hyperbilirubinemia, acute hepatitis, hepatic steatosis, and venoocclusive liver disease have been reported in less than 1% of subjects [10]. In addition, BrinCDV administration was notable for gastrointestinal (GI) side effects [41]. During the first 2 weeks of therapy, diarrhea of any grade occurred in 40% of BrinCDV-treated subjects compared with 25% of subjects in the placebo control group. Treatment was discontinued due to diarrhea in 5% of subjects in the BrinCDV group compared with 1% in the placebo [41].

It is important to keep in mind that administration of probenecid to reduce CDV nephrotoxicity can carry its own adverse effects. In one study, 56% of patients had a probenecid-attributed adverse event, such as fever, chills, nausea, or rash, but only 3 of the 41 patients discontinued treatment because of the reaction [51]. Similar observations were made in the SOCA-ACTG study [50].

7 Clinical Utilization, Dosing, and Administration

7.1 FDA-Approved Indications and Dosing

7.1.1 Cidofovir

HCMV Retinitis in Patients with HIV Infection [49]

Induction dose CDV intravenously 5 mg/kg once weekly for 2 weeks followed by maintenance of 5 mg/kg intravenously once every 2 weeks with concomitant oral probenecid and intravenous hydration to minimize drug-related nephrotoxicity [1, 49].

7.1.2 Brincidofovir

Human Smallpox Disease in Adult and Pediatric Patients Including Neonates [10]

Adult and pediatric dosing is weight based and should be given on days 1 and 8. Pediatric dose is <10 kg; oral suspension 6 mg/kg once weekly for 2 doses, 10 to <48 kg: oral suspension 4 mg/kg once weekly for 2 doses and ≥48 kg: oral suspension or tablet: 200 mg once weekly for 2 doses. Adult dose is 10 to <48 kg: 4 mg/kg once weekly for 2 doses and ≥48 kg: 200 mg once weekly for 2 doses [10].

7.2 NON-FDA Approved Uses and Dosing

7.2.1 Cidofovir

Adenovirus Infection in Hematopoietic Cell Transplant Recipients [54]

CDV is used off-label for the treatment of disseminated, severe, progressive adenovirus disease or preemptive therapy in stem cell recipients [1, 54]. The majority of the published data on the use of CDV for adenovirus disease suffer from major limitations, including small and heterogeneous patient cohorts, variable organ involvement, and retrospective designs [1]. Several dosing options have been evaluated, but typically CDV is used at the FDA-approved dosing or 5 mg/kg once weekly or 1 mg/kg three times per week, all with concomitant oral probenecid [1, 54].

Polyomavirus Infections (BK and JC Virus Infection as an Adjunctive Agent) [1, 54]

Polyomavirus (PyV) disease (JC and BK virus) is rare and limited to immunocompromised individuals; thus, data on the use of CDV in these infections is scant, inconsistent, and based on case reports [1]. For BK-associated hemorrhagic cystitis in hematopoietic cell transplant recipients, CDV is given intravenously at doses from 3 to 5 mg/kg/dose every 1 to 2 weeks with concomitant probenecid. Alternatively, 0.5 to 1.5 mg/kg/dose 1 to 3 times weekly without probenecid has been used. Intravesicular installation of 5 mg/kg/dose (in 60 mL NS) instilled into bladder once weekly and retained for ≥1 h has also been reported. In kidney transplant recipients with BK viremia or nephropathy, lower doses of intravenous CDV 0.25 to 1 mg/kg/dose every 1 to 3 weeks without probenecid are recommended [54].

Progressive multifocal leukoencephalopathy (PML) secondary to JC virus has been attempted to be treated with CDV, but its usefulness is unclear with some studies supporting a therapeutic efficacy and others lacking activity of CDV [1].

HCMV Treatment of Invasive Disease Including Retinitis and Preemptive Therapy in Hematopoietic Cell Recipients [54]

Per package insert the "safety and efficacy of Vistide® has not been established for treatment of other HCMV infections (such as pneumonitis or gastroenteritis), congenital or neonatal HCMV disease, or HCMV disease in non-HIV individuals." [51] However, based on small clinical trials CDV is used as an alternative agent in HCMV infections in other then retinitis in HIV positive patients at the FDA-approved dosing for retinitis [1, 54].

Herpes Simplex Virus Infection, Acyclovir-Resistant [54]

Several small studies and case reports have demonstrated successful treatment with topical or systemic CDV in patients with acyclovir and/or foscarnet-resistant HSV infections [1]. Mucosal infections can be managed with intravenous CDV 5 mg/kg/dose once weekly for 3 weeks, then 5 mg/kg/dose once every 2 weeks for 3 doses, all given with concomitant probenecid [54]. For topical application, CDV 1% gel (extemporaneously prepared) to be applied to affected area once daily for 5 days [54]. It is recommended to observe the patients for 10 days following treatment for relapse/recurrence of disease. Therapy may be repeated if necessary, for up to 6 repetitive cycles. It has also been suggested that duration of 21 to 28 days or longer may be required for patients with HIV [54].

7.2.2 Brincidofovir

A box warning in prescribing information states that BrinCDV is associated with "increased risk for mortality when used for longer duration. An increased incidence of mortality was seen in brincidofovir-treated subjects compared to placebo-treated subjects in a 24-week clinical trial when brincidofovir was evaluated in another disease." [10] Negative results from the phase III study assessing the efficacy of BrinCDV for HCMV prophylaxis in adult stem cell transplant led to discontinuation of the drugs oral clinical development program for HCMV infections [2]. Intravenous formulation of the drug may be potentially less toxic and was being evaluated as a potential treatment for ADV, and other transplant-related viral infections also were halted, and global rights to develop BrinCDV for all non-orthopoxvirus indications was transferred to SymBio Pharmaceuticals Ltd. (Tokyo, Japan) [2, 55]. At the time of this writing, BrinCDV was not available through expanded access. Ongoing phase II dose-confirmation clinical trial of intravenous BrinCDV for pediatric patients with serious adenoviral diseases is being conducted in the United States (ClinicalTrials.gov Identifier: NCT04706923). The efficacy and safety profile of intravenous BrinCDV will be evaluated in a subsequent phase III clinical trial. Due to lack of safety and efficacy, intravenous BrinCDV is currently only available through participation in the clinical trial and is not available separately through an expanded access or compassionate use program [55]. Phase I trial for management of BK virus infections in renal transplant recipients and pre-clinical trial of HCMV infections are also ongoing for intravenous BrinCDV.

7.3 Summary of CDV Clinical Uses

Table 2 summarizes the current clinical uses of CDV either based on approval or available clinical data [1, 56].

8 Special Considerations

8.1 Cidofovir

8.1.1 Pediatric Use, Pregnancy, and Lactation

Safety and effectiveness of cidofovir in children have not been studied [49]. In animal studies, CDV was embryotoxic and maternally toxic in rats and rabbits at relatively low doses. An increased incidence of fetal external, soft tissue, and skeletal anomalies occurred in rabbits at a high dose (1.0 mg/kg/day) [1, 49, 54]. At this dose, CDV was also maternally toxic. There are no adequate and well-controlled studies in pregnant women, and CDV is an FDA category C, used during pregnancy only if the potential benefits justify the potential risk to the fetus [49]. The extent of CDV excretion in human breast milk is not known, thus CDV should not be administered to nursing mothers [49].

8.1.2 Nephrotoxicity: Prevention, Monitoring, and Dosing Considerations

Renal function should be monitored before initiation of CDV therapy and 48 h after each dose [49, 54]. Based on the manufacturer guidelines, initiation of CDV is contraindicated in patients with baseline renal insufficiency (serum creatinine greater than 1.5 mg/dL, creatinine clearance less than or equal to 55 mL/minute, or urine protein greater than 100 mg/dL/greater than or equal to 2+ proteinuria), and should be discontinued in case of elevation of serum creatinine by greater than or equal to 0.5 mg/dL or development of 3+ proteinuria [1]. In patients with a serum creatinine, elevation between 0.3 and 0.4 mg/dL CDV dose should be reduced from 5 to 3 mg/kg [1].

Treatment guidelines have been developed to decrease the risks of nephrotoxicity in patients receiving CDV. Some general recommendations include avoidance of concurrent administration of potentially nephrotoxic drugs and intravascular volume depletion [1]. Before an intravenous CDV infusion, a washout period of at least 7 days is recommended after the use of potentially nephrotoxic agents. Hydration with 1 l of normal saline over 1 to 2 h immediately before intravenous CDV infusion is also recommended. If tolerated, another liter may be administered over 1 to 3 h at the start of or immediately following infusion [49, 54]. In addition, probenecid must be given at a dose of 2 g 3 h before each CDV infusion, and then 1 g 2 h and 8 h after completion of each infusion [54].

In a pharmacokinetic study of CDV in renal insufficiency, a significant correlation was observed between drug clearance and creatinine clearance [57]. The study highlighted the need for dose adjustment in patients with reduced renal function to obtain comparable drug exposure as patients with normal renal function. Although according to the manufacturer CDV is contraindicated in patients with renal impairment data from Brody et al. provides alternative guidelines for CDV dosing. The dosing recommendations are a theoretical dosing strategy to maintain normal serum AUC in patients with renal impairment, not on renal replacement therapy (Table 3). Dosing adjustments are based on 5 mg/kg/dose with induction doses, given weekly and maintenance doses every other week [54, 57]. For patients on hemodialysis, it is recommended to administer standard induction or maintenance dose 2 h before hemodialysis without probenecid. CDV is minimally removed by peritoneal dialysis and patients should receive 0.5 mg/kg/dose without probenecid [54, 57].

Table 3 Alternative dosing for CDV in patients with renal insufficiency

Creatinine clearance (CrCL in mL/minute/kg)	Dose (mg/kg with concomitant probenecid)
≥1.3	5
1–1.2	4
0.8–0.9	3
0.7	2.5
0.5–0.6	2
0.4	1.5
0.2–0.3	1
0.1	0.5

Adapted from UpToDate and Brody [54, 57]

8.2 *Brincidofovir*

8.2.1 Pregnancy and Lactation

Based on findings from animal reproduction studies, BrinCDV may cause fetal harm; however, there are no available human data on the use of BrinCDV in pregnancy [1, 10]. In animal reproduction studies, oral administration of brincidofovir to pregnant rats and rabbits during the period of organogenesis resulted in embryotoxicity and structural malformations. When BrinCDV was administered to lactating rats (4 mg/kg/day or 15 mg/kg twice weekly), the drug was detected in milk but not in the plasma of nursing pups. In addition, due to the lack of data on the presence of BrinCDV in human milk, the effects of the drug on the breastfed infant or on milk production, and the ability of smallpox to be transmitted via direct contact with the breastfeeding infant, breastfeeding is not recommended [41].

8.2.2 Oral Administration

Due to carcinogenic potential of BrinCDV, direct contact with broken or crushed tablets and suspension should be avoided. If contact with skin or mucous membranes occurs, wash thoroughly with soap and water, and rinse eyes thoroughly with water [10]. Tablets should be administered on an empty stomach or with a low-fat meal (approximately 400 calories with approximately 25% calories from fat) and should not be crushed or divided [10, 41]. Suspension should be administered on an empty stomach. Shake oral suspension before use and discard unused portion after two doses. For patients who cannot swallow, oral suspension can be administered by enteral tube (following administration, refill the catheter-tip syringe with 3 mL of water, shake, and administer contents via tube to ensure full dose is administered); flush tube with water before and after enteral administration [10].

8.2.3 Hepatotoxicity: Monitoring

Perform hepatic laboratory testing in all patients before starting BrinCDV and while receiving the drug [10, 41]. Monitor patients who develop abnormal hepatic laboratory tests during therapy for the development of more severe hepatic injury. BrinCDV should be discontinued if ALT levels remain persistently greater than 10 times the upper limit of normal. Do not give the second and final dose of BrinCDV on Day 8 if ALT elevation is accompanied by clinical signs and symptoms of liver inflammation or increasing direct bilirubin, alkaline phosphatase, or International Normalized Ratio [41].

9 Summary and Conclusion

After initial reports of antiviral activities of CDV, the agent was eventually approved as an intravenous formulation for the systemic treatment of HCMV retinitis in patients with HIV infection and AIDS [1]. Early on nephrotoxicity became a well-recognized dose-limiting clinical adverse effect that has prompted research for alternative formulations, dosing, and supportive care strategies. Some general recommendations include addition of probenecid, avoidance of concurrent administration of potentially nephrotoxic drugs, and intravascular volume depletion [1, 54]. In an effort to further minimize toxicity, CDV has been administered locally as intralesional, topical, and intravesical instillation for treatment of DNA virus infections, despite being approved only for intravenous treatment of HCMV retinitis in patients with AIDS [1]. Table 2 summarizes the clinical uses of CDV for local administration.

To overcome side effects, particularly nephrotoxicity of CDV, an alkoxy-alkyl esters of CDV and its cyclic form, cyclic-CDV (c-CDV) was developed [1]. In vitro, the alkoxy-alkyl esters of CDV and c-CDV are significantly more active than the parent compounds against a number of DNA viruses [1, 2]. Brincidofovir is an oral prodrug of CDV in which a natural fatty acid molecule is used as carrier to facilitate drug adsorption in the gastrointestinal tract while avoiding nephrotoxicity [1, 2].

Despite broad in vitro activity of CDV and BrinCDV against many DNA viruses, including adenoviruses, hepadnaviruses, herpesviruses, iridoviruses, papillomaviruses, polyomaviruses, and poxviruses, clinical benefit has not been demonstrated for most, most notably EBV [1]. Outside FDA-approved indication, predominant role of CDV, and BrinCDV if available, remains in management of acyclovir-resistant HSV, HCMV, ADV, and PyV infections in transplant recipients, despite sparse data [54].

Conflict of Interest The author has no conflict of interest.

References

1. M Lindsay Grayson. Kucers' the use of antibiotics: a clinical review of antibacterial, antifungal, antiparasitic, and antiviral drugs. Boca Raton: CRC Press; 2018.
2. Alvarez-Cardona JJ, Whited LK, Chemaly RF. Brincidofovir: understanding its unique profile and potential role against adenovirus and other viral infections. Future Microbiology. 2020 Apr;15(6):389–400.
3. Marty FM, Winston DJ, Chemaly RF, Mullane KM, Shore TB, Papanicolaou GA, et al. A Randomized, Double-Blind, Placebo-Controlled Phase 3 Trial of Oral Brincidofovir for Cytomegalovirus Prophylaxis in Allogeneic Hematopoietic Cell Transplantation. Biology of Blood and Marrow Transplantation. 2019 Feb;25(2):369–81.
4. Andrei G, Snoeck R. Advances and Perspectives in the Management of Varicella-Zoster Virus Infections. Molecules. 2021 Feb 20;26(4):1132.
5. Hostetler KY. Alkoxyalkyl prodrugs of acyclic nucleoside phosphonates enhance oral antiviral activity and reduce toxicity: Current state of the art. Antiviral Research. 2009 May;82(2):A84–98.
6. Mendel DB, Barkhimer DB, Chen MS. Biochemical basis for increased susceptibility to Cidofovir of herpes simplex viruses with altered or deficient thymidine kinase activity. Antimicrobial Agents and Chemotherapy. 1995 Sep;39(9):2120–2.
7. Xiong X, Smith JL, Chen MS. Effect of incorporation of cidofovir into DNA by human cytomegalovirus DNA polymerase on DNA elongation. Antimicrobial Agents and Chemotherapy. 1997 Mar;41(3):594–9.
8. Xiong X, Smith JL, Kim C, Huang E, Chen MS. Kinetic analysis of the interaction of cidofovir diphosphate with human cytomegalovirus DNA polymerase. Biochemical Pharmacology. 1996 Jun;51(11):1563–7.
9. Andrei G, Topalis D, De Schutter T, Snoeck R. Insights into the mechanism of action of cidofovir and other acyclic nucleoside phosphonates against polyoma- and papillomaviruses and non-viral induced neoplasia. Antiviral Research [Internet]. 2015 Feb 1 [cited 2021 Jan 22];114:21–46. Available from: https://www.sciencedirect.com/science/article/pii/S0166354214003052
10. UpToDate [Internet]. http://www.uptodate.com. [cited 2022 Jul 28]. Available from: https://www.uptodate.com/contents/brincidofovir-united-states-availability-limited-to-strategic-national-stockpile-distribution-drug-information?topicRef=8348&source=see_link
11. Williams-Aziz SL, Hartline CB, Harden EA, Daily SL, Prichard MN, Kushner NL, et al. Comparative Activities of Lipid Esters of Cidofovir and Cyclic Cidofovir against Replication of Herpesviruses In Vitro. Antimicrobial Agents and Chemotherapy. 2005 Sep;49(9):3724–33.
12. Andrei G, Snoeck R, Schols D, Goubau P, Desmyter J, Clercq E. Comparative activity of selected antiviral compounds against clinical isolates of human cytomegalovirus. European Journal of Clinical Microbiology & Infectious Diseases. 1991 Dec;10(12):1026–33.
13. Friedrichs C, Neyts J, Gaspar G, Clercq E, Wutzler P. Evaluation of antiviral activity against human herpesvirus 8 (HHV-8) and Epstein–Barr virus (EBV) by a quantitative real-time PCR assay. Antiviral Research. 2004 Jun;62(3):121–3.
14. Coen N, Duraffour S, Naesens L, Krečmerová M, Van den Oord J, Snoeck R, et al. Evaluation of Novel Acyclic Nucleoside Phosphonates against Human and Animal Gammaherpesviruses Revealed an Altered Metabolism of Cyclic Prodrugs upon Epstein-Barr Virus Reactivation in P3HR-1 Cells. Journal of Virology. 2013 Nov 15;87(22):12422–32.
15. Shigeta S, Konno K, Baba M, Yokota T, De Clercq E. Comparative Inhibitory Effects of Nucleoside Analogues on Different Clinical Isolates of Human Cytomegalovirus In Vitro. Journal of Infectious Diseases. 1991 Feb 1;163(2):270–5.
16. Neyts J, Snoeck R, Wutzler P, Cushman M, Klöcking R, Helbig B, et al. Poly(Hydroxy) Carboxylates as Selective Inhibitors of Cytomegalovirus and Herpes Simplex Virus Replication. Antiviral Chemistry and Chemotherapy. 1992 Aug;3(4):215–22.

17. Snoeck R, Sakuma T, De Clercq E, Rosenberg I, Holy A. (S)-1-(3-hydroxy-2-phosphonylmethoxypropyl)cytosine, a potent and selective inhibitor of human cytomegalovirus replication. Antimicrobial Agents and Chemotherapy. 1988 Dec 1;32(12):1839–44.
18. Beadle JR, Hartline C, Aldern KA, Rodriguez N, Harden E, Kern ER, et al. Alkoxyalkyl Esters of Cidofovir and Cyclic Cidofovir Exhibit Multiple-Log Enhancement of Antiviral Activity against Cytomegalovirus and Herpesvirus Replication In Vitro. Antimicrobial Agents and Chemotherapy. 2002 Aug;46(8):2381–6.
19. De Clercq E, Naesens L, De Bolle L, Schols D, Zhang Y, Neyts J. Antiviral agents active against human herpesviruses HHV-6, HHV-7 and HHV-8. Reviews in Medical Virology. 2001;11(6):381–95.
20. Tollefson AE, Spencer JF, Ying B, Buller RML, Wold WSM, Toth K. Cidofovir and brincidofovir reduce the pathology caused by systemic infection with human type 5 adenovirus in immunosuppressed Syrian hamsters, while ribavirin is largely ineffective in this model. Antiviral Research. 2014 Dec;112:38–46.
21. Andrei G, Snoeck R, Piette J, Delvenne P, De Clercq E. Antiproliferative effects of acyclic nucleoside phosphonates on human papillomavirus (HPV)-harboring cell lines compared with HPV-negative cell lines. Oncol Res. 1998;10(10):523-31. PMID: 10338155.
22. Duan J, Paris W, De Marte J, Roopchand D, Fleet T-L, Cordingley MG. Topical effects of cidofovir on cutaneous rabbit warts: treatment regimen and inoculum dependence. Antiviral Research. 2000 May;46(2):135–44.
23. Hostetler KY. Enhanced antiproliferative effects of alkoxyalkyl esters of cidofovir in human cervical cancer cells in vitro. Molecular Cancer Therapeutics. 2006 Jan 1;5(1):156–9.
24. Andrei G, Snoeck R, Vandeputte M, De Clercq E. Activities of various compounds against murine and primate polyomaviruses. Antimicrobial Agents and Chemotherapy. 1997 Mar;41(3):587–93.
25. Topalis D, Lebeau I, Krečmerová M, Andrei G, Snoeck R. Activities of Different Classes of Acyclic Nucleoside Phosphonates against BK Virus in Primary Human Renal Cells. Antimicrobial Agents and Chemotherapy. 2011 May;55(5):1961–7.
26. Andrei G, Fiten P, Froeyen M, Clercq ED, Opdenakker G, Snoeck R. DNA Polymerase Mutations in Drug-Resistant Herpes Simplex Virus Mutants Determine In Vivo Neurovirulence and Drug-Enzyme Interactions. Antiviral Therapy. 2007 Jul;12(5):719–32.
27. Andrei G, Snoeck R, Fiten P, De Clercq E, Opdenakker G, Esnouf R. Resistance of herpes simplex virus type 1 against different phosphonylmethoxyalkyl derivatives of purines and pyrimidines due to specific mutations in the viral DNA polymerase gene. Journal of General Virology. 2000 Mar 1;81(3):639–48.
28. Gilbert C, Boivin G. Human Cytomegalovirus Resistance to Antiviral Drugs. Antimicrobial Agents and Chemotherapy. 2005 Feb 23;49(3):873–83.
29. James SH, Price NB, Hartline CB, Lanier ER, Prichard MN. Selection and Recombinant Phenotyping of a Novel CMX001 and Cidofovir Resistance Mutation in Human Cytomegalovirus. Antimicrobial Agents and Chemotherapy. 2013 Jul;57(7):3321–5.
30. Scott GM, Isaacs MA, Zeng F, Kesson AM, Rawlinson WD. Cytomegalovirus antiviral resistance associated with treatment induced UL97 (protein kinase) and UL54 (DNA polymerase) mutations. Journal of Medical Virology. 2004;74(1):85–93.
31. Springer KL, Chou S, Li S, Giller RH, Quinones R, Shira JE, et al. How Evolution of Mutations Conferring Drug Resistance Affects Viral Dynamics and Clinical Outcomes of Cytomegalovirus-Infected Hematopoietic Cell Transplant Recipients. Journal of Clinical Microbiology. 2005 Jan;43(1):208–13.
32. Hantz S, Cotin S, Borst E, Couvreux A, Salmier A, Garrigue I, et al. Novel DNA polymerase mutations conferring cytomegalovirus resistance: Input of BAC-recombinant phenotyping and 3D model. Antiviral Research. 2013 Apr;98(1):130–4.
33. Cundy KC, Petty BG, Flaherty J, Fisher PE, Polis MA, Wachsman M, et al. Clinical pharmacokinetics of cidofovir in human immunodeficiency virus-infected patients. Antimicrobial Agents and Chemotherapy. 1995 Jun;39(6):1247–52.

34. Painter W, Robertson A, Trost LC, Godkin S, Lampert B, Painter G. First Pharmacokinetic and Safety Study in Humans of the Novel Lipid Antiviral Conjugate CMX001, a Broad-Spectrum Oral Drug Active against Double-Stranded DNA Viruses. Antimicrobial Agents and Chemotherapy. 2012 May;56(5):2726–34.
35. HIGHLIGHTS OF PRESCRIBING INFORMATION These highlights do not include all the information needed to use TEMBEXA ® safely and effectively. See full prescribing information for TEMBEXA. TEMBEXA (brincidofovir) tablets, for oral use TEMBEXA (brincidofovir) oral suspension Initial U.S. Approval: 1996 WARNING: INCREASED RISK FOR MORTALITY WHEN USED FOR LONGER DURATION See full prescribing information for complete boxed warning [Internet]. [cited 2022 Jul 28]. Available from: https://www.chimerix.com/wp-content/uploads/2021/07/USPI-TEMBEXA-Label-July-2021.pdf.
36. Cundy KC. Clinical Pharmacokinetics of the Antiviral Nucleotide Analogues Cidofovir and Adefovir. Clinical Pharmacokinetics. 1999;36(2):127–43.
37. Plosker GL, Noble S. Cidofovir. Drugs. 1999;58(2):325–45.
38. Naiman AN, Roger G, Gagnieu M-C, Bordenave J, Mathaut S, Ayari S, et al. Cidofovir Plasma Assays after Local Injection in Respiratory Papillomatosis. The Laryngoscope. 2004 Jul;114(7):1151–6.
39. Aitken SL, Zhou J, Ghantoji SS, Kontoyiannis DP, Jones RB, Tam VH, et al. Pharmacokinetics and safety of intravesicular cidofovir in allogeneic HSCT recipients. Journal of Antimicrobial Chemotherapy. 2015 Nov 26;71(3):727–30.
40. Ciesla SL, Trahan J, Wan WBrad, Beadle JR, Aldern KA, Painter GR, et al. Esterification of cidofovir with alkoxyalkanols increases oral bioavailability and diminishes drug accumulation in kidney. Antiviral Research. 2003 Aug;59(3):163–71.
41. HIGHLIGHTS OF PRESCRIBING INFORMATION These highlights do not include all the information needed to use TEMBEXA ® safely and effectively. See full prescribing information for TEMBEXA. TEMBEXA (brincidofovir) tablets, for oral use TEMBEXA (brincidofovir) oral suspension Initial U.S. Approval: 1996 WARNING: INCREASED RISK FOR MORTALITY WHEN USED FOR LONGER DURATION See full prescribing information for complete boxed warning [Internet]. Available from: https://www.chimerix.com/wp-content/uploads/2021/07/USPI-TEMBEXA-Label-July-2021.pdf.
42. Lea AP, Bryson HM. Cidofovir. Drugs. 1996 Aug;52(2):225–30.
43. Uwai Y, Ida H, Tsuji Y, Katsura T, Inui K. Renal Transport of Adefovir, Cidofovir, and Tenofovir by SLC22A Family Members (hOAT1, hOAT3, and hOCT2). Pharmaceutical Research. 2007 Feb 15;24(4):811–5.
44. Cihlar T, Lin DC, Pritchard JB, Fuller MD, Mendel DB, Sweet DH. The Antiviral Nucleotide Analogs Cidofovir and Adefovir Are Novel Substrates for Human and Rat Renal Organic Anion Transporter 1. Molecular Pharmacology. 1999 Sep 1;56(3):570–80.
45. Lalezari JP, Stagg RJ, Jaffe HS, Hitchcock MJM, Drew WL. A Preclinical and Clinical Overview of the Nucleotide-Based Antiviral Agent Cidofovir (HPMPC). Advances in Experimental Medicine and Biology. 1996;105–15.
46. Lalezari JP, Drew WL, Glutzer E, James C, Miner D, Flaherty J, et al. (S)-1-[3-Hydroxy-2-(Phosphonylmethoxy)propyl]cytosine (Cidofovir): Results of a Phase I/II Study of a Novel Antiviral Nucleotide Analogue. Journal of Infectious Diseases. 1995 Apr 1;171(4):788–96.
47. Polis MA, Spooner KM, Baird BF, Manischewitz JF, Jaffe HS, Fisher PE, et al. Anticytomegaloviral activity and safety of cidofovir in patients with human immunodeficiency virus infection and cytomegalovirus viruria. Antimicrobial Agents and Chemotherapy. 1995 Apr;39(4):882–6.
48. Bainbridge JWB, Raina J, Shah SM, Ainsworth J, Pinching AJ. Ocular complications of intravenous cidofovir for cytomegalovirus retinitis in patients with AIDS. Eye. 1999 May;13(3):353–6.
49. Cidofovir [package insert]. Foster City, CA: Gilead Sciences, Inc.; 2010.

50. Rhegmatogenous Retinal Detachment in Patients With Cytomegalovirus Retinitis: The Foscarnet-Ganciclovir Cytomegalovirus Retinitis Trial. American Journal of Ophthalmology. 1997 Jul;124(1):61–70.
51. Lalezari JP. Intravenous Cidofovir for Peripheral Cytomegalovirus Retinitis in Patients with AIDS. Annals of Internal Medicine. 1997 Feb 15;126(4):257.
52. Bienvenu B, Martinez F, Devergie A, Rybojad M, Rivet J, Bellenger P, et al. Topical use of cidofovir induced acute renal failure. Transplantation. 2002 Feb;73(4):661–2.
53. Santilli CM, Koozekanani D, Armbrust KR. Title. RETINAL Cases & Brief Reports. 2020 Nov 20;Publish Ahead of Print.
54. UpToDate [Internet]. http://www.uptodate.com. Available from: https://www.uptodate.com/contents/cidofovir-an-overview Accessed, April 18, 2022.
55. https://www.symbiopharma.com/pipeline_e/05.html. Accessed, July 10, 2022.
56. Ogata M, Uchida N, Fukuda T, Ikegame K, Kamimura T, Onizuka M, et al. Clinical practice recommendations for the diagnosis and management of human herpesvirus-6B encephalitis after allogeneic hematopoietic stem cell transplantation: the Japan Society for Hematopoietic Cell Transplantation. Bone Marrow Transplantation. 2019 Nov 19;55(6):1004–13.
57. Brody S, Humphreys M, Gambertoglio J, Schoenfeld P, Cundy K, Aweeka F. Pharmacokinetics of cidofovir in renal insufficiency and in continuous ambulatory peritoneal dialysis or high-flux hemodialysis. Clinical Pharmacology & Therapeutics. 1999 Jan;65(1):21–8.

Antiviral Agents: Ganciclovir/ Valganciclovir

Eric Gaskill

Abstract Ganciclovir was the first antiviral drug noted to have activity against cytomegalovirus (CMV). It has an oral analog, valganciclovir, an L-valine ester of ganciclovir that improves oral bioavailability substantially and allows for equivalent exposure through the oral route. Ganciclovir and valganciclovir display broad anti-herpesvirus activity and represent common options utilized to treat CMV and HHV-6. Both require phosphorylation to an active form inside the virus, with the activated form inhibiting viral DNA polymerase.

Keywords Antiviral · Ganciclovir · Valganciclovir · Cytomegalovirus · CMV · HHV-6

1 Background

Ganciclovir was FDA approved in 1989 and is the first antiviral drug noted to have activity against CMV [1]. Ganciclovir is an acyclic analog of deoxyguanosine with 2-hydroxyl moieties that make it a non-obligate chain terminator. Its mechanism of action first requires phosphorylation by a viral kinase (pUL97 in CMV or similar kinases in other herpesviruses), and then additional phosphorylations which can be performed via human guanylate kinase or dGMP kinase to convert to triphosphate form. After forming ganciclovir triphosphate, the compound competitively inhibits viral DNA polymerase through incorporation into elongating viral DNA [2, 3, 4]. Ganciclovir inhibits DNA polymerase by competitively inhibiting the incorporation

E. Gaskill (✉)
Bone Marrow Transplant/Cellular Immunotherapy Pharmacy Specialist, Moffitt Cancer Center, Florida, USA
e-mail: Eric.gaskill@moffitt.org

C. Somboonwit et al. (eds.), *Global Virology IV: Viral Disease Diagnosis and Treatment Delivery in the 21st Century*,
https://doi.org/10.1007/978-3-031-57369-9_19

of deoxyguanine triphosphate by the viral DNA polymerase and by incorporation into viral DNA, which inhibits chain elongation through its abnormal acyclic structure [3]. This inhibition of viral DNA polymerase is incomplete, and short segments of viral DNA will continue to be synthesized under ganciclovir inhibition [4]. Ganciclovir uptake into herpesvirus-infected cells is increased compared to healthy cells, contributing to its intracellular concentrations and efficacy inside infected cells [5]. Since the first phosphorylation step for ganciclovir requires a viral enzyme catalyzing phosphorylation, it specifically activates in virally infected cells and maintains higher concentrations in them [4]. Ganciclovir is approved for treating CMV retinitis and CMV prophylaxis in transplant patients but has also been used in the treatment of other herpesviruses.

Valganciclovir is an L-valine ester prodrug of ganciclovir with tenfold greater bioavailability than ganciclovir, allowing for comparable drug exposure from an oral route to intravenous ganciclovir [2, 3, 6]. The valine ester is transported from the intestines to the bloodstream and rapidly converted to ganciclovir by liver and intestinal esterases [3]. This prodrug formulation improves the bioavailability of ganciclovir from 6% to 60% in adults [3, 7]. This improved bioavailability allows for similar ganciclovir exposure via taking valganciclovir by mouth to IV ganciclovir with the added convenience of oral dosing [8, 9]. Valganciclovir is FDA approved for the treatment of AIDS-related CMV retinitis and CMV prophylaxis after solid organ transplant and has been utilized for wide-ranging indications for treatment and prophylaxis of many herpesviruses. The structures of these two compounds are shown in Fig. 1.

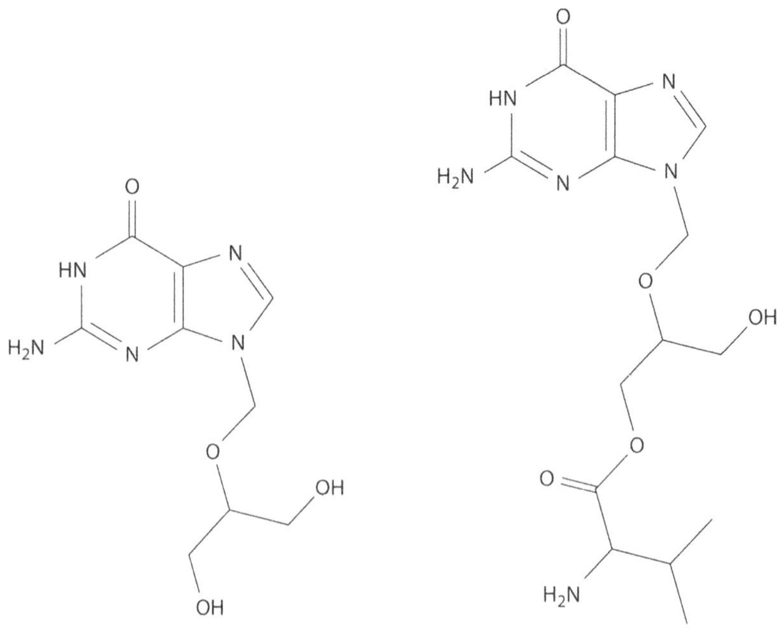

Ganciclovir **Valganciclovir**

Fig. 1 Ganciclovir and valganciclovir chemical structures. Valganciclovir is characterized by the addition of the valine ester to increase bioavailability [1, 2]

2 Spectrum of Activity: In Vitro

Ganciclovir displays activity against several herpesviruses, but no significant clinical utility for other viruses. See Table 1 for general activity, with in-depth discussion of herpesviruses thereafter.

2.1 *Human Herpesviruses*

2.1.1 Herpes Simplex Virus (HSV-1, HSV-2)

In HSV, ganciclovir is first phosphorylated by HSV thymidine kinase, encoded by UL-23 [13]. Resistance has been documented in thymidine kinase–deficient HSV strains and is the most common manifestation of acyclovir and ganciclovir

Table 1 Spectrum of activity

Viral family	Species susceptible	
Adenoviridae	Generally, not active [4] AdV-2 EC_{50} 5.4 μM [10] AdV-19 EC_{50} 7.2 μM [10] AdV-22 EC_{50} 4.5 μM [10] Subgenus D potentially susceptible with much higher ID_{50} compared to herpesviruses [5]	
Herpesviridae	HSV-1	Excellent activity, more potent than acyclovir [4, 5] EC_{50} 0.0025–10 μM [10]
	HSV-2	Excellent activity, more potent than acyclovir [3, 4, 5] 0.016–2.5 μM [10]
	VZV	Excellent activity [3, 4] EC_{50} 0.52–1.3 μM [10]
	EBV	Has activity during active stage, none during latency [3, 5, 11, 12] EC_{50} 4.3 +/− 3.5 μM [5]
	CMV	Excellent activity [4, 5] EC_{50} CMV AD169: 0.04–6.7 μM [10] EC_{50} 0.1–1.6 μg/mL [3]
	HHV-6	Has activity [3, 4, 13] HHV-6A: EC_{50} 2.6–31.9 μM [10] mean 17 [14] HHV-6B: EC_{50} 5.2–68.6 μM [10, 15] mean 4.5 [14]
	HHV-7	Not good activity [13] EC_{50} 61 μM [13]
	HHV-8	Not good activity [13] EC_{50} 0.4–23 μM [10]
	B virus	*Cercopithecine herpesvirus* 1–9 μg/mL [16]
Iridoviridae	No data	
Poxviridae	No activity [10]	
Papillomavirus	No activity [4]	
Polyomavirus	No data	
Influenza	No activity [4]	

resistance [4, 17]. Given the shared thymidine kinase–related resistance mechanism, ganciclovir does not offer much advantage in acyclovir-resistant HSV-1 [11].

2.1.2 Varicella (VZV)

In varicella, ganciclovir is first phosphorylated by VZV thymidine kinase, encoded by ORF-36 [13]. Ganciclovir has good activity against VZV but has no significant advantage over acyclovir derivates due to its increased toxicity.

2.1.3 Epstein–Barr Virus (EBV [HHV-4])

The EBV thymidine kinase, encoded by BXLF-1, possibly participates in the phosphorylation step of ganciclovir in EBV-infected cells, but this enzyme seems to have low affinity for ganciclovir [13]. Latent EBV-infected cells do not typically express the thymidine kinase enzyme, making them non-susceptible to ganciclovir at least while latent [11]. Ganciclovir has been utilized in limited cases of severe active EBV infections [12, 18, 19]. Antiviral prophylaxis has been associated with reduced incidence of EBV-associated post-transplant lymphoproliferative disorders (PTLD) and primary EBV infections, indicating potential clinical benefit [20, 21].

2.1.4 Cytomegalovirus (CMV [HHV-5])

Ganciclovir is activated to a high degree in CMV-infected cells via phosphorylation by the CMV UL97 kinase [3]. Most historical data in vitro utilize testing against CMV strain AD169, in which ganciclovir exhibits EC_{50} values ranging from 0.04 to 6.7 μM [10]. This is a substantially better activity than many other available antiviral compounds such as acyclovir analogs [10], and only antivirals such as letermovir and brincidofovir display more potency in vitro [10].

2.1.5 Other Herpes Viruses [HHV6 – HHV8]

In HHV6, ganciclovir seems to be first phosphorylated by HHV-6 U69 protein kinase, with a lower efficiency than its equivalent enzyme in CMV [13]. An equivalent enzyme also seems to exist in HHV-7, encoded by U69 [13]. The HHV8 thymidine kinase appears to have a low affinity for ganciclovir, leading to decreased activation in infected cells [17]. Ganciclovir seems to have a lower EC_{50} for HHV-6B than HHV-6A in some in vitro testing, but clinical implications are unclear as activity is significantly variable [13].

3 Resistance

3.1 General Herpesviruses

Ganciclovir resistance is not as commonly documented in HSV-1, HSV-2, and VZV, but generally overlaps with acyclovir resistance mutations as they occur most commonly in the viral enzyme that is involved in the initial phosphorylation step of drug activation. If resistance in these viruses is noted, typically, ganciclovir is not a useful alternative. There is little data on ganciclovir resistance in EBV, HHV-7, and HHV-8.

3.2 Cytomegalovirus (CMV)

Ganciclovir is primarily used for CMV infections, so it has the best resistance data available for this virus. The primary mechanism of resistance to ganciclovir and valganciclovir in CMV is through mutations of the UL97 gene encoding pUL97 kinase. Mutations in two major regions render it unable to phosphorylate ganciclovir, preventing activation [3]. UL97 mutations account for 90–95% of ganciclovir resistance in CMV [3, 22]. Mutations in UL97 more often confer lower-level resistance compared to mutations in UL54 [3]. Ganciclovir maintains more potency in CMV compared to acyclovir due to its increased phosphorylation by UL97 in CMV, which has poor affinity for acyclovir and results in low intracellular acyclovir triphosphate levels [4]. Resistance to ganciclovir generally appears in a different area of the UL97 gene than associated with maribavir resistance; thus, cross-resistance is uncommon [17]. The most commonly associated mutations of UL97 occur at codons 460, 520, 592, 594, 595, and 603 [22]. Resistance is found relatively frequently in severely immunocompromised patients (5–10% of solid organ transplant (SOT) recipients on prolonged antiviral therapy) [17, 23]. Low-level resistance (< fivefold) such as M460I, C592, and L595W are potentially overcome via dose escalation to 7.5–10 mg/kg Q12 or 1.5-2x dose valganciclovir, but anything beyond that would demand a change to an alternative agent [11, 22, 24, 25].

Mutations in UL54 (CMV DNA polymerase) offer another site for ganciclovir resistance. Mutations in UL54 may confer cross-resistance to other antivirals such as foscarnet and cidofovir, as their target for activity is the DNA polymerase itself [3]. Mutations in UL54 often confer high-level resistance to ganciclovir [3]. These mutations sometimes confer cross-resistance to cidofovir and, less commonly, foscarnet as they are clustered through conserved regions encoding exonuclease domains I through III and polymerization domains I through III [11, 22]. Generally, if a UL54 resistance mutation is detected, ganciclovir should be switched for a new appropriate therapy, or if no option is available, combination therapies can be considered [25].

Table 2 and 3 shows many notable mutations in UL97 and UL54 that confer ganciclovir resistance.

Table 2 Mutations in UL97

Significant mutations in pUL97	Ganciclovir EC_{50} ratio (mutant EC_{50}/wtEC_{50})	Reference
F342S	7.8	[26]
K355del	16	[26]
V356G	5.5	[26]
L405p	2.5	[8]
D456N	12	[27]
M460I/V/T	5–9.3	[8, 25]
V466G	3.5–11	[8, 26]
C480F	2.3	[25, 28]
C480R	9	[27]
K493N	4.1	[28]
K513T	4.9	[28]
H520Q	10	[8, 25]
P521L	17	[26]
C592G	2.2–4.9	[25, 28]
Del(591–594)	3–10	[8]
Del(591–607)	6.2	[8]
C592G	2.9	[8]
A594T	2.7	[25]
A594V	8.3	[8]
A594G	13.5	[8]
A594E	3	[8]
L595S/F/W	9.2–15.7	[8, 25]
Del(595–603)	8.4	[8]
E596G	2.3	[8]
L600	1.9	[8]
Del(601–603)	11	[8]
C603W/R/S	1.9–8.3	[8, 25]
C607F	1.9	[8]
C607Y	12.5	[8]
Y617del	10	[27]
F342Y + H411Y	5.9	[29]

bold = common

3.3 HHV6

Ganciclovir has been used clinically in HHV-6 infections, but some debate exists if it is an equal empiric therapeutic choice to foscarnet and more clinical data is needed to judge. Mutations have been noted at codon 318 in HHV-6 UL69 protein kinase likely involved in ganciclovir phosphorylation in in vitro testing of HHV-6A that can confer high-level resistance (24-fold) [30]. A DNA polymerase mutation at A961V was also noted that did not seem to confer ganciclovir resistance [30]. Five

Table 3 Mutations in UL54

Significant mutations in UL54	Ganciclovir EC_{50} ratio (mutant EC_{50}/wtEC_{50})	Reference
D301N	2.6	[8, 25]
N408D/K	4.2–4.9	[8, 25]
N410K	2.9	[8, 25]
F412C/V/A/E	4.2–6.5	[8, 25]
D413A	6.5	[8, 25]
L501I	6	[8, 25]
T503I	2.9	[8, 25]
L501I	6	[8, 25]
K513E/N	5–6	[8, 25]
L516R	2.1	[8, 25]
T521I	3.1	[8]
P522A/S	3–3.1	[8, 25]
L545S	3.5	[8, 25]
D588N	3.8	[8, 25]
E756K	3.5	[8, 25]
L776M	2.5	[8, 25]
V787L	2.4	[8, 25]
L802M	1.1–3.5	[8]
A809V	2.6	[8, 25]
V812	2.5	[8, 25]
T813	2.5	[8, 25]
T821I	4.5	[8, 25]
A834P	5.4	[8, 25]
T838A	1.8	[8, 25]
G841A	3.2	[8, 25]
Del(981–982)	8.3	[8]
A987G	5.3	[8, 25]

samples of HHV-6B were tested in the same experiment without findings of this same mutation [30]. A case report of resistant HHV-6B has been noted, with a combination of mutations in U38 (HHV6-6 DNA polymerase) and/or U69 that conferred a >250-fold increased EC50 to ganciclovir [31]. Overall, ganciclovir is less efficiently phosphorylated in HHV-6 infected cells than CMV, leading to comparatively less efficacy and higher EC_{50} values [31, 15]. More information about HHV-6 resistance is needed to inform clinical decisions.

4 Pharmacokinetics(PK)/Pharmacodynamics(PD)/Dosing

The pharmacokinetic, pharmacodynamic, and dosing parameters for ganciclovir and valganciclovir are provided in Tables 4 and 5.

Table 4 Ganciclovir

Dosage	Treatment/induction of CMV: 5 mg/kg every 12 h IV [1] Maintenance: 5 mg/kg daily IV [1] Higher doses 7.5–10 mg/kg every 12 h IV studied in low-level resistance [25]		
Renal Dysfunction [1]	**Creatinine clearance (mL/min)**	**Induction dosing**	**Maintenance dosing**
	>70	5 mg/kg Q12	5 mg/kg daily
	50–69	2.5 mg/kg Q12	2.5 mg/kg daily
	25–49	2.5 mg/kg daily	1.25 mg/kg daily
	10–24	1.25 mg/kg daily	0.625 mg/kg daily
	<10	1.25 mg/kg 3 times/week post-HD	0.625 mg/kg 3 times/week post-HD
	Significantly removed by dialysis ~50% per session [5, 32]		
Hepatic dysfunction	No dose adjustments recommended, not studied		
Interactions	Zidovudine, didanosine reduce ganciclovir AUC, may increase toxicity [1]		
	Probenecid increases ganciclovir AUC 53 +/− 91% [1]		
	Imipenem-cilastatin increased seizure risk [1]		
	Maribavir inhibits activation of ganciclovir, reducing efficacy [25]		
	Mycophenolate may increase hematologic toxicity of ganciclovir [1]		
	Trimethoprim reduces ganciclovir clearance by 16.3% [1]		
	Nephrotoxic medications may increase ganciclovir toxicity by increasing exposure [1]		
Population PK	No PK studies in obesity, but as volume of distribution (Vd) is low and the drug is primarily excreted in urine unchanged, consideration for dosing with adjusted or ideal body weight is recommended by some experts along with appropriate renal adjustment		
	High variability in critically ill and role of PK monitoring debated [33]		
PK [1, 6]	**IV Ganciclovir: General peak concentration after 5 mg/kg dose over 1 h = 8.3 μg/mL** [4] **9.46 μg/mL** [1]		
	Typical troughs after 5 mg/kg dosing 12 h later are <0.5 μg/mL [4] Typical area under curve (AUC) 26.5 mcg h/L		
	Oral ganciclovir: In one study of 1000 mg given TID, Cmax 1.2 μg/mL and min 0.2 μg/mL achieved [1]		
Absorption	Oral ganciclovir poorly absorbed, 6% bioavailable [1]		
Distribution	Vd 33-45 L [5]		
	Minimal plasma protein binding 1–2% [5]		
	One study showed that at a serum concentration of 2.2 μg/mL (3.5–3.8 h after an IV dose of 2.5 mg/kg in adults), a cerebrospinal fluid concentration of 0.7 μg/mL was obtained [1, 34] CSF concentrations 31–67% of plasma in IV ganciclovir [5, 34] Subretinal fluid exceeded plasma concentrations [5]		

(continued)

Table 4 (continued)

Metabolism	Cmax of ganciclovir occurring 2–3 h after administration, half-life 2 to 4 h [3]
Elimination	Almost 100% in urine with elimination t1/2 of 2–4 h in normal renal function [5] >90% of administered dose unchanged in urine [34] Half-life increases to 30–40 h in severe renal dysfunction [5] T1/2 13 h in intravitreal fluid after direct administration [5]

Table 5 Valganciclovir

Dosage	Treatment of CMV: 900 mg twice daily [6] Prophylaxis/maintenance: 900 mg daily [6]		
Renal dysfunction [5, 23]	**Creatinine clearance (mL/min)**	**Induction**	**Maintenance**
	≥60	900 mg Q12 h	900 mg daily
	40–59	450 mg Q12 hs	450 mg daily
	25–39	450 mg daily	450 mg every other day
	10–24	450 mg every other day	450 mg twice weekly
	<10	200 mg 3 times weekly post HD	100 mg 3 times weekly post HD
	Significantly removed by dialysis [5]		
Hepatic dysfunction	See ganciclovir		
Interactions	See ganciclovir		
Population PK	Valganciclovir Cmax is 40% lower than ganciclovir [35]		
PK	Cmax 5.61 μg/mL [6]		
	Comparable AUC from 900 mg PO to ganciclovir 5 mg/kg = 29.1 mcg h/L [6]		
Absorption	Bioavailability 59.4% [9] Cmax is 40% lower than ganciclovir but AUC is similar [35] High fat meals increase the AUC by 24–56% [3] Rapidly converted to active moiety ganciclovir, no other metabolites detected [9]		
Distribution	See ganciclovir		
Metabolism	The mean half-life of valganciclovir is <1 h after administration with rapid conversion from prodrug, subsequent ganciclovir active moeity has a half-life 4.08 h [2] The mean half-life of prodrug valganciclovir is <1 h after administration, with Cmax of ganciclovir occurring 2–3 h after administration [3]		
Elimination	Almost 100% in urine with elimination of 2–4 h in normal renal function [5] >90% of administered dose unchanged in urine [34] Half-life increases to 30–40 h in severe renal dysfunction [5] Half-life 13 h in intravitreal fluid after direct administration [5]		

5 Adverse Events

The primary limiting toxicity of ganciclovir and valganciclovir is hematologic, with neutropenia and thrombocytopenia being common (10–73% of patients, more commonly in hematopoietic cell transplant (HCT) than SOT) and increasing in incidence depending on the intensity and length of treatment [2, 36]. Other adverse events occasionally reported include abdominal pain, diarrhea, nausea, increased infection risk, less commonly renal and hepatic dysfunction, and rarely CNS adverse events [1, 5, 37].

5.1 Hematologic Adverse Reactions

In ganciclovir prophylaxis trials in allogeneic HCT, the most common adverse event was neutropenia, which occurred in 27–71% of patients at a median of 35 days [38]. For treatment of CMV in SOT, a large trial of valganciclovir vs ganciclovir found rates of leukopenia 11.6% vs 12.1% and anemia 12.2% vs 7% [7]. Following a 5 mg/kg dose of ganciclovir, peak plasma levels exceed the IC_{50} for human bone marrow colony-forming cells in vitro, indicating significant potential for bone marrow suppression during therapy [5]. In vitro studies have found that ganciclovir inhibits bone marrow-derived colony-forming cells at an EC_{50} of 0.028–0.7 μg/mL, a concentration achieved during typical ganciclovir therapy [1]. Neutropenia is associated with ganciclovir dose, often occurring before a total cumulative dose of 200 mg/kg has been given [5]. Total incidence of neutropenia is around 40%, with risk not clearly predicted by a neutrophil count at the start of therapy [5]. Suppression usually develops in the first or second week of treatment [35]. Although the typical method of ganciclovir activation involves viral kinases for the first step, this phosphorylation takes place to a smaller degree in healthy cells and seemingly in particular in the bone marrow [17]. Studies analyzing risk factors for neutropenia on ganciclovir have associated this adverse event with creatinine clearance <50 mL/min, serum creatinine ≥2 mg/dL, bilirubin ≥6 mg/dL, and low marrow cellularity [36]. High viral loads, WBC < 6000 cells/mm^3, ANC < 3000 cells/mm^3, and ganciclovir dosage >10 mg/kg/day are also associated [36]. Rare cases of severe agranulocytosis have been noted [39]. One study attempting to create a risk prediction model for neutropenia found that the most relevant factors were baseline ANC < 3854 cells/mm^3, therapy ≥15 days, and HCT [36]. Pre-emptive use of filgrastim may help mitigate the hematologic toxicity of ganciclovir and generally is preferred compared to dose reduction in treating active CMV infections [25]. Reduced incidence of these adverse effects has been noted with oral ganciclovir, likely due to reduced exposure with poor absorption [4]. Certain drug combinations may contribute significantly to the risk of myelosuppression, as it was noted that 80% of patients on a combination of ganciclovir and zidovudine experienced significant toxicity [4].

Thrombocytopenia occurs in approximately 20% of patients, with increased risk in patients with initial platelet count <100,000 [5]. In older PK studies, neutropenia is potentially associated with 5 mg/kg Q8 dosing and peaks >12.76 μg/mL or troughs <0.25 μg/mL [34]. Another study associated thrombocytopenia with ganciclovir doses >12 mg/kg/day and CrCL <20 mL/min [37]. Overall exposure to ganciclovir seems to be associated with this adverse event, but specific PK goals to minimize risk have not been established.

5.2 Reproductive Toxicity/Carcinogenesis

Ganciclovir is a hazardous agent and has animal data indicating that it is potentially teratogenic and carcinogenic and exhibits reproductive toxicity in males and females [1, 5]. In male animal studies, it is associated with aspermatogenesis and loss of fertility in mice and dogs [35, 1]. A study in renal transplant patients found slower recovery of sperm production in males on ganciclovir, but this improved after cessation of therapy [40]. In females, animal studies have shown decreased fertility and increased embryo lethality in mice at exposures slightly higher than human dosing. It is carcinogenic in mice at exposure levels correlated to that experienced by humans at therapeutic dosing [1]. This reproductive toxicity may have a longer duration than during the time of therapy [5]. It is recommended to avoid ganciclovir during pregnancy, avoid pregnancy while taking it, and consider the risks of longer-term therapy [9].

6 Treatment

6.1 CMV

Ganciclovir and valganciclovir are both utilized extensively in treating CMV viremia and disease in both HCT and SOT patients. Dosing between the patient populations has traditionally been similar, with an initial higher dose induction period for several weeks and a maintenance or prophylaxis period thereafter to ensure sufficient suppression of CMV. A comparison trial of valganciclovir and ganciclovir in SOT resulted in similar outcomes (cure rates at day 21 45.1% valganciclovir vs 48.4% ganciclovir), allowing for valganciclovir to become a preferred option in those able to take oral medications [7]. Adverse events between valganciclovir and ganciclovir were similar as well [7]. A trial in HCT patients found that valganciclovir and ganciclovir achieved similar efficacy [41]. A PK study found typically higher exposure with valganciclovir compared to ganciclovir in HCT in patients with GI GVHD grades 0–2. The single patient in the study with Grade III GI GVHD did not absorb oral valganciclovir sufficiently [42]. Another study examining valganciclovir PK in patients with stable GI GVHD of 300–1500 mL of diarrhea per day found that medication absorption was still relatively equivalent to IV ganciclovir but did not include patients with unstable diarrhea [43]. Pre-emptive treatment of CMV viremia with ganciclovir or valganciclovir has been shown to reduce incidence of CMV disease and has lowered CMV disease rates drastically when used appropriately [44].

6.2 CMV Retinitis

Ganciclovir has been utilized in the treatment of CMV retinitis, initially in HIV/AIDS when trials found it achieved comparable results to foscarnet in resolving infections [4, 45]. Relapse after treatment is very common, and maintenance therapy likely reduces this risk [4]. CMV retinitis is treated similarly in transplant patients, with the general recommendation to start a systemic therapy rather than a direct ocular therapy. Intravitreal therapy, often ganciclovir, may be used in conjunction with systemic treatment to achieve higher drug levels in the ocular compartment or when systemic therapy is not available [46]. Risks of intravitreal ganciclovir include vitreous hemorrhage, retinal detachment, and endophthalmitis [47].

6.3 HHV-6

Ganciclovir has been utilized in the treatment of HHV-6-related encephalitis with some success in case series and single arm trials. It is unclear what the appropriate treatment and treatment threshold to use ganciclovir in HHV-6 is without larger studies, but in some case series, treatment of HHV-6 that included ganciclovir as an option resulted in possible increased survival after HHV-6 infections [48, 49]. Cases of resistance to ganciclovir have been noted but are not common [31, 50]. One study comparing ganciclovir and foscarnet use in HHV-6 encephalitis indicated that foscarnet might be more effective, potentially due to increased CNS penetration, and thus their main finding was to ensure that sufficient dosage of either agent was used to achieve good CNS concentrations [51, 52].

7 Prophylaxis

7.1 CMV

In several early trials, IV ganciclovir was utilized as CMV prophylaxis for 120 days in HCT, heart, and liver transplantation with success in reducing the risk of CMV disease [4]. In a HCT trial of patients who had CMV noted in bronchoalveolar lavage fluid, ganciclovir 5 mg/kg twice daily given for 2 weeks then reduced to a maintenance dose was effective at reducing the risk of CMV pneumonia or death from 14% to 5% by week 120 [49]. In a meta-analysis of allo-HCT CMV prophylaxis trials, the only medication associated with a lessened risk of CMV disease was ganciclovir. [38] This benefit was largely derived from studies before the pre-emptive era of CMV treatment and reflect the benefit of usage of ganciclovir vs placebo [38]. Ganciclovir was not shown to improve survival [38]. Ganciclovir prevention in SOT generally confers reduced risk of CMV disease, but no significant benefit in terms of graft rejection and survival [44].

Valganciclovir reduces and delays CMV disease in renal-transplanted recipients and is associated with reduced biopsy-proven rejection [44]. In heart transplant recipients, ganciclovir effectively reduces CMV disease [44]. In lung transplant recipients, it reduces CMV disease [44]. General recommendation includes consideration of long-term valganciclovir prophylaxis in SOT, and 90 days of prophylaxis should be considered in high-risk renal transplant [44]. Equivalent efficacy was found between oral ganciclovir and valganciclovir in a large study of CMV-mismatched transplant recipients, but there was a noted uptick in CMV disease in liver transplant patients on valganciclovir compared to other transplant types [23, 53]. Some studies have utilized lower dose valganciclovir prophylaxis (450 mg daily) in attempts to reduce costs and improve tolerability, with outcomes potentially similar in intermediate-risk groups [23]. A trial in renal transplants examining the duration of valganciclovir of 3 months vs. 6 months found lessened viremia in the longer prophylaxis group and a trend toward less graft rejection. Hematologic adverse events were increased in the longer prophylaxis group (leukopenia 38% at day 200 vs. 26% at day 100) [54]. For SOT recipients, depending on risk and organ, prophylaxis against CMV with valganciclovir is recommended for at least 3–6 months and often longer in lung transplants [23].

8 Special Considerations

8.1 Combination Therapy

Combination therapy with other therapies has been attempted in cases and case series, most typically with foscarnet. Combined with foscarnet, in vitro effects against HSV1 and HSV2 are synergistic, reducing virus titers by over 50-fold more than expected if they were additive [5]. Ganciclovir and foscarnet display in vitro synergy against CMV as well [55]. Case series using a combination of ganciclovir and foscarnet at less than recommended doses of each has resulted in some clinical success, and some guidelines recommend consideration of combination therapy in resistant CMV [25, 56]. Combinations with newer agents such as letermovir have also resulted in potential clinical benefit, but the timing and dosing of this combination still need to be validated in a well-designed study [25]. Ganciclovir activation is inhibited by maribavir, so they should not be combined [25]. Combination therapy of ganciclovir and foscarnet is recommended by some experts for severe HHV-6 infections, but it also lacks strong comparative data [57, 58].

8.2 Renal Replacement Therapies

Since ganciclovir has been studied in dialysis, clear dosing recommendations exist for it in stable dialysis patients. It is typically about 50% removed by dialysis and can be dosed three times weekly after dialysis sessions [5, 32]. In other types of renal replacement therapy, it is also significantly removed, but appropriate dosing is more in question, and the risks of both over- and under-dosing patients who are critically ill are significant. Significant variability of kinetics has been shown in critically ill patients that may make consideration of monitoring ganciclovir levels reasonable [33]. One study using PK modeling predicts that 2.5 mg/kg/day should be adequate dosing for anuric patients on CVVHDF, but in the patients this model was created with, the actual dosing utilized was 5 mg/kg/day and was tolerated by most patients, so the more aggressive dosing may be safe to utilize [59].

9 Summary and Future Direction

Ganciclovir and now valganciclovir have been the mainstays for anti-CMV therapy for several decades and are well established for most indications. Areas for future development may include individualized PK monitoring, for which the benefit of or appropriate monitoring parameters is yet to be fully established. The other primary question arising for ganciclovir and valganciclovir is their place among emerging new therapies that may offer advantages in some scenarios or allow the opportunity for better-tolerated combinations that may prove beneficial over current standards. Perhaps the most notable recent trial is the AURORA trial where maribavir failed to meet non-inferiority against valganciclovir in pre-emptive therapy for CMV, leaving first line options for CMV unchanged for the near future (NCT02927067). It remains to be seen if another agent or a differently designed trial of maribavir will be successful against valganciclovir. Additionally, what effects the now-widespread usage of letermovir prophylaxis will have on valganciclovir usage is an open question.

Conflict of Interest The author has no conflict of interest.

References

1. Ganciclovir [package insert]. Lenoir, NC: Exela Pharmaceuticals, Inc.; 2017.
2. Britt WJ, Prichard MN. New therapies for human cytomegalovirus infections. Antiviral Res. 2018;159:153–174.
3. Marshall BC, Koch WC. Antivirals for cytomegalovirus infection in neonates and infants: focus on pharmacokinetics, formulations, dosing, and adverse events. Paediatr Drugs. 2009;11(5):309–321.
4. Crumpacker CS. Ganciclovir. N Engl J Med. 1996;335(10):721–729.
5. Faulds D, Heel RC. Ganciclovir. A review of its antiviral activity, pharmacokinetic properties and therapeutic efficacy in cytomegalovirus infections. Drugs. 1990;39(4):597–638.
6. Valcyte [package insert]. South San Francisco, CA: Genetech, Inc.;2010
7. Asberg A, Humar A, Rollag H, et al. Oral valganciclovir is noninferior to intravenous ganciclovir for the treatment of cytomegalovirus disease in solid organ transplant recipients. Am J Transplant. 2007;7(9):2106–2113.
8. Lurain NS, Chou S. Antiviral drug resistance of human cytomegalovirus. Clin Microbiol Rev. 2010;23(4):689–712.
9. Pescovitz MD, Rabkin J, Merion RM, et al. Valganciclovir results in improved oral absorption of ganciclovir in liver transplant recipients. Antimicrob Agents Chemother. 2000;44(10):2811–2815.
10. Chemaly RF, Hill JA, Voigt S, Peggs KS. In vitro comparison of currently available and investigational antiviral agents against pathogenic human double-stranded DNA viruses: A systematic literature review. Antiviral Res. 2019;163:50–58.
11. Kodad S, Sabry W, Elsayed A, Goubran H, Elemary M. Viral infections in allogeneic hematopoietic stem cell transplant recipients: literature review. OBM Transplantation. 2019;3(3):1–22.
12. Adams LA, Deboer B, Jeffrey G, Marley R, Garas G. Ganciclovir and the treatment of Epstein-Barr virus hepatitis. J Gastroenterol Hepatol. 2006;21(11):1758–1760.
13. De Clercq E, Naesens L, De Bolle L, Schols D, Zhang Y, Neyts J. Antiviral agents active against human herpesviruses HHV-6, HHV-7 and HHV-8. Rev Med Virol. 2001;11(6):381–395.
14. Prichard MN, Whitley RJ. The development of new therapies for human herpesvirus 6. Curr Opin Virol. 2014;9:148–153.
15. De Bolle L, Manichanh C, Agut H, De Clercq E, Naesens L. Human herpesvirus 6 DNA polymerase: enzymatic parameters, sensitivity to ganciclovir and determination of the role of the A961V mutation in HHV-6 ganciclovir resistance. Antiviral Res. 2004;64(1):17–25.
16. Cohen JI, Davenport DS, Stewart JA, et al. Recommendations for prevention of and therapy for exposure to B virus (Cercopithecine herpesvirus 1). Clin Infect Dis. 2002;35(10):1191–1203.
17. Coen N, Duraffour S, Topalis D, Snoeck R, Andrei G. Spectrum of activity and mechanisms of resistance of various nucleoside derivatives against gammaherpesviruses. Antimicrob Agents Chemother. 2014;58(12):7312–7323.
18. Cho J, Bhavsar AR, Cho DH, Sattarova V, Abel AS, Zadroga R. Successful treatment of epstein-barr virus-induced necrotizing retinitis with intravitreous ganciclovir, foscarnet, and methotrexate. COP. 2022;13:116–123.
19. Bossolasco S, Falk KI, Ponzoni M, et al. Ganciclovir is associated with low or undetectable Epstein-Barr virus DNA load in cerebrospinal fluid of patients with HIV-related primary central nervous system lymphoma. Clin Infect Dis. 2006;42(4):e21–25.
20. Malouf MA, Chhajed PN, Hopkins P, Plit M, Turner J, Glanville AR. Anti-viral prophylaxis reduces the incidence of lymphoproliferative disease in lung transplant recipients. J Heart Lung Transplant. 2002;21(5):547–554.
21. Höcker B, Böhm S, Fickenscher H, et al. (Val-)Ganciclovir prophylaxis reduces Epstein-Barr virus primary infection in pediatric renal transplantation. Transpl Int. 2012;25(7):723–731.
22. El Chaer F, Shah DP, Chemaly RF. How I treat resistant cytomegalovirus infection in hematopoietic cell transplantation recipients. Blood. 2016;128(23):2624–2636.

23. Kotton CN, Kumar D, Caliendo AM, et al. The third international consensus guidelines on the management of cytomegalovirus in solid-organ transplantation. Transplantation. 2018;102(6):900–931.
24. Gracia-Ahufinger I, Gutiérrez-Aroca J, Cordero E, et al. Use of high-dose ganciclovir for the treatment of cytomegalovirus replication in solid organ transplant patients with ganciclovir resistance-inducing mutations. Transplantation. 2013;95(8):1015–1020.
25. Yong MK, Shigle TL, Kim YJ, Carpenter PA, Chemaly RF, Papanicolaou GA. American Society for Transplantation and Cellular Therapy Series: #4 – Cytomegalovirus treatment and management of resistant or refractory infections after hematopoietic cell transplantation. Transplantation and Cellular Therapy, Official Publication of the American Society for Transplantation and Cellular Therapy. 2021;27(12):957–967.
26. Chou S, Ercolani RJ, Marousek G, Bowlin TL. Cytomegalovirus UL97 kinase catalytic domain mutations that confer multidrug resistance. *Antimicrob Agents Chemother*. 2013;57(7):3375–3379.
27. Piret J, Boivin G. Clinical development of letermovir and maribavir: Overview of human cytomegalovirus drug resistance. *Antiviral Res*. 2019;163:91–105.
28. Chou S, Song K, Wu J, Bo T, Crumpacker C. Drug resistance mutations and associated phenotypes detected in clinical trials of maribavir for treatment of cytomegalovirus infection. *J Infect Dis*. Published online July 29, 2020:jiaa462.
29. Chou S, Wu J, Song K, Bo T. Novel UL97 drug resistance mutations identified at baseline in a clinical trial of maribavir for resistant or refractory cytomegalovirus infection. *Antiviral Res*. 2019;172:104616.
30. Manichanh C, Olivier-Aubron C, Lagarde JP, et al. Selection of the same mutation in the U69 protein kinase gene of human herpesvirus-6 after prolonged exposure to ganciclovir in vitro and in vivo. Journal of General Virology. 82(11):2767–2776.
31. Isegawa Y, Hara J, Amo K, et al. Human herpesvirus 6 ganciclovir-resistant strain with amino acid substitutions associated with the death of an allogeneic stem cell transplant recipient. J Clin Virol. 2009;44(1):15–19.
32. Lake KD, Fletcher CV, Love KR, Brown DC, Joyce LD, Pritzker MR. Ganciclovir pharmacokinetics during renal impairment. Antimicrob Agents Chemother. 1988;32(12):1899–1900.
33. Krens SD, Hodiamont CJ, Juffermans NP, Mathôt RAA, van Hest RM. Population pharmacokinetics of ganciclovir in critically ill patients. Ther Drug Monit. 2020;42(2):295–301.
34. Fletcher C, Sawchuk R, Chinnock B, de Miranda P, Balfour HH. Human pharmacokinetics of the antiviral drug DHPG. Clin Pharmacol Ther. 1986;40(3):281–286.
35. Upadhyayula S, Michaels MG. Ganciclovir, foscarnet, and cidofovir: antiviral drugs not just for cytomegalovirus. J Pediatric Infect Dis Soc. 2013;2(3):286–290.
36. Imai S, Yamada T, Kasashi K, Ishiguro N, Kobayashi M, Iseki K. Construction of a flow chart-like risk prediction model of ganciclovir-induced neutropaenia including severity grade: A data mining approach using decision tree. J Clin Pharm Ther. 2019;44(5):726–734.
37. Selby PR, Shakib S, Peake SL, et al. A systematic review of the clinical pharmacokinetics, pharmacodynamics and toxicodynamics of ganciclovir/valganciclovir in allogeneic haematopoietic stem cell transplant patients. Clin Pharmacokinet. 2021;60(6):727–739.
38. Gagelmann N, Ljungman P, Styczynski J, Kröger N. Comparative efficacy and safety of different antiviral agents for cytomegalovirus prophylaxis in allogeneic hematopoietic cell transplantation: a systematic review and meta-analysis. Biol Blood Marrow Transplant. 2018;24(10):2101–2109.
39. Barkam C, Kamal H, Dammann E, et al. Improving safety of preemptive therapy with oral valganciclovir for cytomegalovirus infection after allogeneic hematopoietic stem cell transplantation. Bone Marrow Res. 2012;2012:874601.
40. McLeroth P, Paduch DA, Abt M, Hughes R, Moore S, Mudie N. Effects of valganciclovir on spermatogenesis in renal transplant patients—results of a multicenter prospective nonrandomized study. Transpl Int. 2020;33(3):310–320.

41. van der Heiden PLJ, Kalpoe JS, Barge RM, Willemze R, Kroes ACM, Schippers EF. Oral valganciclovir as pre-emptive therapy has similar efficacy on cytomegalovirus DNA load reduction as intravenous ganciclovir in allogeneic stem cell transplantation recipients. Bone Marrow Transplant. 2006;37(7):693–698.
42. Einsele H, Reusser P, Bornhäuser M, et al. Oral valganciclovir leads to higher exposure to ganciclovir than intravenous ganciclovir in patients following allogeneic stem cell transplantation. Blood. 2006;107(7):3002–3008.
43. Winston DJ, Baden LR, Gabriel DA, et al. Pharmacokinetics of ganciclovir after oral valganciclovir versus intravenous ganciclovir in allogeneic stem cell transplant patients with graft-versus-host disease of the gastrointestinal tract. Biol Blood Marrow Transplant. 2006;12(6):635–640.
44. Slifkin M, Doron S, Snydman DR. Viral prophylaxis in organ transplant patients. Drugs. 2004;64(24):2763–2792.
45. Studies of Ocular Complications of AIDS Research Group, AIDS Clinical Trials Group. Mortality in patients with the acquired immunodeficiency syndrome treated with either foscarnet or ganciclovir for cytomegalovirus retinitis. N Engl J Med. 1992;326(4):213–220.
46. Teoh SC, Ou X, Lim TH. Intravitreal ganciclovir maintenance injection for cytomegalovirus retinitis: efficacy of a low-volume, intermediate-dose regimen. Ophthalmology. 2012;119(3):588–595.
47. Holland GN. AIDS and ophthalmology: the first quarter century. Am J Ophthalmol. 2008;145(3):397–408.
48. Betts BC, Young JAH, Ustun C, Cao Q, Weisdorf DJ. Human herpesvirus 6 infection after hematopoietic cell transplantation: is routine surveillance necessary? Biol Blood Marrow Transplant. 2011;17(10):1562–1568.
49. Schmidt GM, Horak DA, Niland JC, Duncan SR, Forman SJ, Zaia JA. A randomized, controlled trial of prophylactic ganciclovir for cytomegalovirus pulmonary infection in recipients of allogeneic bone marrow transplants. New England Journal of Medicine. 1991;324(15):1005–1011.
50. Bounaadja L, Piret J, Goyette N, Boivin G. Analysis of HHV-6 mutations in solid organ transplant recipients at the onset of cytomegalovirus disease and following treatment with intravenous ganciclovir or oral valganciclovir. J Clin Virol. 2013;58(1):279–282.
51. Ogata M, Oshima K, Ikebe T, et al. Clinical characteristics and outcome of human herpesvirus-6 encephalitis after allogeneic hematopoietic stem cell transplantation. Bone Marrow Transplant. 2017;52(11):1563–1570.
52. Ward KN, Hill JA, Hubacek P, et al. Guidelines from the 2017 European Conference on Infections in Leukaemia for management of HHV-6 infection in patients with hematologic malignancies and after hematopoietic stem cell transplantation. Haematologica. 2019;104(11):2155–2163.
53. Paya C, Humar A, Dominguez E, et al. Efficacy and safety of valganciclovir vs. oral ganciclovir for prevention of cytomegalovirus disease in solid organ transplant recipients. Am J Transplant. 2004;4(4):611–620.
54. Humar A, Lebranchu Y, Vincenti F, et al. The efficacy and safety of 200 days valganciclovir cytomegalovirus prophylaxis in high-risk kidney transplant recipients. Am J Transplant. 2010;10(5):1228–1237.
55. Manischewitz JF, Quinnan GV, Lane HC, Wittek AE. Synergistic effect of ganciclovir and foscarnet on cytomegalovirus replication in vitro. Antimicrob Agents Chemother. 1990;34(2):373–375.
56. Mylonakis E, Kallas WM, Fishman JA. Combination antiviral therapy for ganciclovir-resistant cytomegalovirus infection in solid-organ transplant recipients. Clin Infect Dis. 2002;34(10):1337–1341.
57. Ogata M, Uchida N, Fukuda T, et al. Clinical practice recommendations for the diagnosis and management of human herpesvirus-6B encephalitis after allogeneic hematopoietic stem cell transplantation: the Japan Society for Hematopoietic Cell Transplantation. Bone Marrow Transplant. 2020;55(6):1004–1013.

58. Toomey D, Phan TL, Nguyen V, Phan TT, Ogata M. Retrospective case analysis of antiviral therapies for HHV-6 encephalitis after hematopoietic stem cell transplantation. Transpl Infect Dis. 2021;23(1):e13443.
59. Horvatits T, Kitzberger R, Drolz A, et al. Pharmacokinetics of ganciclovir during continuous venovenous hemodiafiltration in critically ill patients. Antimicrob Agents Chemother. 2014;58(1):94–101.

Antiviral Agents: Letermovir

Eric Gaskill

Abstract Letermovir is the only FDA-approved agent for cytomegalovirus (CMV) prophylaxis in hematopoietic stem cell transplant (HSCT) patients. It was FDA approved in 2017 for CMV-seropositive HSCT recipients and has become a standard of care. Letermovir targets the CMV-terminase complex, preventing proper packaging of viral DNA and resulting in non-viricidal inhibition of replication. Letermovir is the only agent approved for this molecular target.

Keywords Letermovir · Cytomegalovirus · CMV · Antiviral · Resistance · Hematopoietic stem cell transplant · HSCT · Solid organ transplant · SOT

1 Background

Letermovir was FDA-approved in November 2017 for prophylaxis of CMV in seropositive HSCT recipients [1, 2]. It targets the CMV-terminase complex, preventing normal packaging of CMV DNA into capsids to release for viral propagation [3]. Letermovir, a 3,4-dihydro-quinazoline-4-yl-acetic acid derivative, was the first anti-CMV compound approved with a mechanism independent of targeting DNA polymerase [4, 5].

The CMV-terminase complex is encoded by CMV genes UL56, UL51, and UL89 [6]. CMV genomic replication involves multiple DNA concatemers linked head-to-tail, then cleaved by DNA terminase into single genomes to package into capsids for release [5]. Letermovir inhibits the cleavage of DNA concatemers into

E. Gaskill (✉)
Bone Marrow Transplant/Cellular Immunotherapy Pharmacy Specialist, Moffitt Cancer Center, Tampa, FL, USA
e-mail: Eric.gaskill@moffitt.org

C. Somboonwit et al. (eds.), *Global Virology IV: Viral Disease Diagnosis and Treatment Delivery in the 21st Century*,
https://doi.org/10.1007/978-3-031-57369-9_20

Fig. 1 Letermovir structure [2]

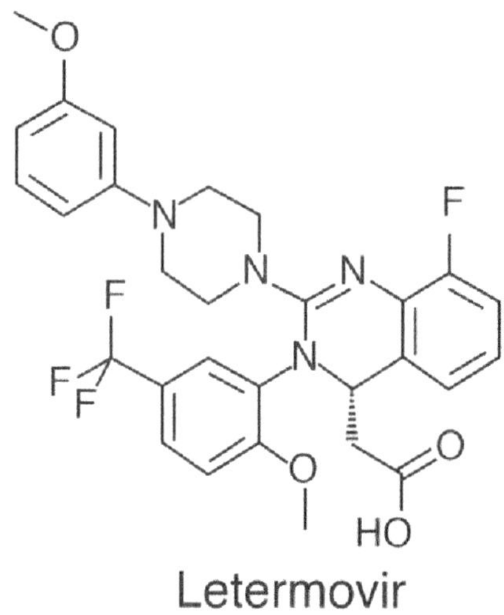

individual viral subunits via inhibition of pUL56, thus preventing the DNA from being utilized to viably replicate [6, 7]. The pUL56 subunit also binds pUL104 to form the entrance through which viral DNA enters the capsid and the *pac* site, which helps signal the viral DNA packaging process [8]. This mechanism of action is unique to CMV, and letermovir has not shown activity against any other herpesviruses. [6, 7] This mechanism is appealing as a target because no human DNA equivalent target exists, leading to less likelihood of off-target effects [5]. Homologues to the DNA terminase polymerase complex have been identified in some bacteriophages and other herpesviruses, but there is no equivalent in mammalian cells [8] (Fig. 1).

2 Spectrum of Activity: In Vitro

Letermovir only demonstrates activity against the CMV DNA terminase complex and has not been shown to inhibit other viruses, as shown in Table 1 [7].

3 Human Herpesvirusescytomegalovirus (CMV [HHV-5])

Letermovir has potent in vitro activity against CMV, displaying an EC_{50} of 4.5 nM compared to 2 μM with ganciclovir, making it roughly 400-fold more potent [4]. It displays a steep dose-response curve, and the EC_{50} and EC_{90} values are very close to each other. In assays looking at letermovir in the face of various viral loads, the EC_{50} of letermovir rose threefold across a 333-fold increase in viral inoculum, while the comparator ganciclovir rose sixfold across this range. Against the AD169

Table 1 Spectrum of activity [7]

Viral family	Species susceptible
Adenoviridae	No activity [7]
Herpesviridae	CMV at EC_{50} of 4.5 nM [7] No activity against HSV-1, HSV-2, VZV, HHV-6A, EBV [7]
Hepadnavirus	No activity [7]
Orthomyxovirus	No activity [7]
Flavivirus	No activity [7]

ganciclovir-resistant strain, letermovir maintained normal EC_{50}. When removed from the culture medium, CMV strains rapidly resume replication as soon as 48 h later showing that the blockade of viral replication by letermovir is reversible [4]. Letermovir displays a high selectivity index of over 15,000, with no activity on non-human CMV strains, human DNA, or other herpesviruses. [4].

Letermovir did not inhibit CMV DNA replication or synthesis compared to placebo in in vitro assays. However, it did fully stop the release of intact viral particles, confirming its effects on a late stage of viral replication. Intact viral yields at 72 and 96 h post-infection, while inoculated with letermovir and ganciclovir, resulted in essentially the same low viral yields at these time points [9]. Further study of viral capsid packaging in models indicates that letermovir resulted in the release of subviral dense particles of concatemeric DNA that are not viable [9].

4 Resistance

Letermovir does not inhibit CMV DNA replication directly, so it is unclear what level of CMV DNA detection should be considered significant in plasma to judge the failure of therapy. One in vitro study demonstrated that whole-cell lysate assays should not be used as intracellular CMV DNA levels remain significant while on letermovir therapy [10]. In one study analyzing whole blood and plasma samples of 26 patients on letermovir with CMV viremia with <10,000 copies, none of the samples contained viable CMV, indicating that this level of viremia was a marker of abortive infection [11]. Letermovir was tested against many samples of CMV with mutations to maribavir, ganciclovir, and foscarnet and was found to retain good activity against most samples [12]. In 80 clinical samples from patients without prior exposure to letermovir, 2.5% of samples expressed letermovir resistance mutations (both R246C) [13].

In clinical trials of letermovir, resistance mutations were uncommon and usually mapped to UL56 codons 25 and 229–369, with 4 mutations noted in the letermovir phase III trial in HSCT (V236M, C325W, E237G, R369T) [3, 6, 8, 14]. Most mutations have been noted in UL56, but in vitro mutations have been noted in both UL51 and UL89 that may have clinical relevance [6]. When true letermovir resistance occurs, it does not seem to be associated with a loss of viral fitness in general.

Testing of mutations at UL56 codons 231, 236, 237, 244, 257, 261, 325, and 329 did not reduce viral fitness compared to wild-type CMV, even with high levels of resistance to letermovir. [15] In a single-center retrospective of 42 thoracic transplant (heart/lung) patients receiving letermovir prophylaxis for a median of 280 days, only 1 patient experienced a clinically significant CMV infection, although transient low-level viremia occurred in 15 episodes [16]. Characterized mutations in letermovir appear in Table 2 below.

One concern about letermovir has been a possible lower barrier to genetic resistance compared to other anti-CMV agents. In an in-vitro experiment using an error-prone DNA polymerase-containing strain of CMV, the virus was inoculated and incubated with escalating concentrations of letermovir and screened for mutations. This experiment manifested a wide range of mutations of UL56, but none were detected in UL51 or UL89. Mutations were detected in all letermovir experiments within five passages at a median concentration of 8 nM, while mutations were not detected with foscarnet until 11–20 passages at a median concentration of 650 μM [15]. This rate of evolution of mutations seems to be faster than other anti-CMV compounds but may primarily be an indicator of the partial effects against viral fitness that are conferred by partial resistance to other CMV therapies. Given these results, concerns have persisted about using letermovir against high inoculums of CMV. [15] In a post hoc analysis of the letermovir phase III trial, patients who were treated with CMV viremia at randomization still had clinical benefits from letermovir prophylaxis but had numerically higher rates of CMV reactivation than those who did not have detectable DNA at randomization [17].

5 Pharmacokinetics (PK)/Pharmacodynamics (PD)/Dosing

Dosage	Oral or IV 480 mg daily 240 mg when combined with cyclosporine IV in hydroxypropyl beta-cyclodextrin
Renal Dysfunction	No dosage adjustments needed for CrCL >10 mL/min [2, 22] No data in CrCL <10 mL/min [2, 22]
	Concentrations were similar on and off dialysis and continuous renal replacement in 2 patients [23]
Hepatic Dysfunction	No dosage adjustment required for mild-moderate hepatic impairment (Child-Pugh A or B) Not recommended in severe hepatic impairment [2] AUC (Area under curve) increased 1.59-fold in Child-Pugh B, and 3.82-fold in Child-Pugh C [2, 24]

Table 2 Letermovir resistance mutations in CMV

Significant mutations		Letermovir EC_{50} ratio (mutant $EC_{50}/wtEC_{50}$)	Reference
UL51	P91S	2.1	[18]
UL56	C25F	5.4	[19]
	S229F	1.8	[18]
	V231A	2.1	[15]
	V231L	5–28	[15, 18–20]
	N232Y	17	[21]
	V236A	2.9	[19]
	V236L	14	[15]
	V236M	19–49	[18–20]
	E237D	10–16	[15, 21]
	L241P	96–266	[9, 15, 18, 20]
	T244K	3.3	[15]
	L254F	3.2	[18]
	L257F	8.6–14	[18, 19]
	L257I	4.9	[15]
	K258E	14	[21]
	F261C	4.4	[15]
	F261L	2.5–2.8	[15, 21]
	Y321C	4.6	[15]
	C325F	>3000	[15]
	C325R	>3000	[15]
	C325Y	>3000	[15, 19, 20]
	C325W	9300	[3, 19]
	L328V	1.9	[19]
	M329T	4.4–4.5	[15, 21]
	A365S	2	[19]
	N368D	2	[18]
	R369G	44	[20]
	R369M	13–23	[18, 20]
	R369S	38–81	[9, 20]
	C25F + V231L	46	[19]
	T244K, F261L	8.2	[15]
	E237D, T244K, F261L	104	[15]
	V236I, L257I	260	[15]
	V236M, L257I, M329T	>3000	[15]
	S229F, L254F, L257I	54	[18]
UL89	N320H	1.8	[21]
	N329S	2	[21]
	D344E	1.6–1.8	[18, 21]
	T350M	2.8	[21]

(continued)

Table 2 (continued)

Significant mutations		Letermovir EC_{50} ratio (mutant EC_{50}/wtEC_{50})	Reference
Combo	E237D (UL56), D344E (UL89)	34	[21]
	F261L (UL56), D344E (UL89)	4.8	[21]
	M329T (UL56), D344E (UL89)	8	[21]
	Q204R (UL56), D344E (UL89)	2.5	[21]
	P91S (UL51), S229F (UL56)	6.3	[18]
	P91S (UL51), V236M (UL56)	28	[18]
	P91S (UL51), R369M (UL56)	126	[18]
	P91S (UL51), S229F, L254F, L257I (UL56)	290	[18]
	P91S (UL51), D344E (UL89)	4	[18]

Interactions—perpetrator	Moderate CYP3A4 inhibitor [2, 25] Increased midazolam AUC by 49% [25] Increases tacrolimus and cyclosporine levels [25–32] Increases sirolimus levels [31, 33]
	Inhibitor of P-gp, OATP1B1/3 transporter [2]
	CYP2C9/19 induction Can lower voriconazole, phenytoin, omeprazole, pantoprazole levels [2] Induction of voriconazole metabolism is clinically significant and warrants caution, and extended therapeutic monitoring [28, 34–36]
	May increase levels of HMG-CoA reductase inhibitors, particularly atorvastatin, pitavastatin, and simvastatin [2, 31]
	Contraindicated with ergot alkaloids, pimozide [2]
Interactions—victim	Levels increased with combined with cyclosporine, which it also increases levels of [6, 31] AUC doubled when combined with cyclosporine, t1/2 unchanged
	Vulnerable to enzyme induction by rifampin, caution in combination [37]
	Substrate of OATP1B1/3 [2, 38]
Population PK	Intra and interindividual variation minimal, standard doses result in troughs >EC_{90} [6]

PK	Maximum concentration 13,000 ng/mL [2]
	AUC 71,500 ng*h/mL [2] in healthy, AUC 34,000–100,000 ng*h/mL in HSCT [2]
Absorption	94% bioavailability in healthy, 35% in HSCT without cyclosporine but increased to 85% with cyclosporine [2, 6, 38] Crushing—No PK data available Cmax 1.5 h after oral admin [8]
Food effects	Food intake lowered rate of absorption and reduced Cmax by 24%, AUC not affected [25], no significant effects [2]
Distribution	99% protein bound [2, 8] Volume of distribution 45.5 L [2]

Metabolism	UGT1A1/1A3 minor contribution [2]
	Half-life of approximately 10 h [2, 8] Time to steady state of 9–10 days [2]
Elimination	Excretion 93% in feces, 70% as unchanged drug [2, 6] OABP1B1/3 involved in elimination [2]

6 Adverse Events

In its Phase III trial in HSCT, letermovir had no adverse events overtly different from placebo. Side effects that were noted to be elevated compared to placebo include nausea (26 vs. 23.4%), vomiting (19 vs. 14%), peripheral edema (14.5 vs. 9.4%), cough (14.2 vs. 10.4%), headache (13.9 vs. 9.4%), and abdominal pain (11.8 vs. 9.4%) [3]. Cardiac, gastrointestinal, and renal adverse events did not increase across increasing AUCs in pharmacokinetic data gathered during the Phase III trial [39]. Significant renal and myelosuppressive effects have not been noted [2, 6].

Cardiac event rates are noted to be slightly higher than placebo comparators (13 vs. 6%), with elevations in tachycardia (4 vs 2%) and atrial fibrillation (3 vs 1%) compared to placebo. There have not been any associations with major cardiac events [2]. At least one case of drug-induced hepatitis has been reported [40].

Myelosuppression has not been associated with letermovir use. In the available data from its Phase III trial, three cases of neutropenia were reported that were judged as potentially attributable [1]. Compared to placebo, letermovir patients had equal to slightly lower rates of neutropenia, anemia, and thrombocytopenia [2]. In a study of thoracic transplants, letermovir was discontinued in two patients for suspected myelosuppression, but in the full trial cohort mean white blood cells count and hemoglobin increased, while median platelet counts remained stable, indicating letermovir was not likely causative [16].

7 Clinical Utilization, Dosing, and Administration

7.1 FDA Approved Indications

Letermovir is FDA-approved for prevention of CMV reactivation in CMV seropositive allogeneic HSCT recipients through 100 days post-transplant, as well as for potential extension to 200 days post-HSCT, and for prevention of CMV in high risk renal transplants [1, 2]. It was approved for HSCT use based on the results of a phase III trial of letermovir vs. placebo starting between Day 0 and Day +28 (median start Day +9) and continued until Day +100. For this outcome of clinically significant CMV reactivation, letermovir reduced the incidence significantly compared to placebo at week 24 (37.5% vs. 60.6%) [3]. The difference between the groups was driven by increased viremia in the placebo group; rates of CMV disease were similar and low (1.5% for letermovir, 1.8% for placebo). In an exploratory outcome of all-cause mortality, letermovir performed significantly better than placebo at week 24 but not at other time points [3]. Post-hoc analysis of the trial indicates that the mortality benefit of letermovir is largely derived from delaying CMV reactivation past the early transplant period. Mortality on letermovir with or without CMV reactivation was similar to patients not given prophylaxis and who did not have CMV reactivation, while patients who did have CMV reactivation and were not on prophylaxis had substantially worse mortality [41]. Additionally, patients with higher risk of CMV reactivation (HLA mismatches, cord blood transplants, haploidentical transplants, ex vivo T cell depletion, or GVHD) derived more benefit from prophylaxis. [3] A trial examining the benefit of extending prophylaxis to day +200 has been completed, resulting in the extension of the label to include up to 200 days post-HSCT.

7.2 *NON-FDA Approved Uses*

7.2.1 Prophylaxis in SOT

Prophylaxis of CMV reactivation in solid organ transplants is an area of clinical exploration. Results from a Phase III trial in high CMV risk renal transplants of letermovir vs valganciclovir prophylaxis are expected soon (NCT03443869), but in the meantime, there is limited data available from institutional series as its use has grown since becoming available on the market. One single-center study of 31 patients who converted from valganciclovir to letermovir prophylaxis found equivalent rates of CMV breakthrough between groups (8.7% letermovir, 13.5% ganciclovir) [42]. Another institutional study examined letermovir primary and secondary prophylaxis in thoracic transplant patients and found only one (2.4%) episode of CMV breakthrough but noted transient low-level viremia (<450 IU/mL) in 35.7% of patients that resolved without intervention [16]. Another case series looked at the usage of letermovir to replace ganciclovir in 7 SOT patients with improvement in lymphopenia after conversion as well as detection of CMV-specific T cells, leading the authors to believe letermovir assisted in the promotion of CMV-specific cellular immunity. [43] A case of letermovir as primary prophylaxis in small bowel transplant resulted in effective prophylaxis, but lower letermovir levels were noted in this setting while the patient had an ileostomy [44]. Letermovir was recently granted FDA approval for use in prevention of CMV in donor CMV+/recipient CMV- renal transplants based on the successful phase 3 trial versus valganciclovir [1].

7.2.2 Secondary Prophylaxis

Data is published from a compassionate use program of letermovir used as secondary prophylaxis in HSCT. In this 80-patient series, letermovir was successful and reduced CMV infections better than historical rates, with only four patients experiencing breakthrough infection [45]. Case series in SOT patients have shown potential, with one showing very few failures and another showing several, potentially based on their different assessments of low-level viremia [16, 46].

7.2.3 Treatment of CMV

Limited data is available on the use of letermovir for CMV treatment. A phase 2 trial of 27 patients compared letermovir 80 mg/day for 14 days vs valganciclovir in HSCT/SOT patients [47]. All patients in the trial had a significant decrease in CMV-DNA after treatment, with no significant differences between letermovir and valganciclovir at 14 days. It was noted that CMV-DMV response was seen after 4 days in the valganciclovir group but 11–15 days in the letermovir group, which was attributed to letermovir's non-viricidal mechanism that does not halt DNA production [47].

In a phase III trial, patients that started letermovir but were found to have baseline CMV viremia were analyzed (n = 48) [17]. Median CMV at letermovir initiation was 448 copies/mL, and no patients had over 1000 copies/mL [17]. In these patients, letermovir still reduced the rate of clinically significant CMV reactivation (31.3% vs 77.3%) compared to placebo at week 14 despite the initial low-level viremia [17]. At week 24, several letermovir patients reactivated CMV in the time between letermovir cessation and final assessment (an additional 12.5% beyond the 31.3% that had reactivated by week 14). The success of letermovir beginning during periods of low-level viremia demonstrate potential further applicability as treatment [17].

Several case series and case reports exist for letermovir as a treatment option for HSCT and SOT patients. One review compiled 47 cases of SOT and HSCT patients and assessed responses with patients receiving letermovir 480 mg daily to 960 mg daily [48]. In the 30 patients with viremia, letermovir was successful as treatment and seemed to achieve response in patients irrespective of viral loads, but in the 17 CMV disease patients, letermovir failure was more predominant in patients who had a CMV viral load of >1000 IU/mL (n = 3, 50%) [48]. Other case series examine the viability of letermovir for treatment in thoracic SOT [49–53], kidney/pancreatic transplants [54–57], and liver transplants [58, 59]. A small case series is available on letermovir used for CMV retinitis, with mixed success [60]. An ongoing Phase II trial is evaluating letermovir as a salvage treatment for CMV infection or disease after the failure of available therapies (NCT03728426).

8 Special Considerations

8.1 Pediatric Use and Pregnancy

There is no adequate data available on pregnancy safety in humans. Embryo-fetal toxicity was observed in rats and rabbits at higher AUC than recommended for human dosing, noting skeletal malformations, spontaneous abortion, and umbilical cord shortening. No embryo-fetal toxicity was noted in rabbits at up to double the recommended human dose, but effects were noted in rats at this exposure level [2]. It is unknown if letermovir is present in human breast milk or affects production. It was noted to be present in the milk of lactating rats [2].

Data on pediatric use is limited, and the label does not recommend it for this use [2]. Several small studies in pediatric use have been published, including pre-emptive use of letermovir in a 17-year-old [61], use in a 2-year-old with therapeutic monitoring of letermovir levels failing to achieve a successful outcome [62], and treatment in a 14-year-old with success but later emergence of resistance after continuing letermovir as secondary prophylaxis. [63] A pediatric pharmacokinetic study is currently recruiting (NCT03940586).

8.2 Low Level Viremia

In clinical trials, typically very low thresholds of CMV viremia were chosen to stop prophylaxis and initiate pre-emptive treatment. It is not fully clear what the appropriate level of viremia should be to switch to treatment, given that letermovir does not inhibit viral replication and transient viremia has been noted in several studies without clinical consequence [10, 17]. Whole blood assays may show inaccurately high levels of CMV DNA, but there is also evidence that lower-level viremia in plasma samples may not be indicative of active infection [10]. One study aiming to distinguish detection of DNA from viable infection found that in 77 patients given letermovir prophylaxis, 26 (34.7%) developed CMV viremia via whole blood and 21 (28%) developed viremia via plasma samples, yet after digestion of unpackaged DNA with DNAse, none of the samples were positive, indicating they were all non-viable CMV particles [11]. Risk factors for breakthrough CMV on letermovir may include higher grade GI GVHD or a positive CMV viremia at the time of initiation [64].

8.3 *CMV Specific Immune Reconstitution*

Since letermovir does not fully inhibit viral replication, it has been suggested that it may allow the promotion of some CMV-specific immune reconstitution during periods of prophylaxis. Data on this subject is somewhat contradictory, with 2 SOT papers indicating promotion of CMV-specific immune reconstitution by switching to letermovir, but two papers in the HSCT population indicating letermovir may delay CMV-specific immune reconstitution [43, 65–67]. Despite these delays in CMV-specific immune reconstitution in HSCT, delaying CMV reactivation until after Day +100 seems to benefit patients by delaying CMV viremia beyond the early transplant period [41]. Related to immune reconstitution, the effects of letermovir on the risk of chronic graft vs host disease is a developing topic, with competing papers indicating it may affect early immune reconstitution in ways that either reduce or increase cGVHD. In 30 haploidentical transplants, letermovir was associated with an increased risk of cGVHD, while a larger study of 210 patients receiving predominantly post-transplant cyclophosphamide-based regimens from a mix of donors showed decreased cGVHD in patients receiving letermovir. [68, 69]

8.3.1 Combination Therapy

Possible synergy or at least additive effects in vitro have been suggested between letermovir and all other available CMV therapies [70, 71]. A case report of combined valganciclovir and letermovir maintenance in a liver transplant with ganciclovir resistance was successful [59]. Eight renal transplant patients received letermovir combined with valganciclovir for refractory CMV, but this combination did not result in clearance of viremia in any of the patients until immunosuppression was reduced [72]. In another case series, letermovir, ganciclovir, and CMV-specific IVIG were combined in 3 renal transplant patients with successful outcomes [73]. Letermovir combined with maribavir was successful in treating CMV in a HSCT patient who was resistant to all other therapies [74].

9 Summary and Future Direction

Letermovir has a growing role in both HSCT and SOT, with now-well-established data supporting its role in primary prophylaxis in HSCT. Many other clinical areas where letermovir may have a role have trials recently completed, including an official Phase III trial in renal transplants, extending prophylaxis beyond D + 100 in HSCT, and small trials utilizing letermovir as a treatment option. Given the favorable tolerability profile of letermovir compared to other available anti-CMV therapies, these areas of research will continue, and over time we will gain more precise knowledge on letermovir's effect on long term outcomes in both HSCT and SOT.

Conflicts of Interest The author has no conflicts of interest.

References

1. U.S. Food and Drug Administration, Center for Drug Evaluation and Research. Prevymis FDA Application Multi-Discipline Review, Nov 8, 2017. Retrieved July 1, 2024, from https://www.accessdata.fda.gov/drugsatfda_docs/nda/2017/209939Orig1s000,209940Orig1s000SumR.pdf
2. Prevymis [package insert]. Whitehouse Station, NJ: Merck; 2017.
3. Marty FM, Ljungman P, Chemaly RF, et al. Letermovir prophylaxis for cytomegalovirus in hematopoietic-cell transplantation. *New England Journal of Medicine*. 2017;377(25):2433–2444.
4. Lischka P, Hewlett G, Wunberg T, et al. In vitro and in vivo activities of the novel anticytomegalovirus compound AIC246. *Antimicrob Agents Chemother*. 2010;54(3):1290–1297.
5. Hakki M. Moving past ganciclovir and foscarnet: advances in cmv therapy. *Curr Hematol Malig Rep*. 2020;15(2):90–102.
6. El Helou G, Razonable RR. Letermovir for the prevention of cytomegalovirus infection and disease in transplant recipients: an evidence-based review. Infect Drug Resist. 2019;12:1481–1491.
7. Marschall M, Stamminger T, Urban A, et al. In vitro evaluation of the activities of the novel anticytomegalovirus compound aic246 (Letermovir) against herpesviruses and other human pathogenic viruses. Antimicrob Agents Chemother. 2012;56(2):1135–1137.
8. Piret J, Boivin G. Clinical development of letermovir and maribavir: Overview of human cytomegalovirus drug resistance. Antiviral Res. 2019;163:91–105.
9. Goldner T, Hewlett G, Ettischer N, Ruebsamen-Schaeff H, Zimmermann H, Lischka P. The novel anticytomegalovirus compound aic246 (Letermovir) inhibits human cytomegalovirus replication through a specific antiviral mechanism that involves the viral terminase. J Virol. 2011;85(20):10884–10893.
10. Weinberger S, Steininger C. Reliable quantification of Cytomegalovirus DNAemia in Letermovir treated patients. Antiviral Res. 2022;201:105299.
11. Cassaniti I, Colombo AA, Bernasconi P, et al. Positive HCMV DNAemia in stem cell recipients undergoing letermovir prophylaxis is expression of abortive infection. Am J Transplant. 2021;21(4):1622–1628.
12. Piret J, Goyette N, Boivin G. In vitro activity of letermovir against human cytomegalovirus isolates with different drug susceptibility phenotypes. Antiviral Res. 2022;202:105328.
13. Santos Bravo M, Tilloy V, Plault N, et al. Assessment of ul56 mutations before letermovir therapy in refractory cytomegalovirus transplant recipients. Microbiol Spectr. 2022;10(2):e0019122.
14. Douglas CM, Barnard R, Holder D, et al. Letermovir resistance analysis in a clinical trial of cytomegalovirus prophylaxis for hematopoietic stem cell transplant recipients. J Infect Dis. 2020;221(7):1117–1126.
15. Chou S. Rapid in vitro evolution of human cytomegalovirus ul56 mutations that confer letermovir resistance. Antimicrob Agents Chemother. 2015;59(10):6588–6593.
16. Saullo JL, Baker AW, Snyder LD, et al. Cytomegalovirus prevention in thoracic organ transplantation: A single-center evaluation of letermovir prophylaxis. J Heart Lung Transplant. 2022;41(4):508–515.
17. Marty FM, Ljungman PT, Chemaly RF, et al. Outcomes of patients with detectable CMV DNA at randomization in the phase III trial of letermovir for the prevention of CMV infection in allogeneic hematopoietic cell transplantation. Am J Transplant. 2020;20(6):1703–1711.
18. Chou S. Comparison of cytomegalovirus terminase gene mutations selected after exposure to three distinct inhibitor compounds. Antimicrob Agents Chemother. 2017;61(11):e01325–17.

19. Chou S, Ercolani RJ, Derakhchan K. Antiviral activity of maribavir in combination with other drugs active against human cytomegalovirus. Antiviral Res. 2018;157:128–133.
20. Goldner T, Hempel C, Ruebsamen-Schaeff H, Zimmermann H, Lischka P. Geno- and phenotypic characterization of human cytomegalovirus mutants selected in vitro after letermovir (Aic246) exposure. Antimicrob Agents Chemother. 2014;58(1):610–613.
21. Chou S. A third component of the human cytomegalovirus terminase complex is involved in letermovir resistance. Antiviral Res. 2017;148:1–4.
22. Kropeit D, Scheuenpflug J, Erb-Zohar K, et al. Pharmacokinetics and safety of letermovir, a novel anti-human cytomegalovirus drug, in patients with renal impairment. Br J Clin Pharmacol. 2017;83(9):1944–1953.
23. Isberner N, Van Daele R, Klinker H, et al. Letermovir exposure in transplant patients with end-stage renal disease on renal replacement therapy. J Antimicrob Chemother. 2021;76(12):3322–3325.
24. Kropeit D, McCormick D, Erb-Zohar K, et al. Pharmacokinetics and safety of the anti-human cytomegalovirus drug letermovir in subjects with hepatic impairment. Br J Clin Pharmacol. 2017;83(12):2678–2686.
25. Kropeit D, McCormick D, Erb-Zohar K, Stobernack HP, Zimmermann H, Rübsamen-Schaeff H. Pharmacokinetics and safety of letermovir and midazolam coadministration in healthy subjects. Clin Pharmacol Drug Dev. 2022;11(1):16–24.
26. Marciano KA, Seago K, Dillaman M, Ross KG, Veltri L, Cumpston A. Evaluation of the pharmacokinetic interaction between letermovir and tacrolimus in allogeneic hematopoietic cell transplantation recipients. Transplant Cell Ther. 2022;28(7):371.e1–371.e4.
27. Hedvat J, Choe JY, Salerno DM, et al. Managing the significant drug-drug interaction between tacrolimus and letermovir in solid organ transplant recipients. Clin Transplant. 2021;35(3):e14213.
28. Hikasa S, Shimabukuro S, Osugi Y, et al. Tacrolimus concentration after letermovir initiation in hematopoietic stem cell transplantation recipients receiving voriconazole: a retrospective, observational study. Int J Med Sci. 2020;17(7):859–864.
29. Maples KT, Maloy M, Devlin S, et al. Lack of a significant pharmacokinetic interaction between letermovir and calcineurin inhibitors in allogeneic HCT recipients. Bone Marrow Transplant. 2020;55(8):1687–1689.
30. Kropeit D, von Richter O, Stobernack HP, Rübsamen-Schaeff H, Zimmermann H. Pharmacokinetics and safety of letermovir coadministered with cyclosporine a or tacrolimus in healthy subjects. Clin Pharmacol Drug Dev. 2018;7(1):9–21.
31. McCrea JB, Macha S, Adedoyin A, et al. Pharmacokinetic drug-drug interactions between letermovir and the immunosuppressants cyclosporine, tacrolimus, sirolimus, and mycophenolate mofetil. J Clin Pharmacol. 2019;59(10):1331–1339.
32. Nakashima T, Inamoto Y, Aoki J, et al. Differences in kinetics of tacrolimus concentration after letermovir discontinuation by type of concomitant azole antifungal. Int J Hematol. 2022;115(2):158–162.
33. Xue E, Lorentino F, Clerici D, et al. Coadministration of letermovir and sirolimus in allogeneic hematopoietic cell transplant recipients. Bone Marrow Transplant. 2022;57(1):128–130.
34. Hikasa S, Shimabukuro S, Osugi Y, et al. Decrease in voriconazole concentration-to-dose ratio after letermovir initiation: a retrospective, observational study. Bone Marrow Transplant. 2021;56(4):949–951.
35. Impact of letermovir prophylaxis on voriconazole exposure in allogeneic hematopoietic cell transplant recipients. Transplantation and Cellular Therapy. 2021;27(3):S124.
36. Suetsugu K, Muraki S, Fukumoto J, et al. Effects of letermovir and/or methylprednisolone coadministration on voriconazole pharmacokinetics in hematopoietic stem cell transplantation: a population pharmacokinetic study. Drugs R D. 2021;21(4):419–429.
37. Robbins JA, Menzel K, Lassman M, et al. Acute and chronic effects of rifampin on letermovir suggest transporter inhibition and induction contribute to letermovir pharmacokinetics. Clin Pharmacol Ther. 2022;111(3):664–675.

38. Prohn M, Viberg A, Zhang D, et al. Population pharmacokinetics of letermovir following oral and intravenous administration in healthy participants and allogeneic hematopoietic cell transplantation recipients. CPT Pharmacometrics Syst Pharmacol. 2021;10(3):255–267.
39. Prohn M, Cho CR, Viberg A, et al. Exposure-response analyses of letermovir following oral and intravenous administration in allogeneic hematopoietic cell transplantation recipients. Clin Pharmacol Ther. 2022;111(2):485–495.
40. Geswein L. Letermovir associated hepatic transaminitis: A case report. J Oncol Pharm Pract. 2019;25(5):1235–1238.
41. Ljungman P, Schmitt M, Marty FM, et al. A mortality analysis of letermovir prophylaxis for cytomegalovirus (Cmv) in cmv-seropositive recipients of allogeneic hematopoietic cell transplantation. Clin Infect Dis. 2020;70(8):1525–1533.
42. Winstead RJ, Kumar D, Brown A, et al. Letermovir prophylaxis in solid organ transplant-Assessing CMV breakthrough and tacrolimus drug interaction. Transpl Infect Dis. 2021;23(4):e13570.
43. Jorgenson MR, Kleiboeker H, Garg N, et al. Letermovir conversion after valganciclovir treatment in cytomegalovirus high-risk abdominal solid organ transplant recipients may promote development of cytomegalovirus-specific cell mediated immunity. Transpl Infect Dis. 2022;24(1):e13766.
44. Nguyen A, Finci L, Berney T, et al. Letermovir for cytomegalovirus primary prophylaxis in a multiple abdominal/small bowel transplant recipient. Clin Transplant. 2022;36(5):e14624.
45. Robin C, Thiebaut A, Alain S, et al. Letermovir for secondary prophylaxis of cytomegalovirus infection and disease after allogeneic hematopoietic cell transplantation: results from the french compassionate program. Biol Blood Marrow Transplant. 2020;26(5):978–984.
46. Hofmann E, Sidler D, Dahdal S, et al. Emergence of letermovir resistance in solid organ transplant recipients with ganciclovir resistant cytomegalovirus infection: A case series and review of the literature. Transpl Infect Dis. 2021;23(3):e13515.
47. Stoelben S, Arns W, Renders L, et al. Preemptive treatment of Cytomegalovirus infection in kidney transplant recipients with letermovir: results of a Phase 2a study. Transpl Int. 2014;27(1):77–86.
48. Linder KA, Kovacs C, Mullane KM, et al. Letermovir treatment of cytomegalovirus infection or disease in solid organ and hematopoietic cell transplant recipients. Transpl Infect Dis. 2021;23(4).
49. Aryal S, Katugaha SB, Cochrane A, et al. Single-center experience with use of letermovir for CMV prophylaxis or treatment in thoracic organ transplant recipients. Transpl Infect Dis. 2019;21(6):e13166.
50. Veit T, Munker D, Barton J, et al. Letermovir in lung transplant recipients with cytomegalovirus infection: A retrospective observational study. Am J Transplant. 2021;21(10):3449–3455.
51. Chong PP, Teiber D, Prokesch BC, et al. Letermovir successfully used for secondary prophylaxis in a heart transplant recipient with ganciclovir-resistant cytomegalovirus syndrome (UL97 mutation). Transpl Infect Dis. 2018;20(5):e12965.
52. Imlay HN, Kaul DR. Letermovir and maribavir for the treatment and prevention of cytomegalovirus infection in solid organ and stem cell transplant recipients. Clin Infect Dis. 2021;73(1):156–160.
53. Phoompoung P, Ferreira VH, Tikkanen J, et al. Letermovir as salvage therapy for cytomegalovirus infection in transplant recipients. Transplantation. 2020;104(2):404–409.
54. Ortiz F, Lempinen M, Aaltonen S, Koivuviita N, Helanterä I. Letermovir treatment for CMV infection in kidney and pancreas transplantation: A valuable option for complicated cases. Clin Transplant. 2022;36(2):e14537.
55. Koepf US, Klehr HU, Eis-Huebinger AM, et al. Suppression of cmv infection with letermovir in a kidney transplant patient. Eur J Case Rep Intern Med. 2020;7(7):001622.
56. Pearston AP, Ingemi AI, Ripley K, et al. Successful treatment of ul97 mutation ganciclovir-resistant cytomegalovirus viremia in a renal transplant recipient with letermovir and adjunct hyperimmune cytomegalovirus immunoglobulin: a case report. Transplant Proc. 2021;53(4):1284–1287.

57. Rho E, Näf B, Müller TF, Wüthrich RP, Schachter T, von Moos S. Use of letermovir-valganciclovir combination as a step-down treatment after foscarnet for ganciclovir-resistant CMV infection in kidney transplant recipients. Clin Transplant. 2021;35(11):e14401.
58. Paolucci S, Campanini G, Cassaniti I, et al. Emergence of Letermovir-resistant HCMV UL56 mutant during rescue treatment in a liver transplant recipient with ganciclovir-resistant infection HCMV: a case report. BMC Infect Dis. 2021;21(1):994.
59. Kronig I, Elkrief L, Berney T, Van Delden C, Neofytos D. Combination treatment with letermovir and ganciclovir for maintenance therapy of multidrug-resistant cmv infection in a liver transplant recipient. Transplantation. 2020;104(8):e248–e249.
60. Turner N, Strand A, Grewal DS, et al. Use of letermovir as salvage therapy for drug-resistant cytomegalovirus retinitis. Antimicrob Agents Chemother. 2019;63(3):e02337–18.
61. Chiereghin A, Belotti T, Borgatti EC, et al. Off-label use of letermovir as preemptive anti-cytomegalovirus therapy in a pediatric allogeneic peripheral blood stem cell transplant. Infect Drug Resist. 2021;14:1185–1190.
62. Pérez Marín M, Decosterd LA, Andre P, et al. Compassionate use of letermovir in a 2-year-old immunocompromised child with resistant cytomegalovirus disease. J Pediatric Infect Dis Soc. 2020;9(1):96–99.
63. Kilgore JT, Becken B, Varga MG, et al. Use of letermovir for salvage therapy for resistant cytomegalovirus in a pediatric hematopoietic stem cell transplant recipient. J Pediatric Infect Dis Soc. 2020;9(4):486–489.
64. Royston L, Royston E, Masouridi-Levrat S, Chalandon Y, Van Delden C, Neofytos D. Predictors of breakthrough clinically significant cytomegalovirus infection during letermovir prophylaxis in high-risk hematopoietic cell transplant recipients. Immun Inflamm Dis. 2021;9(3):771–776.
65. Abu-Omar A, Sester M. Ways to boost cellular immunity in solid organ transplant recipients - The case of letermovir. Transpl Infect Dis. 2022;24(1):e13787.
66. Zamora D, Duke ER, Xie H, et al. Cytomegalovirus-specific T-cell reconstitution following letermovir prophylaxis after hematopoietic cell transplantation. Blood. 2021;138(1):34–43.
67. Gabanti E, Borsani O, Colombo AA, et al. Human cytomegalovirus-specific t-cell reconstitution and late-onset cytomegalovirus infection in hematopoietic stem cell transplantation recipients following letermovir prophylaxis. Transplant Cell Ther. 2022;28(4):211.e1–211.e9.
68. Terao T, Matsuoka KI, Narita K, et al. Letermovir administration to prevent cytomegalovirus reactivation is the potential risk of chronic graft-versus-host disease in patients who received haploidentical stem-cell transplantation with post-transplant cyclophosphamide. Front Oncol. 2021;11:666774.
69. Lorentino F, Xue E, Mastaglio S, et al. Letermovir reduces chronic GVHD risk in calcineurin inhibitor-free GVHD prophylaxis after hematopoietic cell transplantation. Blood Adv. 2022;6(10):3053–3057.
70. Chou S, Satterwhite LE, Ercolani RJ. New locus of drug resistance in the human cytomegalovirus ul56 gene revealed by in vitro exposure to letermovir and ganciclovir. Antimicrob Agents Chemother. 2018;62(9):e00922–18.
71. Wildum S, Zimmermann H, Lischka P. In vitro drug combination studies of letermovir (AIC246, MK-8228) with approved anti-human cytomegalovirus (Hcmv) and anti-hiv compounds in inhibition of hcmv and hiv replication. Antimicrob Agents Chemother. 2015;59(6):3140–3148.
72. Jorgenson MR, Descourouez JL, Garg N, et al. The addition of adjunctive letermovir to valganciclovir for refractory cytomegalovirus viremia in kidney transplant recipients. Transpl Infect Dis. 2021;23(4):e13693.
73. Di Cristanziano V, Affeldt P, Trappe M, et al. Combined therapy with intravenous immunoglobulins, letermovir and (Val-)ganciclovir in complicated courses of cmv-infection in transplant recipients. Microorganisms. 2021;9(8):1666.
74. Dickter JK, Ross JA, Zain JM, Tegtmeier BR, Lee BV, Dadwal SS. Letermovir and maribavir for pan-resistant cytomegalovirus infection in a patient with haematologic malignancy: Consideration for combination therapy. J Clin Pharm Ther. 2022;47(5):699–702.

Antiviral Agents: Maribavir

Eric Gaskill

Abstract Maribavir is the only FDA-approved agent for refractory/resistant cytomegalovirus (CMV). It was approved 11/2021 after prior trials of lower doses and in prophylaxis settings did not meet primary endpoints. It has a limited spectrum of activity, generally only targeting CMV but no other viruses other than in vitro activity against EBV. Maribavir targets UL97 protein kinase of CMV and is the only approved agent to target this enzyme.

Keywords Maribavir · Cytomegalovirus · Antiviral

1 Introduction

Maribavir was FDA approved in November 2021 after 25 years of development and trials investigating its activity in prophylaxis and treatment of transplant recipients [1, 2]. Maribavir is an orally available benzimidazole riboside (structure shown in Fig. 1) that specifically inhibits the human CMV protein kinase UL97, inhibiting CMV DNA replication, encapsidation, and nuclear egress [3–5]. Although UL97 is not essential for viral survival, inhibition of UL97 results in severe replication defects [5, 6]. In in vitro models, viral replication is reduced over 100-fold when UL97 activity is abolished [5, 7]. Maribavir also has activity against the Epstein-Barr virus (EBV) enzyme BGLF4, a homolog of HCMV UL97 [7–9]. Maribavir does not require intracellular activation for activity [10].

E. Gaskill (✉)
Bone Marrow Transplant/Cellular Immunotherapy Pharmacy Specialist, Moffitt Cancer Center, Tampa, FL, USA
e-mail: Eric.gaskill@moffitt.org

C. Somboonwit et al. (eds.), *Global Virology IV: Viral Disease Diagnosis and Treatment Delivery in the 21st Century*,
https://doi.org/10.1007/978-3-031-57369-9_21

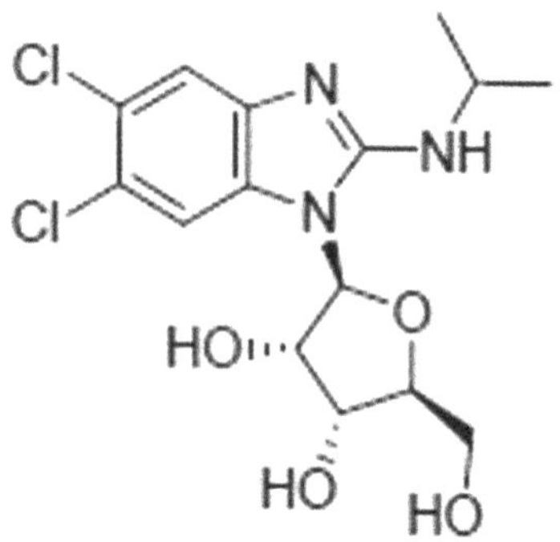

Fig. 1 Maribavir structure [11]

Table 1 Spectrum of activity [12, 13]

Viral family	Species susceptible
Adenoviridae	No data
Herpesviridae	CMV (high potency, clinical data available) EBV (high potency, no animal or human data available) Inactive against: HSV-1,HSV-2, VZV, HHV-6A, HHV-6B, HHV-8
Iridoviridae	No data
Poxviridae	No data
Papillomavirus	No data
Polyomavirus	No data

2 Spectrum of Activity: In Vitro

Maribavir has only clinically been studied against CMV, but it has in vitro data available against other herpesviruses. Given its mechanism of action, it is unlikely to have activity against non-herpesviruses. Available susceptibility data is displayed in Table 1 [12, 13].

3 Human Herpes Viruses

3.1 Epstein–Barr Virus (EBV [HHV-4])

Maribavir is reported to have activity in cell cultures against EBV with EC_{50} values ranging from 0.15 μM to 1.1 μM in an EBV-infected Burkitt lymphoma cell line. Maribavir inhibits phosphorylation of EBV replicative cofactor early antigen D (EA-D) via inhibition of EBV enzyme BGLF4, which is necessary for lytic replication of EBV [8, 9, 14, 15]. Against the P3HR-1 strain of EBV, maribavir displayed similar activity to acyclovir against the virus [16].

3.2 *Cytomegalovirus (CMV [HHV-5])*

Maribavir demonstrated EC_{50} values ranging from 0.3 μM to 14 μM in various cell lines, including ganciclovir and foscarnet-resistant strains [16]. Maribavir demonstrated IC_{50} of 0.04–0.1 μM against HCMV strain AD169 and a range of 0.03–0.132 μM in HCMV isolates [15]. Maribavir lacks activity against murine, rat, or guinea pig CMV strains but does display some activity against rhesus CMV [15, 16]. It displays activity in foscarnet and ganciclovir-resistant isolates, supporting its mechanism of action being independent of DNA polymerase [16]. In the AD169 laboratory HCMV strain, maribavir has an IC_{50} of 0.12 μM compared to 0.53 μM for ganciclovir, displaying increased potency [17]. In a series of 35 HCMV isolates, 9 of which were resistant to ganciclovir, maribavir displayed IC_{50} values <1 μM [19].

Maribavir specifically inhibits pUL97 with IC values of 0.003 μM, and maribavir and its primary 5′-mono and 5′triphosphate derivatives exhibit no significant effect on human CMV DNA polymerase at concentrations of 100 μM. The 5′-monophosphophate derivate does inhibit the incorporation of dNTPs by CMV DNA polymerase delta by 55% [11].

3.2.1 Other Herpes Viruses [HHV6–HHV8]

Maribavir has demonstrated some activity against the HHV-6 U69 kinase, but concentrations required to achieve this inhibition are significantly elevated. For HHV-6A, an EC_{50} of 60 μM was noted, while for HHV-6B, an EC_{50} of >20 has been noted. In vitro data supports that it should have little clinical activity [13].

4 Resistance

Maribavir resistance has been noted via mutations of CMV UL97 and UL27 [19]. The primary resistance mechanism is through UL97 and has generally not overlapped with known ganciclovir resistance mutations and is unrelated to foscarnet or cidofovir resistance [18–23]. The major codons associated are UL97 codons 409, 411, and 480, and in one case series of resistance, 74% of cases had mutations at codons 409 or 411 and 22% at 480 [18, 24, 25]. These mutations are associated with the UL97 ATP binding domain that spans codons 337–480 [20]. As most ganciclovir mutations cluster at codons 460, 520, and 590–607, cross-resistance is uncommon [21]. The dose intensity of maribavir was not associated with resistance rates [25]. A summary of characterized resistance mutations with associated increases in EC_{50} (half maximal effective concentration) is shown in Table 2.

Common resistance mutations conferring significant resistance to maribavir include T409M, H411Y, and C480F [20]. Less commonly noted high-level resistance mutations include V353A, L397R, and H411L/N [26]. Known mutations to

Table 2 Maribavir resistance mutations

Significant mutations in pUL97	Maribavir EC_{50} ratio (mutant EC_{50}/wtEC_{50})	Ganciclovir EC50 ratio (mutant EC_{50}/wtEC_{50})	Reference
L337M	3.4–3.5	1.02	[28]
F342S	18	7.8	[27]
F342Y	4.5	–	[11]
V353A	10–16	1–1.5	[27, 28]
K355del	304	16	[27]
V356G	108	5.5	[27]
L397R	>200	1.6	[30]
T409M	81	0.9	[25, 30]
H411L	69	0.7	[30]
H411N	9	1	[30]
H411Y	12	0.5	[25, 30]
D456N	278	12	[30]
V466G	321	11	[27]
C480F	224	2.3	[20]
C480R	243	9	[30]
K493N	2	4.1	[20]
P497S	2.4	1.9	[20]
K513T	19	4.9	[20]
P521L	428–445	17	[27]
L565V	1.6	1.8	[20]
Y617del	372	10	[30]
V823A	3.6	2.4	[20]
E989D	1.9	3.2	[20]
F342Y + H411Y	56	5.9	[19]
V353A + H411L	227	1	[30]
V353A + H411Y	164	0.7	[30]

maribavir that confer cross-resistance to ganciclovir include F342S (high-level ganciclovir resistance, low-level maribavir resistance) [26, 27]; F342Y [19]; L337M [28]; K359E, K359Q [19]; V466G [29]; C480F [20]; and P521L [29]. The C480F mutation confers high-grade resistance to maribavir, but only low-grade resistance to ganciclovir [20, 23]. The combination of F342Y and H411Y mutations was noted to confer extremely high-level maribavir resistance (EC_{50} increased by 56-fold) [19].

In a post-treatment analysis of mutations in the SOLSTICE trial, UL97 mutations were noted in 58/214 patients with CMV genotyping. Two ganciclovir cross-resistance mutations were noted: C480F in 20 patients (224-fold increase in EC_{50} to maribavir and 2.3-fold increase to ganciclovir) and F342Y in 3 patients (4.5-fold increase in maribavir EC_{50} and six-fold increase in ganciclovir EC_{50}). Overall, 58 maribavir resistance mutations were noted in the trial [1, 3].

Many in vivo mutations conferring low-level resistance to maribavir have been noted in UL27 [5, 19]. These mutations confer only modest resistance but with a higher rate of occurrence. The relation of UL27 to UL97 is unclear currently, but there is likely some modulatory effect on pUL97 that pUL27 possesses [5]. Noted UL27 mutations include R233S, A406V, C415, and W326R [20, 29]. One patient in SOLSTICE had an L193F mutation that conferred 2.6-fold reduced susceptibility [11].

5 Pharmacokinetics (PK)/Pharmacodynamics (PD)/Dosing

Maribavir is available as an oral medication only with relatively reliable target attainment and flat dosing recommendations in most patients. The primary drug interactions of concern relate to antagonism of ganciclovir/valganciclovir and potential effects on common immunosuppression medication levels. Details are provided in Table 3.

6 Adverse Effects

One of the most appealing aspects of maribavir as an option for CMV therapy is its improved tolerability compared to comparators. In the SOLSTICE trial, maribavir was compared against investigator assigned therapy, primarily ganciclovir/valganci clovir and foscarnet. Only 13% of maribavir patients discontinued therapy due to adverse events, as opposed to 32% in the comparator group, demonstrating its improved adverse event profile [3].

6.1 Dysgeusia

The primary tolerability issue that arises with maribavir is dysgeusia (altered taste and/or loss of taste). This was noted in early phase trials, where in dose-ranging studies up to 2400 mg daily, dysgeusia and headache were noted to occur in up to 80% of patients [6]. In subsequent trials, lower doses of maribavir were attempted to reduce dysgeusia and may have been related to insufficient exposure to the drug and reduced efficacy at a prophylaxis dosage of 100 mg twice daily. In SOLSTICE, the dosage was increased to 400 mg twice daily to balance efficacy and toxicity and clinically succeeded with acceptable dysgeusia while maintaining activity. In this trial, dysgeusia rates were 37.2% with a median duration of 42 days, and 89% of patients' dysgeusia resolved after cessation of therapy. Taste alterations have been described as altered taste (bitter or metallic), reduced taste, or loss of taste [3, 6, 11].

Table 3 PK/PD/Dosing

Dosage	400 mg (2 tablets) by mouth twice daily [11]
Renal dysfunction	No dose adjustment needed for mild-severe chronic kidney disease (CKD). Not studied in dialysis. No significant clearance changes across mild to severe CKD [11, 31]
	No data on dialysis removal but is highly protein bound and unlikely to be removed significantly [11]
Hepatic dysfunction	No differences noted in liver dysfunction up to Child-Pugh B [1, 11] 10 subjects with child-Pugh B studied with a modest increase in Cmax, AUC compared to healthy subjects [1]
Interactions—perpetrator	Ganciclovir/valganciclovir—maribavir antagonizes activity via UL97 inhibition [11, 32, 33]
	Maribavir is an inhibitor of 3A4 and may increase concentrations of 3A4 substrates such as tacrolimus, everolimus, sirolimus, cyclosporine [11, 34, 35]
	Maribavir inhibits P-gp and may increase concentration of sensitive substrates such as digoxin [11, 36]
	Maribavir is a weak inhibitor of CYP2C19, P-gp, and potentially UGT1A1 [1, 11]
	Maribavir may increase rosuvastatin levels and increase risk for toxicity [11]
Interactions—victim	Maribavir is a substrate of CYP3A4, CYP1A2, P-gp, UGT [34, 37]
	Ketoconazole increases maribavir AUC by 46–54%, thus it is likely other strong 3A4 inhibitors will increase maribavir levels [1, 36, 38]. Given the wide ranges of doses studied, there is no dose adjustment for this interaction [1]
	Maribavir may have increased likelihood of therapeutic failure if combined with CYP inducing medications such as carbamazepine, phenytoin, phenobarbital, rifampin, rifabutin, and efavirenz. If combined with carbamazepine, increase dosage to 800 mg twice daily. If combined with phenobarbital or phenytoin, increase dosage to 1200 mg twice daily. Combination with rifabutin and rifampin is not recommended [11]
Population PK	No significant PK differences noted by age, race, or weight [11]
	Steady state AUC values in hematopoietic stem cell transplant recipients (HSCT) were noted to be 19–66% higher than in healthy subjects [1]

(continued)

Table 3 (continued)

PK	Linear, Cmax achieved in 1–3 h, t/12 3–5 h [11, 39]
	Free drug <30 ng/mL on 200 mg or less [6]
	Phase 2 dose finding trial—400 mg dosage achieved plasma levels 2.5× higher than 100 mg [40, 41]
Absorption	Rapid absorption with bioavailability 30–40% [11]
Food effects	High fat meals decrease concentration by 30% [6, 11]
Distribution	Volume of distribution (Vd) 27.3 L; 98% plasma protein bound [6, 39]
	CNS/Ocular penetration reduced in animal models [6]
	In monkeys, plasma concentrations 1–13 ug/mL did not result in detectable concentrations in aqueous humor, reaching 4–20% of plasma levels in brain tissue, 1–2% of plasma levels in CSF, and <1% in aqueous humor [17]
Metabolism	Extensively hepatically metabolized. CYP3A4: 70–85%, CYP1A2: 15–30% [10, 11]
	Half-life 4.32 h in transplant recipients [10]
Elimination	Primarily biliary, <3% renal excretion [11, 31, 39]

6.2 *Other Adverse Reactoins*

Many other adverse events of maribavir have been associated with dose and exposure [6]. Headaches were noted in up to 80% of patients in dose-ranging studies, but in SOLSTICE headaches were experienced by only 8.1% of patients on 400 mg twice daily. The other common adverse events experienced during therapy include nausea (21.4%), diarrhea (18.8%), and vomiting (14.1%). No other events were increased compared to comparator therapies in SOLSTICE, and notably avoided the major organ toxicities associated with ganciclovir (bone marrow suppression) and foscarnet (renal toxicity) [3].

7 Clinical Utilization, Dosing, and Administration

7.1 *FDA-Approved Indications and Dosing*

Maribavir is approved for the treatment of resistant or refractory CMV infections in solid organ transplant recipients (SOT) or HSCT patients after failure of ganciclovir, valganciclovir, cidofovir, or foscarnet in adults and pediatrics 12 years of age and older and at least 35 kg [1, 11]. In its Phase III trial for this indication, it achieved superior resolution of CMV compared to investigator assigned therapy when given for 8 weeks in this category of patients [3].

7.2 Non-FDA-Approved Uses and Dosing

Maribavir has been studied in other settings, such as prophylaxis of CMV as well as initial pre-emptive therapy for CMV infections but is not currently approved for these indications.

7.2.1 First-Line Pre-emptive Therapy of CMV in SOT and Allogeneic HSCT

A Phase II trial of maribavir versus valganciclovir utilizing doses ranging from 400 mg to 1200 mg BID of maribavir for 3–12 weeks for pre-emptive treatment of CMV viremia. Maribavir had similar efficacy to valganciclovir at all endpoints and better tolerability [42]. A Phase 3 trial more thoroughly examining efficacy against valganciclovir is in progress with maribavir 400 mg twice daily. The trial has completed, and maribavir did not meet non-inferiority endpoints versus valganciclovir (NCT02927067).

7.2.2 Prophylaxis of CMV in Allogeneic HSCT and SOT

Maribavir was given in Phase 2 trials to prevent CMV at doses ranging from 100 mg to 400 mg twice daily from engraftment up to 12 weeks after transplant [41]. This trial found promising results that led to a Phase 3 trial using 100 mg BID (chosen for similar efficacy and potentially reduced side effect profile compared to higher doses) [40]. In the later Phase 3 trial, the dose of 100 mg twice daily given until week 12 did not reduce CMV infection or disease compared to placebo, and further development in the prophylaxis setting was dropped [40]. A similar trial in liver transplant patients randomized patients to maribavir 100 mg twice daily or ganciclovir 1000 mg three times daily in CMV-seronegative patients with CMV-seropositive donor livers. This trial found good tolerability of maribavir, but it did not achieve non-inferiority compared to ganciclovir [43]. There is no further development of maribavir in this setting.

8 Special Considerations

8.1 Pediatric Use, Pregnancy, and Lactation

There is no data regarding the safety of maribavir in pregnancy or lactation. In animal studies at maribavir exposures less than those used in human trials, embryo-fetal survival was not affected in rabbits but was decreased in rat models. Maribavir is only studied in patients 12 years and older and at least 35 kg based on available human data and PK modeling [11]. Additional pharmacokinetic studies are underway in younger children (NCT05319353).

8.2 CNS Penetration

Patients with CNS involvement of CMV were excluded from maribavir trials [PI]. In primate models, CNS penetration was significantly reduced, causing concern that therapeutic failure in CNS infections is a significant risk. No human data is available. Animal models also indicate that ocular penetration is significantly reduced [17].

8.3 Combination Therapy

Very little human data is available on combining maribavir with other anti-CMV medications, other than the well-published antagonism with ganciclovir [1, 6, 11, 44]. In in vitro experiments, maribavir synergized with sirolimus, cidofovir, foscarnet, and letermovir [44]. One case report exists of a successful combination of letermovir and maribavir in treating a case of CMV with UL54 mutations conferring resistance to ganciclovir, foscarnet, and cidofovir [45].

9 Summary and Future Direction

Maribavir has been shown to be a promising agent against refractory and resistant CMV in HSCT and SOT patients. Its improved tolerability profile gives it a unique place in therapy compared to historical options. With a trial in front line pre-emptive therapy now yielding negative results, the current role of therapy for maribavir will remain as an option primarily for refractory/resistant CMV until further data arises.

Questions remain on many other possible roles of maribavir. Since its prophylaxis trials at lower doses failed, it remains to be seen if maribavir will ever have a role in prophylaxis. Additionally, the currently published data in refractory CMV utilizes maribavir for 8 weeks of therapy. It is still to be determined if shorter durations would remain as effective in this setting. Lastly, maribavir remains unstudied in some possible settings, such as against EBV, or in combination with other agents for resistant CMV.

Conflicts of Interest The author declares no conflict of interest.

References

1. U.S. Food and Drug Administration, Center for Drug Evaluation and Research. Livtencity FDA Application Multi-Discipline Review, Nov 23, 2021. Retrieved Dec 6, 2021, from https://www.accessdata.fda.gov/drugsatfda_docs/nda/2021/215596Orig1s000IntegratedR.pdf
2. Mullard A. FDA approves decades-old maribavir for CMV infection. *Nature Reviews Drug Discovery*. 2021;21(1):9–9.
3. Avery RK, Alain S, Alexander BD, et al. Maribavir for refractory cytomegalovirus infections with or without resistance post-transplant: results from a phase 3 randomized clinical trial. *Clin Infect Dis*. Published online December 2, 2021:ciab988.
4. Marty FM, Ljungman P, Papanicolaou GA, et al. Maribavir prophylaxis for prevention of cytomegalovirus disease in recipients of allogeneic stem-cell transplants: a phase 3, double-blind, placebo-controlled, randomised trial. *Lancet Infect Dis*. 2011;11(4):284–292.
5. Prichard MN. Function of human cytomegalovirus UL97 kinase in viral infection and its inhibition by maribavir. *Rev Med Virol*. 2009;19(4):215–229.
6. Gandhi RG, Kotton CN. Evaluating the safety of maribavir for the treatment of cytomegalovirus. *Ther Clin Risk Manag*. 2022;18:223–232.
7. Wang FZ, Roy D, Gershburg E, Whitehurst CB, Dittmer DP, Pagano JS. Maribavir inhibits epstein-barr virus transcription in addition to viral DNA replication. *J Virol*. 2009;83(23):12108–12117.
8. Whitehurst CB, Sanders MK, Law M, et al. Maribavir inhibits Epstein-Barr virus transcription through the EBV protein kinase. *J Virol*. 2013;87(9):5311–5315.
9. Chemaly RF, Hill JA, Voigt S, Peggs KS. In vitro comparison of currently available and investigational antiviral agents against pathogenic human double-stranded DNA viruses: A systematic literature review. *Antiviral Res*. 2019;163:50–58.
10. Lalezari JP, Aberg JA, Wang LH, et al. Phase i dose escalation trial evaluating the pharmacokinetics, anti-human cytomegalovirus (Hcmv) activity, and safety of 1263w94 in human immunodeficiency virus-infected men with asymptomatic hcmv shedding. *Antimicrob Agents Chemother*. 2002;46(9):2969–2976.
11. Livtencity [package insert]. Lexington, MA: Takeda Pharmaceuticals, Inc.; 2021.
12. Williams SL, Hartline CB, Kushner NL, et al. In vitro activities of benzimidazole D- and L-ribonucleosides against herpesviruses. *Antimicrob Agents Chemother*. 2003;47(7):2186–2192.
13. Prichard MN, Frederick SL, Daily S, et al. Benzimidazole analogs inhibit human herpesvirus 6▿. *Antimicrob Agents Chemother*. 2011;55(5):2442–2445.
14. Zacny VL, Gershburg E, Davis MG, Biron KK, Pagano JS. Inhibition of Epstein-Barr virus replication by a benzimidazole L-riboside: novel antiviral mechanism of 5, 6-dichloro-2-(Isopropylamino)-1-beta-L-ribofuranosyl-1H-benzimidazole. *J Virol*. 1999;73(9):7271–7277.
15. Zacny VL, Gershburg E, Davis MG, Biron KK, Pagano JS. Inhibition of epstein-barr virus replication by a benzimidazole l-riboside: novel antiviral mechanism of 5,6-dichloro-2-(Isopropylamino)-1-β-l-ribofuranosyl-1h-benzimidazole. *J Virol*. 1999;73(9):7271–7277.
16. Williams SL, Hartline CB, Kushner NL, et al. In vitro activities of benzimidazole D- and L-ribonucleosides against herpesviruses. *Antimicrob Agents Chemother*. 2003;47(7):2186–2192.
17. Koszalka GW, Johnson NW, Good SS, et al. Preclinical and toxicology studies of 1263w94, a potent and selective inhibitor of human cytomegalovirus replication. *Antimicrob Agents Chemother*. 2002;46(8):2373–2380.
18. Drew WL, Miner, RC, Marousek GI, et al. Maribavir sensitivity of cytomegalovirus isolates resistant to ganciclovir, cidofovir or foscarnet. *Journal of Clinical Virology*. 2006;37(2):124–127.

19. Chou S, Wu J, Song K, Bo T. Novel UL97 drug resistance mutations identified at baseline in a clinical trial of maribavir for resistant or refractory cytomegalovirus infection. *Antiviral Res.* 2019;172:104616.
20. Chou S, Song K, Wu J, Bo T, Crumpacker C. Drug resistance mutations and associated phenotypes detected in clinical trials of maribavir for treatment of cytomegalovirus infection. *J Infect Dis.* Published online July 29, 2020:jiaa462.
21. Chou S, Watters M, Sinha R, Kleiboeker S. Ganciclovir and maribavir susceptibility phenotypes of cytomegalovirus UL97 ATP binding region mutations detected by expanded genotypic testing. *Antiviral Res.* 2021;193:105139.
22. Bigley TM, Reitsma JM, Terhune SS. Antagonistic relationship between human cytomegalovirus pul27 and pul97 activities during infection. *J Virol.* 2015;89(20):10230–10246.
23. Santos Bravo M, Plault N, Sánchez Palomino S, et al. Phenotype and genotype study of novel c480f maribavir-ganciclovir cross-resistance mutation detected in hematopoietic stem cell and solid organ transplant recipients. *J Infect Dis.* 2021;224(6):1024–1028.
24. Chou S, Wechel LCV, Marousek GI. Cytomegalovirus UL97 kinase mutations that confer maribavir resistance. *J Infect Dis.* 2007;196(1):91–94.
25. Papanicolaou GA, Silveira FP, Langston AA, et al. Maribavir for refractory or resistant cytomegalovirus infections in hematopoietic-cell or solid-organ transplant recipients: a randomized, dose-ranging, double-blind, phase 2 study. *Clin Infect Dis.* 2019;68(8):1255–1264.
26. Hakki M. Moving past ganciclovir and foscarnet: advances in cmv therapy. *Curr Hematol Malig Rep.* 2020;15(2):90–102.
27. Chou S, Ercolani RJ, Marousek G, Bowlin TL. Cytomegalovirus UL97 kinase catalytic domain mutations that confer multidrug resistance. *Antimicrob Agents Chemother.* 2013;57(7):3375–3379.
28. Chou S, Hakki M, Villano S. Effects on maribavir susceptibility of cytomegalovirus UL97 kinase ATP binding region mutations detected after drug exposure in vitro and in vivo. *Antiviral Res.* 2012;95(2):88–92.
29. Yong MK, Shigle TL, Kim YJ, Carpenter PA, Chemaly RF, Papanicolaou GA. American Society for Transplantation and Cellular Therapy Series: #4 – Cytomegalovirus treatment and management of resistant or refractory infections after hematopoietic cell transplantation. *Transplant Cell Ther.* 2021;27(12):957–967.
30. Piret J, Boivin G. Clinical development of letermovir and maribavir: Overview of human cytomegalovirus drug resistance. *Antiviral Res.* 2019;163:91–105.
31. Swan SK, Smith WB, Marbury TC, et al. Pharmacokinetics of maribavir, a novel oral anticytomegalovirus agent, in subjects with varying degrees of renal impairment. *J Clin Pharmacol.* 2007;47(2):209–217.
32. Chou S, Van Wechel LC, Marousek GI. Effect of cell culture conditions on the anticytomegalovirus activity of maribavir. *Antimicrob Agents Chemother.* 2006;50(7):2557–2559.
33. Chou S, Ercolani RJ, Derakhchan K. Antiviral activity of maribavir in combination with other drugs active against human cytomegalovirus. *Antiviral Res.* 2018;157:128–133.
34. Pescovitz MD, Bloom R, Pirsch J, Johnson J, Gelone S, Villano SA. A randomized, double-blind, pharmacokinetic study of oral maribavir with tacrolimus in stable renal transplant recipients. *Am J Transplant.* 2009;9(10):2324–2330.
35. Verdier MC, Patrat-Delon S, Lemaitre F, et al. Suspicion of interaction between maribavir and everolimus in a renal transplant recipient. *Transplantation.* 2014;98(3):e20–21.
36. Goldwater DR, Dougherty C, Schumacher M, Villano SA. Effect of ketoconazole on the pharmacokinetics of maribavir in healthy adults. *Antimicrob Agents Chemother.* 2008;52(5):1794–1798.
37. Summary of maribavir (Shp620) drug–drug interactions based on accumulated clinical and nonclinical data. *Biology of Blood and Marrow Transplantation.* 2019;25(3):S370–S371.
38. Song IH, Ilic K, Wu J Lack of drug–drug interaction between maribavir and voriconazole. Poster presented at: American Transplant Congress (ATC) Meeting May 30, 2020. Poster A-176; 2020.

39. Wang LH, Peck RW, Yin Y, Allanson J, Wiggs R, Wire MB. Phase I safety and pharmacokinetic trials of 1263W94, a novel oral anti-human cytomegalovirus agent, in healthy and human immunodeficiency virus-infected subjects. *Antimicrob Agents Chemother*. 2003;47(4):1334–1342.
40. Marty FM, Boeckh M. Maribavir and human cytomegalovirus-what happened in the clinical trials and why might the drug have failed? *Curr Opin Virol*. 2011;1(6):555–562.
41. Winston DJ, Young JAH, Pullarkat V, et al. Maribavir prophylaxis for prevention of cytomegalovirus infection in allogeneic stem cell transplant recipients: a multicenter, randomized, double-blind, placebo-controlled, dose-ranging study. *Blood*. 2008;111(11):5403–5410.
42. Maertens J, Cordonnier C, Jaksch P, et al. Maribavir for preemptive treatment of cytomegalovirus reactivation. *N Engl J Med*. 2019;381(12):1136–1147.
43. Winston DJ, Saliba F, Blumberg E, et al. Efficacy and safety of maribavir dosed at 100 mg orally twice daily for the prevention of cytomegalovirus disease in liver transplant recipients: a randomized, double-blind, multicenter controlled trial. *Am J Transplant*. 2012;12(11):3021–3030.
44. Chou S, Ercolani RJ, Derakhchan K. Antiviral activity of maribavir in combination with other drugs active against human cytomegalovirus. *Antiviral Res*. 2018;157:128–133.
45. Dickter JK, Ross JA, Zain JM, Tegtmeier BR, Lee BV, Dadwal SS. Letermovir and maribavir for pan-resistant cytomegalovirus infection in a patient with haematologic malignancy: Consideration for combination therapy. *J Clin Pharm Ther*. 2022;47(5):699–702.

Antiviral Agents in the Hematopoietic Stem Cell Transplant Population: Acyclovir, Valacyclovir, Penciclovir, and Famciclovir

Elizabeth DiMaggio

Abstract Over the last five decades, the nucleoside analogues acyclovir and penciclovir, and their respective prodrugs, valacyclovir and famciclovir, have been used to prevent and control viral infections commonly found in the hematopoietic cell transplantation community. Despite all being guanosine analogs, structural differences yield variability in their affinity to viral enzymes and subsequent viral control. What follows is an introduction to each agent, highlighting their indications for use and dosing, pharmacokinetic and pharmacodynamic properties, and clinical data supporting their utilization as both preventative and treatment modalities.

Keywords Antivirals · HSCT · Hematopoietic stem cell transplant · Acyclovir · Penciclovir · Famciclovir · Prophylaxis

1 Background: Acyclovir and Valacyclovir

A second-generation synthetic acyclic nucleoside analogue, acyclovir (ACV) (Fig. 1), was initially studied in clinical trials in the mid-1970s, formally approved in 1982, and continues to be used today [1, 2]. ACV, also known by the chemical formula 9-((2-Hydroxyethoxy)methyl) guanine or acycloguanosine [3], is simply an analogue of the DNA component guanosine, differing only by the substitution of a sugar residue for a 2-hydroxyethoxymethyl acyclic side chain (at position 9) [2, 3]. Additional chemistry properties are described in Table 1.

E. DiMaggio (✉)
Department of Pharmacy, Moffitt Cancer Center, Tampa, FL, USA
e-mail: Elizabeth.DiMaggio@moffitt.org

C. Somboonwit et al. (eds.), *Global Virology IV: Viral Disease Diagnosis and Treatment Delivery in the 21st Century*,
https://doi.org/10.1007/978-3-031-57369-9_22

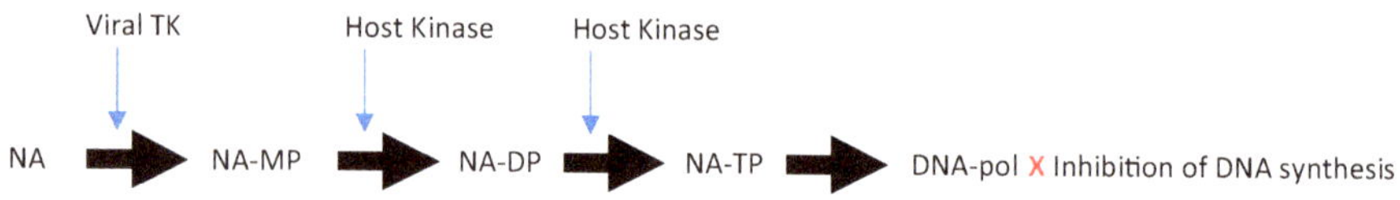

Fig. 1 The mechanism of action of nucleoside analogs such as acyclovir and penciclovir. NA-TP acts as both a substrate and inhibitor of viral DNA-pol. Abbreviations: *DP* diphosphate, *MP* monophosphate, *NA* nucleoside analog, *pol* polymerase, *TK* viral thymidine, *TP* triphosphate. (Reference: Figure adapted from Majewska A, et al. [2])

Table 1 Chemical properties of Acyclovir and Valacyclovir [3, 11]

	Acyclovir	Valacyclovir
Physical description	White solid	Solid
Melting point (°C)	256.5–257	170–172
Water solubility	Poor (0.2% at 25 °C)	174 mg/mL
pKA value(s)	2.27, 9.25	1.9, 7.47, 9.43

Once ACV diffuses into a virally infected cell, it is activated following tri-phosphorylation: first with viral thymidine kinase (TK) enzymes (encoded by genes UL23 in HSV and ORF36 in VZV), next by cellular guanylate kinase, and last by host cellular kinases [2]. Not only does non-phosphorylated ACV inhibit viral TK [4], tri-phosphorylated ACV acts as a substrate of the viral DNA polymerase (encoded by genes UL30/UL42 in HSV and ORF28 in VZV), fusing to the 3′ terminus of the DNA and blocking further chain formation [2].

Overall, ACV is well tolerated, with only mild gastrointestinal (GI) side effects, rash, and headache reported [5, 6]. At higher peak concentrations, reversible kidney failure and neurotoxicity have occurred [6]. It is available topically, intravenously, and orally [2, 6], but ACV's bioavailability is poor at only 15–30% [2]. Its derivative or prodrug, valacyclovir (VACV) (L-valine ester of ACV, Fig. 2), has better bioavailability at approximately 54.2% [7]. VACV's chemical formula is L-valine, 2-[(2-amino-1,6-dihydro-6 382 oxo-9H-purin-9-yl)methoxy]ethyl ester, monohydrochloride. Following oral administration, VACV's quick conversion to ACV via hepatic and intestinal first-pass metabolism yields a threefold to fivefold increase in the concentration of ACV compared to oral administration of the parent compound ACV [7]. See Table 3 for dosing equivalents. As expected, the side effects of VACV are similar to those of ACV [7]. Once VACV is in the systemic circulation, distribution, intracellular PK, metabolism, and excretion of "converted VACV to ACV" are identical to those of ACV [8]. Both drugs are active only in virus-infected cells [2]. In uninfected cells, the drug acts as a poor substrate for TK and DNA polymerase, whereas in viral-infected cells, it is highly selective for viral proteins [9].

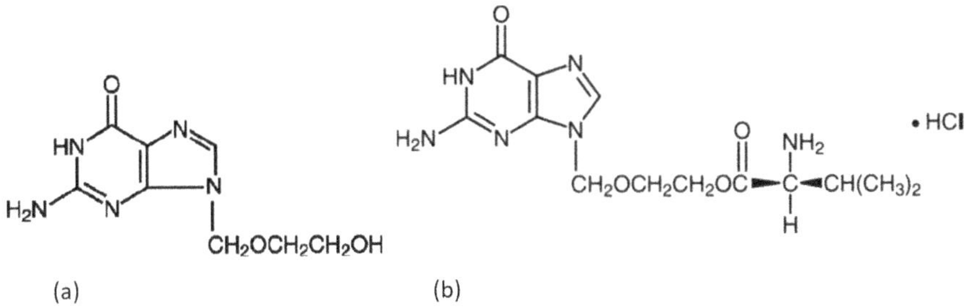

Fig. 2 Structural formulas for (**a**) acyclovir and (**b**) valacyclovir [10]

Table 2 ACV and VACV viral inhibition in vitro [7]

Virus	IC_{50} (μM)
CMV	200
EBV (ACV)	0.2–25
HSV (ACV)	0.1
HSV (VZV)	0.84
VZV (ACV)	1.3–20.6

Abbreviations: *ACV* acyclovir, *CMV* cytomegalovirus, *EBV* Epstein-Barr virus, *HSV* herpes simplex virus, *VACV* valacyclovir, *VZV* varicella zoster virus

2 Spectrum of Activity: In Vitro

VACV and ACV's antiviral activity includes in vitro inhibition against (in order of most susceptible to least) human alphaherpesvirus (HSV) 1 (HSV-1), human alphaherpesvirus 2 (HSV-2), human alphaherpesvirus 3 (varicella-zoster virus, VZV), Epstein-Barr virus (EBV), human betaherpes virus 6 (HHV-6), and cytomegalovirus (CMV) [2, 6]. The IC50 value, or antiviral concentration resulting in 50% inhibition in vitro, has been reported as the following (Table 2):

To put into context, VACV would need to be given at toxic doses for CMV activity to be achieved [12]. There is no *in vitro* activity beyond the Herpesviridae family.

3 Resistance

With the use of antiviral therapy, there remains a potential risk of selecting nucleoside-resistant viruses. This is more common with prolonged drug use in the setting of prophylaxis or suboptimal dosing [2]. Resistance should be suspected in cases of failed clinical resolution or the emergence of new viral lesions despite appropriate treatment duration and dosing, or in cases of disease recurrence [2]. ACV resistance is most often conferred by UL23-thymidine kinase gene mutations (up to 95%) [13], though UL30-viral DNA polymerase gene mutations also occur much less frequently (5–10%) [2]. Unlike DNA polymerase, a functional thymidine kinase is not essential for viral replication, and the mutation perpetuates [13].

Only rarely does the HSV genome develop resistance, owing in part to its robust enzymatic DNA proofreading and replication prevention from potent ACV and other nucleoside antivirals such as penciclovir [9]. Even so, when mutations arise, they may be one of the following: The most common a *TK minus* strain may form following a deletion, insertion, or point mutation of the viral TK gene, giving rise to an early stop codon, which subsequently produces an inactive viral enzyme [9]. Any further viral replication is slowed, permitting the host's immune system to respond [9]. Rare mutations may produce viruses that retain activity and virulence. A *TK partial* has incomplete but functional activity [9]. A *TK altered* detects the antivirals but can continue the phosphorylation steps required for replication. The viral DNA polymerase can also be altered to only utilize host substrates for phosphorylation, prompting viral replication, even in the presence of ACV or penciclovir [9].

In an immunocompetent patient, the prevalence of resistant HSV strains is low (<1%), while in an immunocompromised patient, namely a hematopoietic cell transplant recipient, rates can range 4–36% [13]. Unfortunately, if ACV resistance arises, as in the case of HSV, foscarnet treatment is required [9]. Once symptoms are resolved, the ACV can resume in the case of immunocompromised patients who had previously received ACV prophylaxis (such as hematopoietic cell transplant (HCT) recipients). Further viral reactivation will occur with wild-type viruses, as mutated viruses have been eliminated [9].

Similar to HSV, strains of resistant VZV are rare but more often reported in immunocompromised patients, such as HCT recipients [14, 15]. Though the exact prevalence is unknown, in cases of persistent VZV, mutations possibly contributing to resistance have been detected in up to 27% of cases [14]. In a report by van der Beek and colleagues, within a cohort of 87 hematological patients, of whom 65 (75%) underwent allogeneic HCT, there were 89 episodes of VZV, including 32 persistent. Of these, six patients (27%) were found to have mutations and developed complications of VZV. Again, foscarnet is the treatment of choice for ACV-resistant strains [17], though case reports using either a high-dose continuous ACV infusion [16] or famciclovir (FAM) [17] have shown efficacy.

4 Pharmacokinetic/Pharmacodynamic/Dosing

Pharmacokinetic and pharmacodynamic properties of ACV and its prodrug, VACV, are well established. At equivalent doses, oral ACV demonstrates a lower area under the curve (AUC) than VACV (Table 3) [7]. Peak concentrations of ACV following oral administration of 800 mg five times daily are also lower when compared to VACV dosed at 1000 mg four times daily (Table 3) [7]. The bioavailability of ACV compared with VACV also appears to lag behind VACV (Table 4) [2, 6, 7, 21]. Additional pharmacokinetic properties are outlined in Table 4. Only ACV is available in an intravenous form, which also demonstrates a higher AUC and achieves a

Table 3 Dosing comparison for Acyclovir and Valacyclovir [7]

Acyclovir			Valacyclovir		
Dosage	AUC (h*micrograms/mL)	Cmax (micrograms/mL)	Dosage	AUC (h*micrograms/mL)	Cmax (micrograms/mL)
200 mg PO 5X/day	12	0.8	250 mg PO QID	23	23
800 mg PO 5X/day	24	1.6	1000 mg PO QID	63	63
5 mg/kg every 8 h IV	54	9.8	2000 mg PO QID	109	109

Abbreviations: *AUC* area under the curve, *Cmax* maximum plasma concentration, *h* hour, *IV* intravenously, *PO* per os, *QID* four times daily

higher peak than oral ACV [7]. Thus, to obtain optimal concentrations, either higher doses of oral ACV are needed to be effective or use of the intravenous and VACV forms is warranted. See Table 5 for different indications and common doses for oral and intravenous ACV and oral VACV [18–20]. Dosing strategies also include considerations of renal and hepatic impairment [18–20]. Potential drug interactions via the cytochrome (CYP) P450 pathway for either ACV or VACV appear to be clinically insignificant. Both are substrates of OAT1/3 and OCT1 and very weak inhibitors of CYP1A2 (Table 5) [18–20].

5 Clinical Utilization, Dosing, and Administration: FDA-Approved Indications for Acyclovir and Valacyclovir

FDA-approved indications for ACV and VACV are outlined below in Table 6. Additionally, use of these agents as prophylactic therapy is discussed in the next section.

6 Special Populations: Nucleoside Antiviral Use in HCT Recipients

Since the introduction of acyclovir in 1977, it has been a mainstay of therapy for herpesvirus infections, followed shortly thereafter by valacyclovir. The purpose of this section is to discuss clinical data supporting its use in the transplant population.

The presence of HSV seropositivity is ubiquitous in the general population but typically remains dormant in the dorsal ganglia of immunocompetent patients. In contrast, immunosuppressed HCT recipients undergoing an allogeneic HCT [22] have a relatively high risk of reactivation of 80% [22] to 90% [23], depending on

Table 4 Pharmacokinetic properties of Acyclovir and Valacyclovir [2, 6, 7, 21]

BA / Absorption	Cmax	Half-life	Distribution	Clearance	Elimination	Css max per dose regimen	Time to Css
Passive across small intestine ACV 15–30% VACV PO 54.2%	1.5–2.5 h Doubled in ESRF	Normal renal function (IV) 2–3 h ESRF 20 h HD 6–10 h Continuous ambulatory peritoneal dialysis 13–18 h Neonates 2.5–5 hours Intracellular infected cells: HSV-1, 0.7 h; HSV-2, 1 h; and VZV, 0.8 h	Protein binding 9–33% Vd: 48 (range 22.5–101) /1.73m^2	15.6 L/h/1.73m^2 Dialyzable	Renal 45–79% recovered unchanged in urine	IV 2.5–15.0 mg/kg q8h: 6.7–20.6 mg/L (similar in neonates) 200 mg PO q4h: 0.52–0.56 mg/L 400 mg PO q4h: 0.79–1.22 mg/L	Oral 1–2 d

Abbreviations: *ACV* acyclovir, *BA* bioavailability, *Cmax* maximum plasma concentration, *Css* concentration at steady state, *d* day(s), *ESRF* end stage renal failure, *h* hour(s), *HD* hemodialysis, *IV* intravenous, *PO* per os, *VACV* valacyclovir

Table 5 General dosing guides [18–20]

	Acyclovir[a]	Valacyclovir[a]
Standard dosages *(Adult)* depending on indication (see below)	200–800 mg PO BID to 5X/d 250–500 mg/m^2 IV q8-12h 5–10 mg/kg IV q8h	500–2000 mg PO daily to TID
Prophylaxis in HCT recipients	HSV or VZV: 800 mg PO BID for 30 days post allogeneic HCT CMV (allogeneic HCT only): 500 mg/m^2 IV TID or 800 mg QID	HSV or VZV: 500 mg daily (HSV) or BID (BID) 1 year after HCT (or until off IS if after an allogeneic HCT) VZV (post exposure): 1 g TID ×3 weeks CMV (allogeneic HCT only): 2 g TID to QID
Treatment in HCT recipients	HSV: 250–500 mg/m^2 IV q8h or 400 mg PO q8h VZV: 10 to 15 mg/kg IV q8h for at least 2 days or after all lesions have crusted	
Renal Dysfunction	Reduced with CrCL <30 mL/min or <50 mL/min depending on initial dose / frequency Dialyzable; give after HD (1/3 removed during a 4-h session)	Reduced with CrCL <30 mL/min or <50 mL/min depending on initial dose / frequency Dialyzable; give after HD (1/3 removed during a 4-h session)
Hepatic Impairment	No dose adjustment	No dose adjustment
Interactions	Substrate of OAT1/3, OCT1 Inhibits CYP1A2 (weak) May interact with drugs metabolically impacted by above substrates / enzymes	Substrate of OAT1/3, OCT1 Inhibits CYP1A2 (weak) May interact with drugs metabolically impacted by above substrates / enzymes

Abbreviations: *BID* twice daily, *CMV* cytomegalovirus, *CrCL* creatinine clearance, *d* day(s), *h* hour(s), *HCT* hematopoietic cell transplantation, *HD* hemodialysis, *HSV* herpes simplex virus, *IS* immunosuppression, *IV* intravenous, *PO* per os, *q* every, *QID* four times daily, *TID* three times daily, *VZV* varicella-zoster virus

[a]Dose, frequency, and duration dependent on indication

seropositivity and transplant type. Disease reactivation may present as mucosal lesions in or around the oral and pharyngeal cavity (herpes labialis) (HSV-1) or genital herpes infections (HSV-2), but in more severe forms, it may result in ocular lesions, encephalitis, or death [24]. It has been well established since the early 1980s that prophylaxis in the first year reduces the rate of HSV activation in both autologous and allogenic transplantation recipients [25]. In a large observational study by Erard and colleagues, the incidence of HSV at 2-years (per 1000 person-months) in HCT recipients that received prophylactic ACV for at least 1 year (cohort 2) or longer if on immunosuppression (cohort 3) was improved over those who stopped ACV on engraftment (cohort 1); 3 (95% CI, 2–4), 0, and 36 (95% CI, 32–41), respectively (Table 7). Rates of ACV-resistance at 2 years were also incredibly low, and the drug was well tolerated, strongly supporting the long-term ACV-prophylaxis [25]. The use of VACV as prophylaxis has also been shown to be

Table 6 General *Approved* indications [18, 19]

Acyclovir	Treatment of herpes simplex virus Genital + immunocompetent (initial, PO/IV; recurrent, PO; suppressive, PO) Genital + immunocompromised (initial, recurrent, PO/IV; suppressive, PO) Genital + pregnant (initial, PO; recurrent PO; severe, IV; suppressive, PO) Orolabial (initial, PO/IV; suppressive PO) Mucocutaneous infection in immunocompromised patients (initial, IV; recurrent, IV) Treatment of herpes zoster (shingles) infections (lesion <72 h old) Immunocompetent (PO) Immunocompromised localized (PO) Immunocompromised extensive (IV) Treatment of varicella-zoster (chickenpox) infection within 24 h after appearance of the typical chickenpox rash Immunocompromised severe / complicated (IV) Immunocompromised uncomplicated (PO) Herpes Simplex Encephalitis (IV) Neonatal Herpes Simplex Virus Infection (IV) Varicella-Zoster Infections in Immunocompromised Patients (IV)
Valacyclovir	Adults: Cold sores (herpes labialis); genital herpes (initial episode, recurrent episode, suppressive therapy, reduction of transmission); herpes zoster Pediatric patients: Cold sores (herpes labialis); chicken pox

equally effective and well tolerated compared with ACV in an autologous HCT population [23, 26]. In a cohort of autologous HCT patients who did not take prophylaxis, Dignani et al. reported 45% HSV reactivation compared to less than 3% with VACV or 2% with ACV (Table 7) [26]. Eisen and colleagues also compared the efficacy of ACV with VACV in a small population of HSV-1 seropositive autologous transplant recipients. None of the 120 patients included presented with HSV-1 or HSV-2 positive cultures or any herpes infections [23]. Current 2009 international guidelines recommend ACV or VACV prophylaxis among HSV-seropositive patients to start with conditioning and continue until engraftment or mucositis resolves, whichever is longer, or for approximately 30 days post-allogeneic and autologous HCT recipients [20].

VZV reactivations are present in 17–50% of allogeneic hematopoietic cell transplant recipients [27] and 16–30% [28] of autologous HCT recipients, occurring up to 10 times more than in the general U.S. adult population [29]. Initial VZV infection presents as chickenpox and is typically seen in childhood. Following this primary insult, the virus remains dormant in the dorsal root ganglia of immunocompetent individuals until reactivation is triggered, causing herpes zoster or shingles. VZV disease may be more mild, such as skin inflammation and disseminated painful cutaneous lesions [1], or, in its more severe form, cause more debilitating complications such as postherpetic neuralgia [1, 30], ocular changes such as vision loss [1], tinnitus [1], nerve palsies [1], hepatic disease [30], encephalitis [1], bacterial super infection [1], and increased mortality [1]. In a HCT population, the disease typically occurs most often within the first- or second-year post-hematopoietic cell transplant [28]. Recurrence in this population is also common (up to 16%) [28].

Table 7 Clinical studies supporting HSV and VZV Prophylaxis in an HCT population

Ref	HCT Type	N	Intervention	Primary Endpoint	Disease, no. (%)	Time to event	Outcomes	*P*
[5]	Allo	38[a]	ACV 800 mg PO BID[b]	VZV disease at 1 yr post HCT	2 (5)	Year 1 post HCT	HR 0.16 (0.035, 0.74)	0.006
					8 (21)	Year 2 post HCT	HR 0.52 (0.21, 1.3)	0.15
					11 (29)	Year 3 post HCT	HR 0.65 (0.29, 1.5)	0.40
					14 (37)	Year 5 post HCT	HR 0.76 (0.36, 1.6)	0.46
		39[a]	Placebo[b]		10 (26)	Year 1 post HCT		
					12 (31)	Year 2 post HCT		
					13 (33)	Year 3 post HCT		
					14 (36)	Year 5 post HCT		
[30]	Allo aHCT	720 212[e]	None[c]	VZV disease within 2 yrs post HCT	Allo 178 (25) Auto 45 (21.2)	Allo Md 6 mo (0–24) aHCT Md 5 mo (0–22)		Allo <0.001 aHCT <0.001
	Allo aHCT	851 266[e]	800 mg PO or VACV 500 mg PO BID for 1 yr post HCT		Allo 73 (9) 35 (7.1)	Allo Md 13 mo (0–23) aHCT Md 14 mo (2–24)		
	Allo aHCT	402 184[e]	Same as above for 1 yr post HCT or 6 mo after IS complete[c]		Allo 18 (4)[d]	Allo Md 16 mo (2–23)[d]		
[1]	Allo aHCT	17 20[e]	VACV 1000 mg PO BID from 4 mo through 2 yrs post HCT	VZV disease	3 (11) occurring *BEFORE* VACV initiated ITT: 0	1 day post randomi-zation *BEFORE* VACV initiated ITT: 0		0.21 ITT (0.025)
	Allo aHCT	19 7[e]	Placebo		6 (23) ITT: 6 (23)	Md 278 days (range NR)		
[28]	aHCT	1000[e]	ACV 500 mg PO BID or VACV 500 mg PO BID for 1 yr post HCT	Herpes zoster	194	Md 19 mo (IQR, 14.4–30.7)	CIHZ 21% (95% CI, 18% to 24%)	

(continued)

Table 7 (continued)

Ref	HCT Type	N	Intervention	Primary Endpoint	Disease, no. (%)	Time to event	Outcomes	P
[26]	aHCT	108	VACV 500 mg PO BID from day −3 of HCT until engraftment, day +30, toxicity, or HSV	HSV reactivation	3 (2.7%)			
	aHCT	43	ACV 5 mg/kg IV every 12 hours from day −3 of HCT until engraftment		1 (2%)			
	aHCT	38	None		17 (45%)			
[25]	Allo aHCT	523 154	ACV (250 mg/m² IV every 12 hours) until engraftment then 800 mg BID until day +30	HSV disease within 2 yrs post HCT	1st at 2-years: 31.6% (95% CI 28–35%)			<0.001
	Allo aHCT	688 210	800 mg PO or VACV 500 mg PO BID for 1 yr post HCT		1st at 2-years: 3.9% (95% CI 2.7–5.2%)			<0.001
	Allo aHCT	330 144	Same as above for 1 yr post HCT plus extended use while on IS		1st at 2-years: 0			
[23]	aHCT	60	ACV 600 mg PO q 6 h at start of conditioning until engraftment	Reactivation of HSV-1 and occurrence of mucositis	0			
	aHCT	60	VACV 800 mg PO BID at start of conditioning until engraftment		0			

Abbreviations: *ACV* acyclovir, *aHCT* autologous hematopoietic cell transplantation, *allo* allogeneic, *BID* twice daily, *CI* confidence interval, *CIHZ* cumulative incidence of herpes zoster, *FAM* famciclovir, *HCT* hematopoietic cell transplantation, *HR* hazard ratio, *HSV* herpes simplex virus, *IQR* interquartile range, *IS* immunosuppression, *IV* intravenous, *Md* median, *mo* months, *N* number of participants, *NR* not reported, *PO* per os, *q* every, *VaCV* valacyclovir, *ValGCN* valganciclovir, *VZV* varicella-zoster virus, *yr* year

[a]VZV Serostatus at entry not reported

[b]If seropositive for HSV prior to HCT, prophylactic ACV (250 mg/m² IV every 12 h) was given from day −7 until day +30 after HCT

[c]If seropositive for HSV prior to HCT, ACV given from day −7 until engraftment or resolution of mucositis

[d]aHCT recipients from cohort 2 and 3 were combined

[e]All patient VZV seropositive at study entry

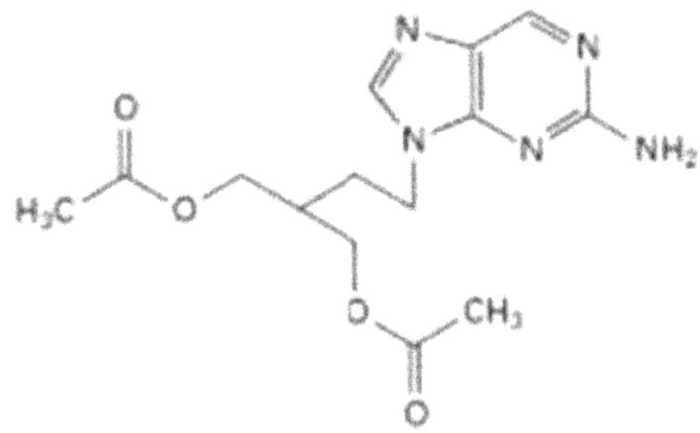

Fig. 3 Structural formula for famciclovir [34]

Table 8 Chemical properties of Penciclovir and Famciclovir [35]

	Penciclovir	Famciclovir
Physical description	White crystalline solid	White to pale yellow solid
Melting point (°C)	275–277	102–104
Water solubility	Soluble at 20 °C: 1.7 mg/mL	Freely soluble (>25% w/v) in water, then rapidly precipitates as the sparingly soluble (2%–3% w/v) monohydrate
pKA value(s)	3.2, 9.4	2.91, 3.83

Prophylaxis with either ACV or VACV not only reduces the risk for VZV infection (Table 7) [1, 5, 28, 30] but has also been shown to improve overall survival in an HCT population [30]. In the mid-1980s, early clinical trials found that use of ACV prophylaxis for 6 months following hematopoietic cell transplantation significantly reduced both HSV and VZV disease incidence in this population [5]. Later studies found that in both allogeneic [5, 30] and autologous HCT recipients [30, 31], antiviral prophylaxis given for 1 year post-HCT significantly reduced VZV disease at 1 year. Following ACV discontinuation, protection was also seen beyond 1 year in both HCT types [30]. Notably, there was a prolonged benefit found in allogeneic HCT recipients who remained on antiviral prophylaxis during concomitant graft versus host disease immunosuppression [30]. A longer duration of 2-years has also been found to be beneficial without resulting in significant rebound or VZV reactivation upon discontinuation of prophylaxis [1]. Current 2009 international guidelines recommend ACV or VACV prophylaxis among VZV-seropositive patients for 1 year post-allogeneic and autologous HCT recipients [20]; however, depending on the patient, different strategies could be considered. In a recent retrospective observational study including 1000 autologous HCT recipients, at least one VZV infection occurred in 21% of patients over a 5-year period, of which 82% of disease occurred in patients not on prophylaxis [28]. Despite patients being prescribed standard prophylaxis for the requisite 1-year period, up to 16% of patients experienced recurrent disease, begging the question of whether longer prophylaxis is warranted [28]. Consequently, the authors noted that continued prophylaxis,

Table 9 PCV viral inhibition *in vitro* [13]

Virus	IC_{50} (μg/ml)
CMV	51
EBV	1.5–3.1
HSV-1	0.04–1.8
HSV-2	0.06–4.4
VZV	1.6–8.0

Abbreviations: *CMV* cytomegalovirus, *EBV* Epstein-Barr virus, *HSV* herpes simplex virus, *PCV* penciclovir, *VZV* varicella zoster virus

especially in the elderly among those on maintenance therapy, may be beneficial [28]. For treatment of HSV or VZV, see Table 5 for dosing guidelines.

7 Background: Penciclovir and Famciclovir

Penciclovir (PCV), also identified by its chemical name 2-amino-9-[4-hydroxy-3-(hydroxymethyl)butyl]-1H-purin-6-one [2], is a guanosine analog [4] in the same family of antiviral agents as acyclovir (ACV) while differing in structure [4]. Subsequently, there is a higher affinity for PCV to herpes simplex virus (HSV) thymidine kinase (TK) but lower affinity for viral DNA polymerase [4].

The bioavailability of PCV is poor. Thus, its prodrug famciclovir (FAM), the diacetyl, 6-deoxy ester of penciclovir, was developed and approved in the mid-1990s (Fig. 3) [2, 21]. Following oral administration, FAM is quickly converted to PCV via hepatic first-pass metabolism; no parent drug remains [21].

Like ACV, FAM is activated for PCV. Once FAM is administered, it is metabolized by the liver to 6-deoxy-penciclovir, then further oxidized to PCN [32]. Once taken in by the viral cell, like ACV, PCN undergoes tri-phosphorylation; first, phosphorylation is with TK enzymes, followed by cellular kinases (Fig. 1) [2, 21]. The triphosphate form inhibits viral DNA polymerase, terminating viral replication [32]. Compared to ACV, PCV has about a 100-fold [12] higher affinity for TK, a longer in vitro intracellular half-life in HSV and varicella zoster virus (VZV)-infected cells, and results in higher accumulation within HSV or VZV [21].

The safety profile of FAM is also similar to that of ACV. The most common side effect reported in studies is headache [33].

Despite its similar properties to ACV, FAM's differs in its bioavailability and pharmacokinetic profile, making it a viable option for activity against Herpesviridae viruses. Additional chemical properties are described in Table 8.

Table 10 Dosing guidelines for famciclovir [21, 34]

Dosage *(Adult)*[a]	Herpes simplex virus (genital): Recurrent 1 g PO once at onset of prodrome then 500 mg BID for 7–10 days Herpes simplex virus (orolabial): Immunocompromised: 500 mg PO BID for 7–10 days Herpes simplex virus prophylaxis: suppressive 250–500 mg PO BID Herpes zoster (shingles): Acute localized: 500 mg PO TID for 7 to 10 days Disseminated, start with intravenous acyclovir then switch to PO famciclovir to complete 10–14 day treatment
Renal Dysfunction	Reduced with CrCL <60 mL/min, <40 mL/min, or <20 mL/min depending on initial dose / frequency Dialyzable; administer after HD session
Hepatic Impairment	Mild to moderate, no dose adjustment Not studied in severe impairment
Interactions: Food	Minimal concern for drug interaction Following FAM administration, if fasting, peak PCV plasma concentrations are attained in under 60 min. If FAM taken with food (within 2 h), this is delayed PCV BA is not impacted by food

Abbreviations: *BA* bioavailability, *BID* twice daily, *CrCL* creatinine clearance, *d* day(s), *h* hour(s), *HD* hemodialysis, *PO* per os, *q* every, *TID* three times daily
[a]Dose, frequency, and duration dependent on indication

Table 11 Pharmacokinetic properties of Penciclovir and Famciclovir [21, 37]

BA/ Absorption	Cmax	Half-life	Distribution	Clearance	Elimination
Delayed by food PCV PO 5% FAM 60–77%	Reduced by food 1.59 mg/L[a] Prolonged in ESRF	Normal renal function 2–2.5 h Prolonged in renal impairment Intracellular infected cells: HSV-1, 10 h; HSV-2, 20 h; and VZV, 9 h	Protein binding <20% Vd: 1.5 L/kg No available data for fluid or tissue distribution	25–30 L/h Dialyzable	Renal, unchanged

Abbreviations: *BA* bioavailability, *ESRF* end stage renal failure, *FAM* famciclovir, *h* hour(s), *HSV* herpes simplex virus, *PCV* penciclovir, *PO* per os, *Vd* volume of distribution, *VZV* varicella zoster virus
[a]For si1ngle oral dose of famciclovir 250 mg

Table 12 General indications (dosing outlined above) [34]

Approved	Herpes simplex virus (genital) Herpes simplex virus (orolabial) - Immunocompetent or immunocompromised Herpes zoster (shingles): - Immunocompetent or immunocompromised

8 Spectrum of Activity: In Vitro

FAM has activity against HSV-1 and HSV-2, and VZV [2]. In vitro inhibitory concentrations of penciclovir are reported as the following (Table 9):

9 Resistance

Little to no data has been reported specific to FAM resistance. Similar to ACV, resistance to HSV may be due to a *TK minus* strain [12]. If the mutation is related to the viral DNA-polymerase, herpesvirus may still be sensitive to PCV, albeit reduced [12]. Thus, the majority of HSV ACV and VZV resistant strains show cross-resistance to FAM [32] though this is not universal [17], and resistance to PCV is less than that to ACV [12].

10 Pharmacokinetic/Pharmacodynamic/Dosing

Renal dysfunction: PCV is predominantly cleared renally. As renal clearance declines, PCV clearance follows. Peak plasma drug concentration (Cmax), area under the concentration-time curve (AUC), and half-life (T ½) all increase [36].

Hepatic impairment: Conversion of FAM to PCN occurs in part by hepatic metabolism, but liver impairment has not been shown to impact the bioavailability of the drug, and dose adjustments are not recommended [21, 32] (Tables 10 and 11).

11 Clinical Utilization, Dosing, and Administration: FDA-Approved Indications

At present, FAM is not routinely recommended as either HSV or VZV prophylaxis for HCT recipients [20, 22], though it can be utilized if necessary. In contrast, in solid organ transplants, when both donor and recipient are cytomegalovirus sero-negative, FAM prophylaxis is an option [38]. Indications for FAM are outlined below in Table 12.

12 Clinical Data

Though there is a paucity of data supporting its use in the HCT population, given its similarities to ACV and the data outside of this population, it can be extrapolated that FAM can be utilized, if necessary, for activity in HSV and VZV [22, 33]. In an immunocompetent patient population with HSV-2 recurrent genital herpes, compared to placebo, FAM was shown to increase the number of aborted lesions (prodromal symptoms without lesion appearance or progression beyond papule stage) while reducing non-aborted lesion healing time [39]. In patients with uncomplicated VZV or herpes zoster (HZ), FAM was found to be equally effective to ACV in treatment with a more convenient dosing strategy [33]. Pott and colleagues conducted a multicenter phase III non-inferiority comparing both agents in treating uncomplicated HZ in immunocompetent adults. Patients were randomized 1:1 to receive either FAM 500 mg orally three times daily ($n = 75$) or ACV 800 mg orally five times daily ($n = 76$) for 7 days, with the primary endpoint being time until complete crusting of all lesions. In the FAM arm, time to full crusting was 14.840 days compared to 15.033 days in the ACV arm ($p = 0.820$); complete cure was achieved in 94.67% and 94.74% (0.985), respectively. Between both treatment arms, signs and symptoms related to HZ disease, including pain, vesicular lesions, loss of sensitivity, burning, and pruritus, were also similar [33]. FAM was also compared to VACV for HZ treatment in a double-blind, multicenter clinical trial to determine the equivalence of efficacy and tolerability. Five-hundred ninety-seven patients were randomized to receive either FAM 500 mg or VACV 1 g thrice daily for 7 days. Overall, there was no difference between the treatment arms in pain related to HZ (hazard ratio, 1.02; 95% confidence interval, 0.84–1.23; $p = 0.84$) or healing of lesions. Side effects were similar, with headache and nausea reported most often in each arm [40]. In immunocompromised patients, FAM can also be considered for VZV and HSV prophylaxis [41]. In a retrospective review assessing the feasibility of FAM to prevent HSV, VZV, or CMV at 1 year following solid organ transplantation, Crouch et al. evaluated 78 patients who received FAM 500 mg daily for 3 months to determine incidence. Overall, no documented cases occurred during the 3-month treatment period, and only one patient developed VZV viral infection at 1-year post SOT [41].

13 Summary

Since the early 1960s, the antiviral armamentarium has continued to expand, all the while improving drug tolerance, ease of administration, bioavailability, and spectrum of activity. Prophylactic use of ACV, VACV, and FAM in the HCT population is well-established in preventing the debilitating complications related to either HSV or VZV. Considerations for future studies may include the development of more potent agents to circumvent the rise of resistance.

Conflict of Interest The author has no conflict of interest.

References

1. Klein A, Miller KB, Sprague K, DesJardin JA, Snydman DR. A randomized, double-blind, placebo-controlled trial of valacyclovir prophylaxis to prevent zoster recurrence from months 4 to 24 after BMT. Bone Marrow Transplant. 2011 Feb;46(2):294–9.
2. Majewska A, Mlynarczyk-Bonikowska B. 40 Years after the Registration of Acyclovir: Do We Need New Anti-Herpetic Drugs? Int J Mol Sci. 2022 Mar 22;23(7):3431.
3. Kłysik K, Pietraszek A, Karewicz A, Nowakowska M. Acyclovir in the Treatment of Herpes Viruses - A Review. Curr Med Chem. 2020;27(24):4118–4137.
4. Sadowski LA, Upadhyay R, Greeley ZW, Margulies BJ. Current Drugs to Treat Infections with Herpes Simplex Viruses-1 and -2. Viruses. 2021 Jun 25;13(7):1228.
5. Boeckh M, Kim HW, Flowers ME, Meyers JD, Bowden RA. Long-term acyclovir for prevention of varicella zoster virus disease after allogeneic hematopoietic cell transplantation--a randomized double-blind placebo-controlled study. Blood. 2006 Mar 1;107(5):1800–5.
6. Wagstaff AJ, Faulds D, Goa KL. Aciclovir. A reappraisal of its antiviral activity, pharmacokinetic properties and therapeutic efficacy. Drugs. 1994 Jan;47(1):153–205.
7. Beutner KR. Valacyclovir: a review of its antiviral activity, pharmacokinetic properties, and clinical efficacy. Antiviral Res. 1995 Dec;28(4):281–90.
8. MacDougall C, Guglielmo BJ. Pharmacokinetics of valaciclovir. J Antimicrob Chemother. 2004 Jun;53(6):899–901.
9. Griffiths PD. A perspective on antiviral resistance. J Clin Virol. 2009 Sep;46(1):3–8.
10. Acyclovir and Valacyclovir at DailyMed (accessed on July 1, 2022 at https://dailymed.nlm.nih.gov/dailymed/index.cfm)
11. Valacyclovir at PubChem (accessed on July 1, 2022 at https://pubchem.ncbi.nlm.nih.gov)
12. Field HJ, Vere Hodge RA. Recent developments in anti-herpesvirus drugs. Br Med Bull. 2013;106:213–49.
13. Piret J, Boivin G. Resistance of herpes simplex viruses to nucleoside analogues: mechanisms, prevalence, and management. Antimicrob Agents Chemother. 2011 Feb;55(2):459–72.
14. van der Beek MT, Vermont CL, Bredius RG, Marijt EW, van der Blij-de Brouwer CS, Kroes AC, Claas EC, Vossen AC. Persistence and antiviral resistance of varicella zoster virus in hematological patients. Clin Infect Dis. 2013 Feb;56(3):335–43.
15. Hatchette T, Tipples GA, Peters G, Alsuwaidi A, Zhou J, Mailman TL. Foscarnet salvage therapy for acyclovir-resistant varicella zoster: report of a novel thymidine kinase mutation and review of the literature. Pediatr Infect Dis J. 2008 Jan;27(1):75–7.
16. Kim JH, Schaenman JM, Ho DY, Brown JM. Treatment of acyclovir-resistant herpes simplex virus with continuous infusion of high-dose acyclovir in hematopoietic cell transplant patients. Biol Blood Marrow Transplant. 2011 Feb;17(2):259–64.
17. Oka T, Hishizawa M, Yamashita K, Shiraki K, Takaori-Kondo A. Successful treatment with famciclovir for varicella zoster virus infection resistant to acyclovir. J Infect Chemother. 2021 May;27(5):755–758.
18. Zovirax Manufacturer's label (accessed on July 1, 2022 at https://gskpro.com/content/dam/global/hcpportal/en_BW/PI/Zovirax-IV-GDS24.pdf
19. Valtrex Manufacturer's label (accessed on July 1, 2022 at http://www.accessdata.fda.gov/drugsatfda_docs/label/2008/020487s014lbl.pdf on July 1, 2022);
20. Tomblyn M, Chiller T, Einsele H, Gress R, Sepkowitz K, Storek J, Wingard JR, Young JA, Boeckh MJ; Center for International Blood and Marrow Research; National Marrow Donor program; European Blood and Marrow Transplant Group; American Society of Blood and Marrow Transplantation; Canadian Blood and Marrow Transplant Group; Infectious Diseases Society of America; Society for Healthcare Epidemiology of America; Association of Medical

Microbiology and Infectious Disease Canada; Centers for Disease Control and Prevention. Guidelines for preventing infectious complications among hematopoietic cell transplantation recipients: a global perspective. Biol Blood Marrow Transplant. 2009 Oct;15(10):1143–238.
21. Gill KS, Wood MJ. The clinical pharmacokinetics of famciclovir. Clin Pharmacokinet. 1996 Jul;31(1):1–8.
22. Ullmann AJ, Schmidt-Hieber M, Bertz H, Heinz WJ, Kiehl M, Krüger W, Mousset S, Neuburger S, Neumann S, Penack O, Silling G, Vehreschild JJ, Einsele H, Maschmeyer G; Infectious Diseases Working Party of the German Society for Hematology and Medical Oncology (AGIHO/DGHO) and the DAG-KBT (German Working Group for Blood and Marrow Transplantation). Infectious diseases in allogeneic haematopoietic stem cell transplantation: prevention and prophylaxis strategy guidelines 2016. Ann Hematol. 2016 Sep;95(9):1435–55.
23. Eisen D, Essell J, Broun ER, Sigmund D, DeVoe M. Clinical utility of oral valacyclovir compared with oral acyclovir for the prevention of herpes simplex virus mucositis following autologous bone marrow transplantation or stem cell rescue therapy. Bone Marrow Transplant. 2003 Jan;31(1):51–5.
24. Pires de Mello CP, Bloom DC, Paixão IC. Herpes simplex virus type-1: replication, latency, reactivation and its antiviral targets. Antivir Ther. 2016;21(4):277–86.
25. Erard V, Wald A, Corey L, Leisenring WM, Boeckh M. Use of long-term suppressive acyclovir after hematopoietic stem-cell transplantation: impact on herpes simplex virus (HSV) disease and drug-resistant HSV disease. J Infect Dis. 2007 Jul 15;196(2):266–70.
26. Dignani MC, Mykietiuk A, Michelet M, Intile D, Mammana L, Desmery P, Milone G, Pavlovsky S. Valacyclovir prophylaxis for the prevention of Herpes simplex virus reactivation in recipients of progenitor cells transplantation. Bone Marrow Transplant. 2002 Feb;29(3):263–7.
27. Truong Q, Veltri L, Kanate AS, Hu Y, Craig M, Hamadani M, Cumpston A. Impact of the duration of antiviral prophylaxis on rates of varicella-zoster virus reactivation disease in autologous hematopoietic cell transplantation recipients. Ann Hematol. 2014 Apr;93(4):677–82.
28. Sahoo F, Hill JA, Xie H, Leisenring W, Yi J, Goyal S, Kimball LE, Lee I, Seo S, Davis C, Pergam SA, Flowers ME, Liaw KL, Holmberg L, Boeckh M. Herpes Zoster in Autologous Hematopoietic Cell Transplant Recipients in the Era of Acyclovir or Valacyclovir Prophylaxis and Novel Treatment and Maintenance Therapies. Biol Blood Marrow Transplant. 2017 Mar;23(3):505–511.
29. McKay SL, Guo A, Pergam SA, Dooling K. Herpes Zoster Risk in Immunocompromised Adults in the United States: A Systematic Review. Clin Infect Dis. 2020 Oct 23;71(7):e125–e134.
30. Erard V, Guthrie KA, Varley C, Heugel J, Wald A, Flowers ME, Corey L, Boeckh M. One-year acyclovir prophylaxis for preventing varicella-zoster virus disease after hematopoietic cell transplantation: no evidence of rebound varicella-zoster virus disease after drug discontinuation. Blood. 2007 Oct 15;110(8):3071–7
31. Zhang D, Weiss T, Feng Y, Finelli L. Duration of Antiviral Prophylaxis and Risk of Herpes Zoster among Patients Receiving Autologous Hematopoietic Stem Cell Transplants: A Retrospective, Observational Study. Adv Ther. 2017 Jul;34(7):1610–1621.
32. Cirelli R, Herne K, McCrary M, Lee P, Tyring SK. Famciclovir: review of clinical efficacy and safety. Antiviral Res. 1996 Mar;29(2–3):141–51.
33. Pott Junior H, de Oliveira MFB, Gambero S, Amazonas RB. Randomized clinical trial of famciclovir or acyclovir for the treatment of herpes zoster in adults. Int J Infect Dis. 2018 Jul;72:11–15.
34. Famciclovir at DailyMed (accessed on July 1, 2022 at https://dailymed.nlm.nih.gov/dailymed/index.cfm)
35. Famciclovir at PubChem (accessed on July 1, 2022 at https://pubchem.ncbi.nlm.nih.gov/ on July 1, 2022.
36. Boike SC, Pue MA, Freed MI, Audet PR, Fairless A, Ilson BE, Zariffa N, Jorkasky DK. Pharmacokinetics of famciclovir in subjects with varying degrees of renal impairment. Clin Pharmacol Ther. 1994 Apr;55(4):418–26.

37. Pue MA, Pratt SK, Fairless AJ, Fowles S, Laroche J, Georgiou P, Prince W. Linear pharmacokinetics of penciclovir following administration of single oral doses of famciclovir 125, 250, 500 and 750 mg to healthy volunteers. J Antimicrob Chemother. 1994 Jan;33(1):119–27.
38. Kotton CN, Kumar D, Caliendo AM, Huprikar S, Chou S, Danziger-Isakov L, Humar A; The Transplantation Society International CMV Consensus Group. The Third International Consensus Guidelines on the Management of Cytomegalovirus in Solid-organ Transplantation. Transplantation. 2018 Jun;102(6):900–931.
39. Aoki FY, Tyring S, Diaz-Mitoma F, Gross G, Gao J, Hamed K. Single-day, patient-initiated famciclovir therapy for recurrent genital herpes: a randomized, double-blind, placebo-controlled trial. Clin Infect Dis. 2006 Jan 1;42(1):8–13.
40. Tyring SK, Beutner KR, Tucker BA, Anderson WC, Crooks RJ. Antiviral therapy for herpes zoster: randomized, controlled clinical trial of valacyclovir and famciclovir therapy in immunocompetent patients 50 years and older. Arch Fam Med. 2000 Sep–Oct;9(9):863–9.
41. Crouch A, Le M, Rogers C, Shao S, Kotton C. Evaluation of low dose famciclovir as herpes simplex virus and varicella zoster virus prophylaxis in cytomegalovirus low-risk solid organ transplant recipients. Transpl Infect Dis. 2021 Oct;23(5):e13711.

Antiviral Agents in the Hematopoietic Stem Cell Transplant Population: Foscarnet

Elizabeth DiMaggio

Abstract Over the past three decades, the antiviral agent foscarnet has been used in the prevention and treatment of several viruses, including herpes simplex infections, varicella zoster virus, cytomegalovirus, and human herpesvirus 6. Foscarnet is an essential tool against viral infections owing to its strong activity against drug-resistant organisms; however, its administration can be difficult and requires close monitoring for electrolyte abnormalities and renal toxicity. What follows is a description of its mechanism of action, pharmacokinetics, and side effects, as well as a review of the clinical data supporting its use in the hematopoietic cell transplant population.

Keywords Antiviral · Foscarnet · Herpes simplex · Varicella zoster · Cytomegalovirus · CMV · HHV-6

1 Background

Foscarnet sodium (FOS), or phosphonoformic acid, or trisodium salt (Na3CO5P•6 H20), was first approved in 1991 for the treatment of cytomegalovirus retinitis (Fig. 1) [1]. An analog to pyrophosphate, FOS works as a selective and reversible inhibitor of the pyrophosphate binding site on the viral DNA polymerase, thereby preventing nucleotide integration and viral DNA elongation; activation or phosphorylation by thymidine kinase or other enzymes is not required (Fig. 2) [2]. Concentrations 100 times higher are necessary for mammalian DNA polymerase inhibition [3]. FOS is virustatic [4], and if discontinued before the infection is cleared, viral replication will resume. Therefore, maintenance therapy is essential to

E. DiMaggio (✉)
Department of Pharmacy, Moffitt Cancer Center, Tampa, FL, USA
e-mail: Elizabeth.DiMaggio@moffitt.org

C. Somboonwit et al. (eds.), *Global Virology IV: Viral Disease Diagnosis and Treatment Delivery in the 21st Century*,
https://doi.org/10.1007/978-3-031-57369-9_23

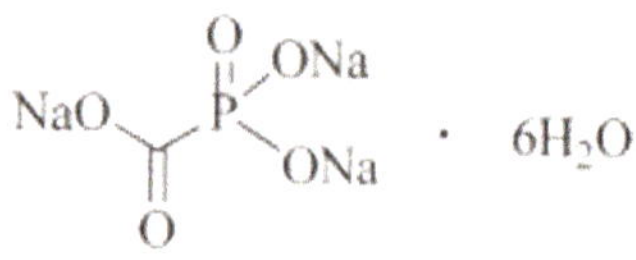

Fig. 1 Structural formula for foscarnet sodium [1]

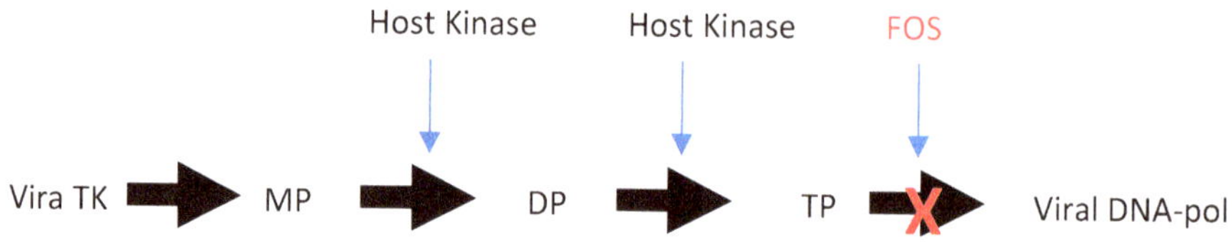

Fig. 2 The mechanism of action of foscarnet. FOS is an inhibitor of viral DNA-pol. Abbreviations: *DP* diphosphate, *FOS* foscarnet, *MP* monophosphate, *NA* nucleoside analog, *pol* polymerase, *TK* viral thymidine, *TP* triphosphate. (Reference: Figure adapted from Majewska A, et al. [2])

Table 1 Chemical properties of foscarnet [7]

	Foscarnet
Physical description	White, crystalline powder
Melting point (°C)	88.06
Water solubility	Complete
pKA value(s)	3.13

permit the body's immune system to fully eradicate the infection [4, 5]. Chemical properties of FOS are described in Table 1.

Common (>25% incidence) adverse effects of FOS include fever, nausea and vomiting, anemia, diarrhea, headache, and renal impairment, and electrolyte imbalances (calcium, potassium, magnesium, and phosphorus) [4, 5]. Less common side effects include seizures, loin pain [4], or genital ulceration [5]; the latter thought to be related to local irritation from unchanged drug in the urine. Personal hygiene is also recommended, and resolution occurs with cessation of the agent [4, 5]. Renal impairment, resulting from acute tubular necrosis, can lead to two- to threefold increases in serum creatinine and possibly hemodialysis [5]. In patients undergoing hematopoietic cell transplantation (HCT), Foster and colleagues found that those exposed to FOS at 6 and 12 months had significantly lower median eGFRs compared to those who did not receive the agent [6]. Renal dysfunction can be avoided with dose reductions based on serum creatinine levels, intermittent infusions, adequate pre- and post-hydration, and discontinuing concomitant nephrotoxic agents [5].

2 Spectrum of Activity: In Vitro

FOS has activity against all known herpes viruses and hepatitis B virus via inhibition of DNA polymerase and against human immunodeficiency virus (HIV) via noncompetitive reversible inhibition to the HIV reverse transcriptase [3]. The IC_{50}

Table 2 Foscarnet inhibition of viral replication *in vitro* [4]

Virus	IC_{50} (μM)
CMV	50–800 (mean = 269)
GCV resistant CMV	190
EBV	<500
HHV-6	49
HIV-1	11–32
HSV-1 and -2	10–130
HSV-TK negative mutant	67
HSV-DNA polymerase mutants	5–443
VZV	48–90

Abbreviations: *CMV* cytomegalovirus, *EBV* Epstein-Barr virus, *GCV* ganciclovir, *HHV* human herpes virus, *HIV* human immunodeficiency virus, *HSV* herpes simplex virus, *TK* thymidine kinase, *VZV* varicella zoster virus

value, or antiviral concentration resulting in 50% inhibition in vitro, has been reported as the following (Table 2):

3 Resistance

Because FOS is a pyrophosphate analog that does not require thymidine kinase phosphorylation and activation, resistance is associated with mutations in the DNA polymerase, namely, the pUL30 in HSV or the pUL54 in cytomegalovirus (CMV) [4].

Specific to CMV, mutations are commonly located on pUL54 DNA polymerase residues found in terminal domain 1 and terminal 2 (residues 555–600), palm, and finger domains (residues 696–981) (Table 3) [8, 9].

Incidence of viral resistance varies but is associated with prolonged FOS use and subtherapeutic doses [10] resulting in a three- to fivefold decrease in FOS susceptibility [8]. In a population of patients with AIDS receiving FOS for CMV retinitis, resistance at 6 and 9 months was 26% and 37%, respectively [11]. Little data is available describing FOS resistant HSV strains and is mostly associated with AIDS patients (Table 4) [12].

4 Pharmacokinetic/Pharmacodynamic

Intravenous infusion is the preferred route of administration due to poor oral absorption, and both intermittent and continuous infusion have been investigated [3, 5]. To avoid toxicity and increase convenience, intermittent infusion is considered standard [14]. FOS has a short first and second phase elimination half-life (1.4 and

Table 3 Mutations of DNA polymerase amino acid substitutions in cell culture [1, 9, 12, 13]

CMV pUL54	T419M, T552N, S585A, F595I, Q807A, M844T/V, V946L
HSV 1 pUL30	Y577H, E597D, A605V, L702H, V714M, L774F, L788M, D780N, L782I, P797T, L802F, V813M, V817M, Y818C, T821M, R842S, S889A, F891C, V892M, D907V, A910V, SRA914-916LCV, V958L, R959H

Abbreviations: *CMV* cytomegalovirus, *HSV* herpes simplex virus

Table 4 Mutations specific to the DNA polymerase amino acid substitutions relevant to clinical practice [1, 9, 12, 13]

CMV pUL54	N495K, Q578H/L, D588 E/ N, T700A, V715M, E756D/K/Q, L773V, L776M, V781I, V787L, L802M, A809V, V812L, T813S, T821I, A834P, T838A, G841A/S, del 981982
HSV 1 pUL30	S599L, D672N, R700G, V715G, A719T/V, S724N, E798K, G841C/S, A910T, Y941H
HSV2 p UL30	A724T, S725G, S729N, Q732R, L783M, D785N, T844I, L850I, D912V

Abbreviations: *CMV* cytomegalovirus, *HSV* herpes simplex virus

6.8 h, respectively) but can deposit in the bone and cartilage, contributing to a prolonged terminal phase elimination half-life [3, 5]. The majority of the drug (83–90%) is excreted unchanged in the urine [3, 5]. With impaired renal function, clearance is decreased while half-life is increased prompting dose adjustment relative to serum creatinine. The pharmacokinetic properties are outlined below in Table 5. Of note, FOS given concomitantly with other nephrotoxic medication may exacerbate renal injury or impairment and should be avoided; otherwise, other drug interactions do not appear to be clinically relevant.

5 Clinical Utilization, Dosing, and Administration

5.1 Indications

FOS is indicated for treatment of CMV retinitis in patients with acquired immunodeficiency syndrome (AIDS) and treatment of acyclovir-resistant HSV infections (Table 6); however, in the HCT population, it is commonly used for either CMV or human herpes virus 6 (HHV6) treatment. In a report by Foster et al. over a 14-year period, FOS exposure occurred in 447 HCT recipients of which 58% were treated for CMV while 31% were treated for HHV6 [6]. Common doses for intravenous FOS for both indications are outlined in Table 7.

Dosing strategies must take into consideration renal impairment. Foscarnet dosages require adjustment based on renal function (Table 8) [1]. Administration is not recommended if CrCL (mL/min/kg) is <0.4. Only limited data available for patients on hemodialysis [18, 19].

Table 5 Pharmacokinetic properties of FOS following induction therapy in AIDS patients with CMV retinitis [1]

	FOS dose	
	60 mg/kg IV every 8 h	90 mg/kg IV every 12 h
C_{max} at ss (μM)	589 ± 192 (24)	623 ± 132 (19)
C_{trough} at ss (μM)	114 ± 91 (24)	63 ± 57 (17)
Vd (l/kg)	0.41 ± 0.13 (12)	0.52 ± 0.20 (18)
Plasma T ½ (h)	4.0 ± 2.0 (24)	3.3 ± 1.4 (18)
Systemic clearance (L/h)	6.2 ± 2.1 (24)	7.1 ± 2.7 (18)
Renal clearance (L/h)	5.6 ± 1.9 (5)	6.4 ± 2.5 (13)
CSF: plasma ratio	*50 mg/kg IV every 8 h* *Samples taken 3 h after completion of 1-h infusion* 0.69 ± 0.19(9)	*90 mg/kg IV every 12 h* *Samples taken 1 h after completion of 2-h infusion* 0.66 ± 0.11(5)

Values expressed as mean ± S.D. (number of subjects studied) for each parameter
Abbreviations: *AIDS* acquired immunodeficiency syndrome, C_{max} maximum plasma concentration, *CMV* cytomegalovirus, *CSF* cerebrospinal fluid, C_{trough} minimum plasma concentration, *h* hour(s), *IV* intravenous, *L* liter, *ss* steady state, *T½* half-life, *Vd* volume of distribution

Table 6 General label indications [1]

Approved	Treatment of CMV retinitis in patients with AIDS Treatment of acyclovir-resistant mucocutaneous HSV infections in immunocompromised patients

Abbreviations: *AIDS* acquired immunodeficiency syndrome, *HSV* herpes simplex virus

Table 7 Dosing guideline for foscarnet in CMV or HHV6 [1, 15–17]

CMV Preemptive therapy		CMV disease (Including retinitis)		HHV6		ACV-resistant HSV patients	
Dosage	Duration	Dosage	Duration	Dosage	Duration	Dosage	Duration
60 mg/kg BID	>=2 weeks (induction)	60 mg/kg TID	>3 weeks (induction)	90 mg/kg BID Or 60 mg/kg TID	>3 weeks and until clearance demonstrated	40 mg/kg TID or BID	2–3 weeks or until healed
90 mg/kg QD	Maintenance until PCR negativity can be considered	60 mg/kg BID or 90 mg/kg QD	Maintenance can be considered				

Abbreviations: *ACV* acyclovir, *BID* twice daily, *CMV* cytomegalovirus, *HHV6* human herpesvirus 6, *HSV* herpes simplex virus, *QD* daily, *TID* three times daily

Table 8 Dosing guideline for foscarnet in renally impaired patients [1]

	Induction HSV: Equivalent to		Induction CMV: Equivalent to		Maintenance CMV: Equivalent to	
CrCl (mL/min/kg)	80 mg/kg/day total	120 mg/kg/day total	60 mg/kg q8h	90 mg/kg q12h	90 mg/kg/daily	120 mg/kg/day
>1.4	40 Q12h	40 Q8h	60 Q8h	90 Q12h	90 Q24h	120 Q24h
>1.0–1.4	30 Q12h	30 Q8h	45 Q8h	70 Q12h	70 Q24h	90 Q24h
>0.8–1.0	20 Q12h	35 Q12h	50 Q12h	50 Q12h	50 Q24h	65 Q24h
>0.6–0.8	35 Q24h	25 Q12h	40 Q12h	80 Q24h	80 Q48h	105 Q48h
>0.5–0.6	25 Q24h	40 Q24h	60 Q24h	60 Q24h	60 Q48h	80 Q48h
>0.4–0.5	20 Q24h	35 Q24h	50 Q24h	50 Q24h	50 Q48h	65 Q48h
<0.4	NR	NR	NR	NR	NR	NR

Abbreviations: *CMV* cytomegalovirus, *CrCl* creatinine clearance, *HSV* herpes simplex virus, *NR* not recommended

Table 9 Foscarnet use in HCT recipients [13, 15]

CMV Pre-emptive therapy	Second-line if ganciclovir or valganciclovir cannot be used
CMV Treatment of disease	Second-line, used as a single-agent if ganciclovir cannot be used Second-line, combined with ganciclovir at full-dose
CMV R/R	If UL97 mutation detected with high level resistance to ganciclovir If UL54 mutation detected but is resistant to ganciclovir and cidofovir only If UL54 mutation conferring resistance to ganciclovir, foscarnet, and cidofovir, give combined with ganciclovir If UL56, UL89, UL51 mutations resistant to letermovir (can give as alternative to ganciclovir) No known mutations, give second-line following ganciclovir
HHV-6B encephalitis	IV foscarnet or ganciclovir is recommended (section dictated by toxicity and tolerance)

Abbreviations: *CMV* cytomegalovirus, *HHV-6B* human herpesvirus 6B, *IV* intravenous, *R/R* resistant and refractory

6 Special Populations: Foscarnet Use in HCT Recipients

Infections from CMV, a member of the Herpesviridae family, result in major complications in HCT recipients, leading to increased morbidity and mortality [15]. Seropositive patients are at the highest risk for development of CMV viremia, with incidence as high as 80%; subsequent multiorgan disease will develop in one-third of these patients [4]. At present, FOS is recommended for use against CMV as noted in Table 9 [13, 15] and with supporting data described in Table 10 [20, 22].

Table 10 Clinical data on foscarnet for use in CMV viremia and disease [20–22]

Study	Patients	Foscarnet	Viremia or Disease	Outcomes
Asakura M, et al.	65 MRD	Median 88 mg/kg (range, 7–216) for 20 days	Gastroenteritis ($n = 46$), pneumonia ($n = 12$), retinitis ($n = 2$), hepatitis ($n = 1$), encephalitis ($n = 1$), cystitis ($n = 1$), pneumonia + retinitis + gastroenteritis ($n = 2$)	8% symptom resolution ($n = 5$) 44% symptom improvement ($n = 28$) 31% no change ($n = 20$) 15% worsened ($n = 10$) 2% no symptoms/unevaluable ($n = 2$) More effective if ganciclovir not used first ($P = 0.1$)
Chang J, et al.	10 MRD	100 mg/kg in 3 divided doses every 8 h	Detection of CMV viremia was evidence of infection	3 patients discontinued therapy early (neurocortical toxicity). 7 patients completed therapy: $n = 5$ previous GCV-failures (CMV clearance at median of 11 days (range, 5–19) and CMV-free for a median of 7 months (range, 1- >14) 2 FOS-primary treatment (CMV-free for >3 and >13 mo post therapy)
Metafuni E, et al.	97 (30 UD with ATG; 67 haplo)	Median 165 mg/Kg (range 50–200)	Positive pp6 CMV antigenemia or positive qPCR	32 patients received foscarnet as 2nd line or 3 as 3rd line due to lack of response to previous GCV or VGCV treatment and/or cytopenia. Time from HCT to CMV infection (median): 33 days in UD, 27 days in haplo. CI of response overall was 87%. Median time to clearance of CMV was 7 days (range, 1–79). Response was higher in haplo arm (84 vs 59%, $p = 0.01$) but time to clear CMV was also shorter (9 vs 16 days, $p = 0.07$)

Abbreviations: *ATG* antithymocyte globulin, *CMV* cytomegalovirus, *GCV* ganciclovir, *haplo* haploidentical donors, *HCT* hematopoietic cell transplantation, *hr* hour(s), *mo* month(s), *MRD* matched-related donor, *qPCR* quantitative polymerase chain reaction, *UD* unrelated donor, *VGCV* valganciclovir

6.1 For CMV: Foscarnet Use for Pre-emptive Therapy

Ganciclovir is considered first-line in pre-emptive therapy against CMV; however, FOS has also been shown to yield similar results and may be deemed appropriate if ganciclovir toxicity, namely marrow failure or drug resistance, is of concern. In a prospective randomized multicenter trial comparing the two agents in matched-related and unrelated transplants, event-free survival at day 180, CMV disease, and overall mortality were similar between the two arms [23]. Growth factors were administered more frequently in the ganciclovir arm, owing to higher rates of severe neutropenia. Rates of impaired renal function were similar.

6.2 For CMV: Foscarnet for Treatment

Ganciclovir is also considered first-line therapy in the treatment of CMV, though alternatively, FOS can also be used. There is a paucity of literature for FOS in the setting of disease treatment, though small studies report on its efficacy in clearing CMV viremia as well.

HHV-6 is another member of the Herpesviridae family, commonly infecting the population but remaining dormant in hematopoietic reservoirs [24]. In allogeneic HCT transplant recipients on immunosuppression, HHV-6 reactivation occurs in up to 70% of patients and has also been found to be associated with an increased mortality [16, 24]. HHV-6 reactivation has been associated with myelitis, lung disease, bone marrow suppression, graft failure, gastrointestinal disease, liver disease, and myocarditis, though causation is only moderate at best [17, 25]. Contrary to this, fever and rash are more commonly detected, and HHV-6 has been found to have a causal relationship with the encephalitis [25], which can ultimately lead to serious CNS complications, including death [26]. The incidence of encephalitis has been reported as up to 2.3% in an HCT population and is more commonly seen in patients receiving HCTs from cord donors [27]. Several small studies have attempted to determine possible preventative strategies against HHV-6 encephalitis using FOS (Table 11) [28–31]. Though toxicity was determined to be mild and manageable, these methods were found to be ineffective, one theory being FOS's poor penetration of meninges when noninflamed [31]. At best, the benefit of FOS preemptive therapy was associated with HHV-6 encephalitis patients presenting with more mild cases compared to those not on preemptive therapy [31]. Currently, routine HHV-6 DNA screening for preemptive or prophylactic therapy is not recommended; however, FOS is recommended for treatment of HHV-6 encephalitis following HCT (Table 12) [16, 17].

6.3 *For HHV-6: Foscarnet for Pre-emptive Therapy*

Table 11 Clinical data on foscarnet for use in prevention of HHV-6 [28–31]

Study	Patients	Foscarnet	Adverse effects/ Toxicity	Outcomes
Ishiyama K, et al.	$N = 20$ UCB ($n = 15$) FOS ($n = 8$) Haplo ($n = 5$) FOS ($n = 0$)	90 mg/kg/day: started on day plasma HHV-6 DNA >5 × 10^2 copies/mL. Increased to 180 mg/kg/day when plasma HHV-6 DNA copy number increased to >1 × 10^2/mL or when symptoms of encephalitis appeared FOS discontinued when the plasma HHV-6 DNA was negative ×3 consecutive tests	UCB: AEs grade >3 in 7/8 FOS group and 4/7 No FOS group FOS AEs: electrolyte abnormalities, grade 2 renal dysfunction, LFT rise, skin rash	Haplo: 0 developed HHV-6 (thus did not receive FOS) UCB FOS group: 12 developed HHV-6 DNAemia (of which 7 were treated). Limbic encephalitis developed in 1 FOS patient 4 deaths or drop-outs (of which 2 developed HHV-6 DNAemia not requiring treatment); causes of death were VOD, PAH, TMA, bacteremia.
Ishiyama K, et al.	$N = 10$ FOS prophylaxis RD ($n = 2$), UD ($n = 6$) UCB ($n = 2$) $N = 10$ No FOS prophylaxis UD ($n = 10$)	90 mg/kg/d beginning day 7 to 21 post HCT (to day 25 in UCB)	Drop out d/t AEs ($n = 2$): Death from bacterial pneumonia ($n = 1$); low GFR ($n = 1$). Requiring FOS dose reduction due to creatinine increases ($n = 2$). Grade 3 AEs occurred in 9/10 FOS patients and 7/10 control patients. Renal dysfunction occurred less in FOS group ($n = 2$) compared to control ($n = 5$). Disturbance of consciousness was seen in the control group only ($n = 4$).	FOS group: 3/10 developed HHV-6 DNAemia between day 18–23 post HCT (during FOS administration ($n = 2$); 3-days after FOS cessation ($n = 1$)). No FOS patients developed neurological symptoms during the test period. Control group: Neurological symptoms detected ($n = 4$). Of the 4, 3 were positive for HHV-6 DNA in their plasma, 2 diagnosed w/HHV-6 encephalitis

(continued)

Table 11 (continued)

Study	Patients	Foscarnet	Adverse effects/ Toxicity	Outcomes
Ogata M, et al.	$N = 118$ FOS UD ($n = 57$) UCB ($n = 10$) Control No FOS UD ($n = 40$) UCB ($n = 11$)	50 mg/kg/d for 10 days from day of neutrophil engraftment (if not engrafted by day 18, FOS was started)	FOS group: early drop-out due to increased serum creatinine ($n = 3$ of 67), nausea ($n = 3$), headache ($n = 1$), electrolyte loss ($n = 1$); median FOS days was 6 (range, 2–9). FOS AEs grade ≥2 included nausea (30.3%), vomiting (13.4%), serum creatinine elevations (4.5%), magnesium loss (6%), calcium loss (4.5%), potassium loss (6%)	No reduction in high-level HHV-6 Reactivation (≥10^4 copies/mL) between FOS (19.4%) and control (33.8%); $p = 0.095$ HHV-6 encephalitis: FOS: $n = 3$ Control: $n = 5$ (2 developed and 3 possibly developed)
Ogata M, et al.	$N = 120$ FOS UCB ($n = 57$) Control No FOS UCB ($n = 63$)	90 mg/kg/d for 21 days from day from day 7 to 27 post HCT	FOS group: early drop-out due to HHV-6 encephalitis ($n = 5$), increased serum creatinine ($n = 3$), infection (not HHV-6, $n = 3$), TMA ($n = 1$), PIR ($n = 1$); median FOS days was 21 (range, 2–21). FOS AEs grade ≥3 at incidence >10% included encephalitis infection (12.3%), febrile neutropenia (18%) nausea (23%), hypokalemia (49%)	High-level HHV-6 Reactivation (≥10^4 copies/mL) in the FOS group was 18.3% compared to the control (57.3%); $p < 0.001$ HHV-6 encephalitis: FOS: $n = 7$ (5 during prophylaxis) Control: $n = 3$ $P = 0.14$

Abbreviations: *AE* adverse event, *d* day(s), *FOS* foscarnet, *GFR* glomerular filtration rate, *Haplo* haploidentical, *HCT* hematopoietic cell transplantation, *PAH* pulmonary alveolar hemorrhage, *PIR* pre-engraftment immune reaction, *RD* related donor, *TMA* thrombotic alveolar hemorrhage, *UCB* umbilical cord blood, *UD* unrelated donors, *VOD* veno-occlusive disease

6.4 *For HHV-6: Foscarnet for Treatment*

Table 12 Clinical data on foscarnet for use in Treatment of HHV-6 encephalitis [26, 27]

Study	Patients/ HCT type	Diagnostic testing and symptoms	Foscarnet	Outcomes
Zerr D, et al.	11 MUD, MRD, MMRD (4/11 UCB)	Positive in CSF (n = 11); HHV-6 encephalitis (n = 8)	FOS 50–180 mg/kg/d for 4–40 d with GCV[a] (n = 7) or without (n = 2) GCV monotherapy (n = 2)	Of 9 FOS-treated patients: favorable response/death not reported (n = 3); death (n = 5); no change (n = 1)
Ogata M, et al.	145 MRD, MMRD, MUD, MMUD, UCB	CNS dysfunction and absence of other causes, positive qPCR in CSF	FOS dose in monotherapy: FOS <90 mg/kg/d (n = 18), 90–179 mg/kg/d (n = 28), ≥180 mg/kg/d (n = 28)	Data available for 133 patients: 123 patients received foscarnet or ganciclovir monotherapy (response rates of neurological symptoms after starting antiviral therapy were 83.8% (FOS monotherapy) and 71.4% (GCV monotherapy); 10 patients FOS + GCV[a] *FOS 1st line*: OS day 30 and day 100: 88.3% (95% CI: 79.4–93.6%) and 60.8% (95% CI: 49.7–70.2%). *FOS not 1st line*: OS day 30 and day 100: 67.0% (95% CI: 52.3–78.0%) and 49.3% (95% CI: 35.0–62.0%) (P = 0.14)

Abbreviations: *CNS* central nervous system, *CSF* cerebrospinal fluid, *d* day(s), *FOS* foscarnet, *GCV* ganciclovir, *HCT* hematopoietic cell transplantation, *MMRD* mismatched related donor, *MMUD* mismatched unrelated donor, *MRD* matched related donor, *MUD* matched unrelated donors, *OS* overall survival, *qPCR* quantitative polymerase chain reaction, *UCB* umbilical cord blood

[a]Treated in combination with ganciclovir

7 Summary

Though FDA-approved indications for FOS are limited, it is a valuable agent in the armamentarium against CMV and HHV-6. Advantages of the drug include its broad coverage in the setting of high viral resistance and limited potential drug interactions. Disadvantages include its inconvenience of administration (intravenous), renal toxicity, and concomitant hydration requirements. With close observation, this toxicity is highly manageable helping to ensure safe and effective administration.

References

1. Foscarnet at DailyMed. Accessed from DailyMed at https://dailymed.nlm.nih.gov/dailymed/ on July 1, 2022.
2. Majewska A, Mlynarczyk-Bonikowska B. 40 Years after the Registration of Acyclovir: Do We Need New Anti-Herpetic Drugs? Int J Mol Sci. 2022 Mar 22;23(7):3431.
3. Wagstaff AJ, Bryson HM. Foscarnet. A reappraisal of its antiviral activity, pharmacokinetic properties and therapeutic use in immunocompromised patients with viral infections. Drugs. 1994 Aug;48(2):199–226.
4. Bacigalupo A, Boyd A, Slipper J, Curtis J, Clissold S. Foscarnet in the management of cytomegalovirus infections in hematopoietic stem cell transplant patients. Expert Rev Anti Infect Ther. 2012 Nov;10(11):1249–64.
5. Chrisp P, Clissold SP. Foscarnet. A review of its antiviral activity, pharmacokinetic properties and therapeutic use in immunocompromised patients with cytomegalovirus retinitis. Drugs. 1991 Jan;41(1):104–29.
6. Foster GG, Grant MJ, Thomas SM, Cameron B, Raiff D, Corbet K, Loitsch G, Ferreri C, Horwitz M. Treatment with Foscarnet after Allogeneic Hematopoietic Cell Transplant (Allo-HCT) Is Associated with Long-Term Loss of Renal Function. Biol Blood Marrow Transplant. 2020 Sep;26(9):1597–1606.
7. Foscarnet at PubChem. Accessed at https://pubchem.ncbi.nlm.nih.gov/ on July 1, 2022.
8. Chou S. Foscarnet resistance mutations mapping to atypical domains of the cytomegalovirus DNA polymerase gene. Antiviral Res. 2017 Feb;138:57–60.
9. Chou S. Advances in the genotypic diagnosis of cytomegalovirus antiviral drug resistance. Antiviral Res. 2020 Apr;176:104711.
10. El Chaer F, Shah DP, Chemaly RF. How I treat resistant cytomegalovirus infection in hematopoietic cell transplantation recipients. Blood. 2016 Dec 8;128(23):2624–2636
11. Jabs DA, Enger C, Forman M, Dunn JP. Incidence of foscarnet resistance and cidofovir resistance in patients treated for cytomegalovirus retinitis. The Cytomegalovirus Retinitis and Viral Resistance Study Group. Antimicrob Agents Chemother. 1998 Sep;42(9):2240–4.
12. Piret J, Boivin G. Resistance of herpes simplex viruses to nucleoside analogues: mechanisms, prevalence, and management. Antimicrob Agents Chemother. 2011 Feb;55(2):459–72.
13. Yong MK, Shigle TL, Kim YJ, Carpenter PA, Chemaly RF, Papanicolaou GA. American Society for Transplantation and Cellular Therapy Series: #4 - Cytomegalovirus treatment and management of resistant or refractory infections after hematopoietic cell transplantation. Transplant Cell Ther. 2021 Dec;27(12):957–967.
14. Domingo W, Nguyen IT, Johnsrud JJ, Brown JW. Continuous-Infusion Foscarnet Facilitates Administration in Hematopoietic Stem Cell Transplantation Patients. Transplant Cell Ther. 2021 Jul;27(7):622.e1–622.e5.
15. Ljungman P, de la Camara R, Robin C, Crocchiolo R, Einsele H, Hill JA, Hubacek P, Navarro D, Cordonnier C, Ward KN; 2017 European Conference on Infections in Leukaemia group. Guidelines for the management of cytomegalovirus infection in patients with haematological malignancies and after stem cell transplantation from the 2017 European Conference on Infections in Leukaemia (ECIL 7). Lancet Infect Dis. 2019 Aug;19(8):e260–e272.
16. Wang X, Patel SA, Haddadin M, Cerny J. Post-allogeneic hematopoietic stem cell transplantation viral reactivations and viremias: a focused review on human herpesvirus-6, BK virus and adenovirus. Ther Adv Infect Dis. 2021 May 24;8:20499361211018027.
17. Ward KN, Hill JA, Hubacek P, de la Camara R, Crocchiolo R, Einsele H, Navarro D, Robin C, Cordonnier C, Ljungman P; 2017 European Conference on Infections in Leukaemia (ECIL). Guidelines from the 2017 European Conference on Infections in Leukaemia for management of HHV-6 infection in patients with hematologic malignancies and after hematopoietic stem cell transplantation. Haematologica. 2019 Nov;104(11):2155–2163.

18. Kang GW, Hong HL, Lee IH, Ahn KS, Kim JD, Choi DL. Ganciclovir and Foscarnet Therapy of Cytomegalovirus-Associated Meningoencephalitis in a Hemodialysis patient with Liver Transplantation: Case Report. Transplant Proc. 2016 May;48(4):1208–11.
19. Sam R, Patel SB, Popli A, Leehey DJ, Gambertoglio JG, Ing TS. Removal of Foscarnet by hemodialysis using dialysate-side values. Int J Artif Organs. 2000 Mar;23(3):165–7.
20. Asakura M, Ikegame K, Yoshihara S, Taniguchi S, Mori T, Etoh T, Takami A, Yoshida T, Fukuda T, Hatanaka K, Kanamori H, Yujiri T, Atsuta Y, Sakamaki H, Suzuki R, Ogawa H. Use of foscarnet for cytomegalovirus infection after allogeneic hematopoietic stem cell transplantation from a related donor. Int J Hematol. 2010 Sep;92(2):351–9.
21. Chang J, Powles R, Singhal S, Paton N, Jameson B, Treleaven J, Mehta J. Foscarnet therapy for cytomegalovirus infection after allogeneic bone marrow transplantation. Clin Infect Dis. 1996 Mar;22(3):583–4.
22. Metafuni E, Chiusolo P, Sica S, Laurenti L, Bregante S, Van Lint MT, Dominietto A, Angelucci E, Bacigalupo A. Foscarnet treatment of cytomegalovirus infection in haploidentical or unrelated donor transplants. Bone Marrow Transplant. 2018 Dec;53(12):1560–1567.
23. Reusser P, Einsele H, Lee J, Volin L, Rovira M, Engelhard D, Finke J, Cordonnier C, Link H, Ljungman P; Infectious Diseases Working Party of the European Group for Blood and Marrow Transplantation. Randomized multicenter trial of foscarnet versus ganciclovir for preemptive therapy of cytomegalovirus infection after allogeneic stem cell transplantation. Blood. 2002 Feb 15;99(4):1159–64.
24. Zerr DM, Boeckh M, Delaney C, Martin PJ, Xie H, Adler AL, Huang ML, Corey L, Leisenring WM. HHV-6 reactivation and associated sequelae after hematopoietic cell transplantation. Biol Blood Marrow Transplant. 2012 Nov;18(11):1700–8.
25. Ogata M, Uchida N, Fukuda T, Ikegame K, Kamimura T, Onizuka M, Kato K, Kobayashi H, Sasahara Y, Sawa M, Sawada A, Hasegawa D, Masuko M, Miyamoto T, Okamoto S. Clinical practice recommendations for the diagnosis and management of human herpesvirus-6B encephalitis after allogeneic hematopoietic stem cell transplantation: the Japan Society for Hematopoietic Cell Transplantation. Bone Marrow Transplant. 2020 Jun;55(6):1004–1013.
26. Zerr DM, Gupta D, Huang ML, Carter R, Corey L. Effect of antivirals on human herpesvirus 6 replication in hematopoietic stem cell transplant recipients. Clin Infect Dis. 2002 Feb 1;34(3):309–17.
27. Ogata M, Oshima K, Ikebe T, Takano K, Kanamori H, Kondo T, Ueda Y, Mori T, Hashimoto H, Ogawa H, Eto T, Ueki T, Miyamoto T, Ichinohe T, Atsuta Y, Fukuda T. Clinical characteristics and outcome of human herpesvirus-6 encephalitis after allogeneic hematopoietic stem cell transplantation. Bone Marrow Transplant. 2017 Nov;52(11):1563–1570.
28. Ishiyama K, Katagiri T, Hoshino T, Yoshida T, Yamaguchi M, Nakao S. Preemptive therapy of human herpesvirus-6 encephalitis with foscarnet sodium for high-risk patients after hematopoietic SCT. Bone Marrow Transplant. 2011 Jun;46(6):863–9.
29. Ishiyama K, Katagiri T, Ohata K, Hosokawa K, Kondo Y, Yamazaki H, Takami A, Nakao S. Safety of pre-engraftment prophylactic foscarnet administration after allogeneic stem cell transplantation. Transpl Infect Dis. 2012 Feb;14(1):33–9.
30. Ogata M, Satou T, Inoue Y, Takano K, Ikebe T, Ando T, Ikewaki J, Kohno K, Nishida A, Saburi M, Miyazaki Y, Ohtsuka E, Saburi Y, Fukuda T, Kadota J. Foscarnet against human herpesvirus (HHV)-6 reactivation after allo-SCT: breakthrough HHV-6 encephalitis following antiviral prophylaxis. Bone Marrow Transplant. 2013 Feb;48(2):257–64.
31. Ogata M, Takano K, Moriuchi Y, Kondo T, Ueki T, Nakano N, Mori T, Uoshima N, Nagafuji K, Yamasaki S, Shibasaki Y, Sakai R, Kato K, Choi I, Jo Y, Eto T, Kako S, Oshima K, Fukuda T. Effects of Prophylactic Foscarnet on Human Herpesvirus-6 Reactivation and Encephalitis in Cord Blood Transplant Recipients: A Prospective Multicenter Trial with an Historical Control Group. Biol Blood Marrow Transplant. 2018 Jun;24(6):1264–1273.

Radioimmunotherapy-An Overview

Komal Rana, Sibi Oommen, and Sanjay Bharati

Abstract Cancer is one of the leading causes of morbidity and death worldwide. Carcinogenesis is a process where normal cells turn malignant through a series of evolutionary events. There are numerous agents whose exposure can lead to the initiation of carcinogenesis in humans or animals, for example, cancer causing chemicals, pollutants, heavy metals, alfatoxins, viruses, bacteria, parasites, etc. Hodgkin's disease (HD) as well as non-Hodgkin lymphoma, Burkitt's lymphoma, and nasopharyngeal carcinoma are some of the cancers developed due to the infection caused by various oncogenic DNA or RNA viruses. Chemotherapy, radiation therapy, immunotherapy, surgical excision, etc. are some of the existing treatment modalities to counter tumors. Radioimmunotherapy (RIT) is an exciting new approach to treating cancer. It involves the conjugation of a therapeutic radionuclide with a monoclonal antibody (mAb), where the ablative properties of therapeutic radionuclides are used to kill cancer cells. The damage to the cancer cells is either through the α, β, or auger electrons emitted by these radionuclides. The choice of radionuclide is based on the type of tumor. This technique is well-established and accepted for hematological malignancies, and so far, products like ^{131}I-tositumomab (Bexxar) and ^{90}Y-ibritumomab tiuxetan (Zevalin) are commercially available for non-Hodgkin lymphomas (NHLs). However, the application of RIT to solid tumors is still challenging, and different approaches are being adopted as well as investigated to increase its therapeutic index in solid tumors.

K. Rana
Manipal GoK Bioincubator, MUTBI, Manipal, KA, India

S. Oommen · S. Bharati (✉)
Department of Nuclear Medicine, Manipal College of Health Professions, Manipal Academy of Higher Education, Manipal, KA, India
e-mail: sibi.oommen@manipal.edu; sanjay.bharati@manipal.edu

C. Somboonwit et al. (eds.), *Global Virology IV: Viral Disease Diagnosis and Treatment Delivery in the 21st Century*,
https://doi.org/10.1007/978-3-031-57369-9_24

Keywords Epstein–Barr virus · Human papilloma virus · Human T-lymphotropic virus · Therapeutic monoclonal antibody · Radioimmunoconjugate · Radioisotopes · Theranostic agents · Targeted radiation therapy · Pre-targeted radioimmunotherapy · Compartmental radioimmunotherapy · Combination radioimmunotherapy

1 Introduction

Cancer is among the leading cause of death worldwide with a death toll of 99,58,133 in the year 2020 [1]. Cancer is an evolutionary process, where, in its evolutionary journey, normal cells acquire certain hallmark characteristics such as independent proliferation, non-response to anti-proliferative signals, evasion of apoptosis, excessively high replicative potential, angiogenesis, and metastasis [2, 3]. The processes of attaining these hallmarks are also referred to as "carcinogenesis." However, before commencing the evolutionary journey, normal cells must undergo a process known as "initiation." During the initiation stage of carcinogenesis, normal cells undergo specific genetic modifications, which help them further progress in the journey [3]. There are numerous agents, exposure to which can lead to the initiation of carcinogenesis [4]. Some important examples include chemical carcinogens and viruses. Chemical carcinogens cause cancer in humans or animals. There are numerous such chemicals to which people are repeatedly exposed during daily activities [5, 6]. Oncogenic DNA and RNA viruses are known to be associated with human cancers. Among them, human papilloma virus (HPV-16, HPV-18), hepatitis B virus (HBV), hepatitis C virus (HCV), human T-lymphotropic virus (HTLV), and Kaposi's sarcoma-associated herpesvirus (HHV-8), etc. are the most common oncogenic viruses [7, 8]. Additionally, Epstein–Barr virus (EBV) is associated with the development of Hodgkin's disease (HD) as well as non-Hodgkin lymphoma (NHL), especially in immunosuppressed individuals [9]. Burkitt's lymphoma and nasopharyngeal carcinoma have also shown correlations with exposure to EBV [10, 11].

Continual research in this field has led to breakthrough discoveries and the development of new modalities of cancer treatment. Chemotherapy and external radiation therapy were conventional methods for cancer treatment; however, in recent years, improved tumor targeting methods have been developed based on our understanding of molecular differences between normal and cancer cells [12]. The combination of nuclear medicine and immunology has led to the introduction of radioimmunotherapy (RIT), a novel cancer treatment approach where a radionuclide is chemically conjugated with an antibody to treat cancer [13]. The specificity of radioimmunoconjugates towards cancer cells is governed by the antibody molecule, whereas the attached radionuclide primarily renders a toxic cell-killing effect.

The idea of using antibodies to deliver therapeutic radiation doses to cancer cells materialized after the introduction of hybridoma technology to produce monoclonal antibodies (mAbs) [14]. Clinically, RIT is a widely accepted treatment, especially for leukemias and lymphomas [15]. The radiolabeled antibody, after injection, distributes in the blood and is recognized by the receptors present on the cancer cells [16]. The binding of antibodies to these receptors allows radionuclides to deliver targeted toxic radiation doses to cancer cells resulting in their death [17]. Therefore, RIT can specifically deliver radiation doses to the diseased tissue while sparing the normal tissue. RIT also offers the advantage of being able to simultaneously image and treat the tumor. The imaging in RIT can be achieved by employing theranostic radionuclides and imaging modalities such as SPECT or PET [18]. The simultaneous imaging allows easy and efficient calculation of overall uptake of radiolabeled therapeutic antibodies by the tumor tissue, which helps in the choosing/establishing dose for therapy as well as prediction of therapeutic response of the RIT.

2 Radioimmunotherapy and Immunotherapy

Immunotherapy employs chemotherapeutic mAbs for cancer treatment. They are capable of specifically targeting receptors present in cancer cells. These mAbs are known to exhibit their effect via direct cellular action or indirect modulation of cell signaling. Direct cellular action is exerted by receptor blocking or ligand blocking, wherein mAbs can affect cell survival or apoptotic cascades [19]. Indirect mechanisms include the modulation of immune system responses such as antibody-dependent cell-mediated cytotoxicity (ADCC), antibody-dependent cellular phagocytosis (ADCP), complement-mediated cytotoxicity, etc. [20–22].

For RIT, radionuclides are conjugated to the mAbs, which enable the possibility of both immunological and radiobiological types of cell killing within the tumor microenvironment [23] (Fig. 1).

RIT provides a platform for the delivery of low-dose radioactivity to the tumor tissue as well as the flexibility to utilize a wide range of available radioisotopes, generally used in nuclear medicine protocols. The tumor ablative properties of α- and β- radiation are different, and RIT provides flexibility to use them according to the tumor type. In cases of larger tumors, mAbs do not distribute homogenously within the whole tumor mass, which could be due to heterogeneity of tumor-associated antigen expression, heterogeneous blood perfusion, hindered diffusion in the interstitium, high avidity of antibodies, etc. Therefore, for such tumors, β-emitting radionuclides are preferred as they have larger range of action and thus overcome the problem of heterogenous antibody distribution [24]. Additionally, RIT is also capable of imaging tumor tissue, which can help in planning tumor ablation therapy [25–27]. Radio-imaging can help in obtaining information such as

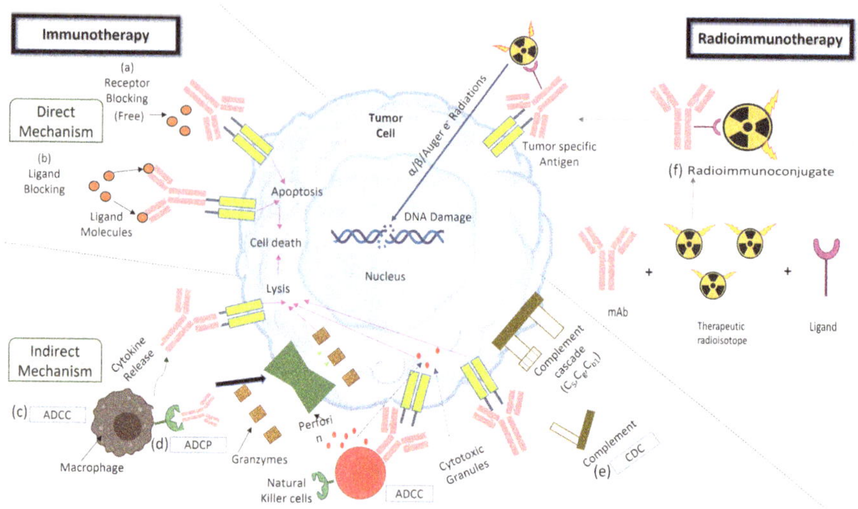

Fig. 1 Mechanism of action of immunotherapy and radioimmunotherapy. Immunotherapy works through two mechanisms: direct mechanisms and indirect mechanisms. Direct mechanisms include (**a**) Receptor blocking, where mAbs can directly get attached to a receptor and inhibit tumor cell growth, division, and angiogenesis. (**b**) Ligand blocking: mAbs get attached to the target cells via some ligand and inhibit various signaling processes such as tumor growth and angiogenesis. Indirect mechanisms include (**c**) Antibody-dependent cellular toxicity (ADCC), where mAb or fragments bind to their target antigen and initiate cell death by involving cells of immune systems such as natural killer cells or macrophages. (**d**) Antibody-dependent cellular phagocytosis (ADCP); mAbs recognize the target tumor cells, get opsonized, and then activate immune cells such as macrophages, which ultimately initiate tumor cell lysis via phagocytic action. (**e**) Complement-dependent cytotoxicity (CDC); mAbs interact with the components of the complement system and initiate cell lysis within the tumor microenvironment. CDC-mediated cell lysis involves the interaction of Fc region of mAb with components of the complement cascade, such as C1q, C5, C9, etc. resulting in the formation of membrane attack complex responsible for disrupting the cell membrane of target cells. (**f**) Radioimmunoconjugate targets tumor-specific antigens present on malignant cells, delivering ionizing energy to the targeted cell, resulting in its death

tumor targeting of radioimmunoconjugate, tumor size, density of targeting molecules, and the effects of the administered radiation on the tumor tissue [28].

3 Antibody Engineering and Its Effect on Immunotherapy

The selection of tumor cells over normal body cells is of utmost importance for successful anticancer therapy. The common route of antibody administration in RIT is through intravenous injection. After entering the bloodstream, radiolabeled antibodies reach the tumor site through blood circulation. After crossing the blood vessel, these molecules are distributed in the interstitial sites and then diffuse through the extracellular matrix, reaching the target tumor cells and binding to their surface antigens. Diffusion through the extracellular matrix is mainly governed by the size and binding affinity of the antibodies [29]. Indirectly, clearance of radiolabeled antibodies from the systemic circulation reduces overall tumor uptake, as diffusion is the function of concentration gradient [30].

Since, antibodies are engineered to be retained in the blood, their concentration builds up in the tumor tissue over a span of 1 or 2 days. However, this is challenging for imaging because of elevated background activity, which requires several days for clearance. To address this issue, modifications in the size and binding properties of antibodies are considered central to the success of RIT.

The applicability of large antibody molecules in RIT is restricted as they remain for a longer time in circulation [31]. This increases an unwanted dose to normal tissue and raises the chances of radiation-induced toxicity. Therefore, smaller portions of antibodies such as single-chain Fv (scFv), which are only variable heavy (V_H) or variable light (V_L) chain domains linked by peptide or disulfide bridges, are tested in RIT. These antibody fragments are smaller in size and thus remain for shorter durations in circulation, as well as showing improvement in renal excretion [32]. However, the tumor uptake of these antibodies is poor as compared to the whole antibody molecule procedure [32]. This problem is resolved by the use of tetrameric molecules ($[sc(Fv)_2]_2$). The uptake of tetramers by tumors is similar to that of whole antibodies, but clearance from the blood is faster [33, 34]. The other options include domain deleted antibodies such as $(scFv\text{-}CH_3)_2$ or small immunoprotein molecules that have reincorporated immunoglobulin-derived constant regions, V_H and V_L domains (scFv fragments). These molecules exhibited superior tumor therapy characteristics as compared to whole antibodies [35].

Another important consideration when selecting antibodies for RIT is their affinity and avidity [36]. *Affinity* is the measure of the strength of binding of the antigen and antibody. It reflects the strength of non-covalent bonds formed between antigen and antibody at a particular binding site [37]. Since, antigen and antibodies are multivalent, which means that they have more than one binding site, then the measure of total binding strength is called *avidity*. This means that for an antigen-antibody interaction, avidity can be equal to or greater than affinity [38]. Although, binding to the target antigen is of utmost importance for tumoral uptake, however, this may also contribute to the heterogenous intra-tumoral distribution of antibodies [39]. As per "binding-site barrier effect" the higher the affinity of the antibody to the antigen, the greater the restriction to its diffusion/penetration into the tumor mass [40].

Antibodies penetrate the interior of tumors in a sequential fashion, diffusing from one layer of cells to the next. In this process, some of the antibodies bind to the antigens present on the surface of cells in external layers, and the unbound antibodies move to the next layer [41]. Therefore, the concentration of antibodies should not be a limiting factor and must be greater than needed for saturation binding. Moreover, the bound antibodies may hinder the movement of unbound antibodies towards deeper layers of the tumor. The larger molecules have higher binding site barrier effect as compared to smaller molecules of similar affinities. This may be due to the higher possibility of molecular crowding with larger molecules [38]. It is observed that affinity can be enhanced to a certain threshold range, beyond which no further increase or restricted tumor uptake is observed [42].

4 Choice of Radionuclide for Radioimmunotherapy

One of the important parameters for the success of radioimmunotherapy is the correct choice of the radionuclide. The choice is based on the factors such as physical and chemical properties. Additionally, the availability and biodistribution of radionuclides are also important considerations [43]. The physical characteristics that are most critical include the type of emission and half-life of the radioisotope. Radioisotopes undergo decay and, in this process, emit radiation such as α, β, and γ radiations, etc. These radiations are utilized for different applications in nuclear medicine [44]. γ-radiations have a very high penetrating power and can be detected outside the body from the source located within the body. Therefore, radioisotopes that emit γ-radiations find applications in diagnostic nuclear medicine, where they are usually employed for the diagnosis of disease [45]. The β and α radiations are less penetrating but have high linear energy transfer (lose their maximum energy in a small area) and are therefore suitable for tumor ablation [46]. Some radioisotopes produce both β or α and γ-radiations. Such radioisotopes, along with the tumor ablation effects, also provided scope for the personalized dosimetry evaluation for RIT [47].

In contrast to radiotherapy, where tumor tissue is exposed to a large radiation flux for a short duration of time, RIT continuously delivers ionizing radiation to the tumor tissue [48]. These radiations can be further classified as lower LET (linear energy transfer) for β-emitters and higher LET radiations for α-particles. α-particles have low range and high killing efficiency as compared to the β radiations which have comparatively longer range but lower cell killing efficiency. Both β and α radiations kill tumor cells in one of the following ways (Fig. 2):

1. Direct hit, in this case, a radiolabeled antibody binds to antigen present on the tumor cell and induces direct damage to cellular DNA, leading to the death of this cell.
2. Crossfire irradiation is a kind of off-target effect that induces cell death in the cells present in the vicinity of cells that are targeted by the radiolabeled antibody.
3. Bystander effect is also an off-target effect that is mediated via the release of death effectors/signals by irradiated cells to other non-irradiated cells.
4. Abscopal effect is another off-target effect that involves the immune system's response to ionizing radiation, causing the cell death of far distant metastatic cells following irradiation of primary metastatic tumors.

The major contributors to RIT-induced cell death are "direct hit" and "crossfire irradiation [49]. Conjugation of antibodies with β emitting radionuclides is widely used in RIT [50]. As compared to α radiations, β radiations have lower-LET, which means that they have a longer tissue range and can irradiate a larger portion of

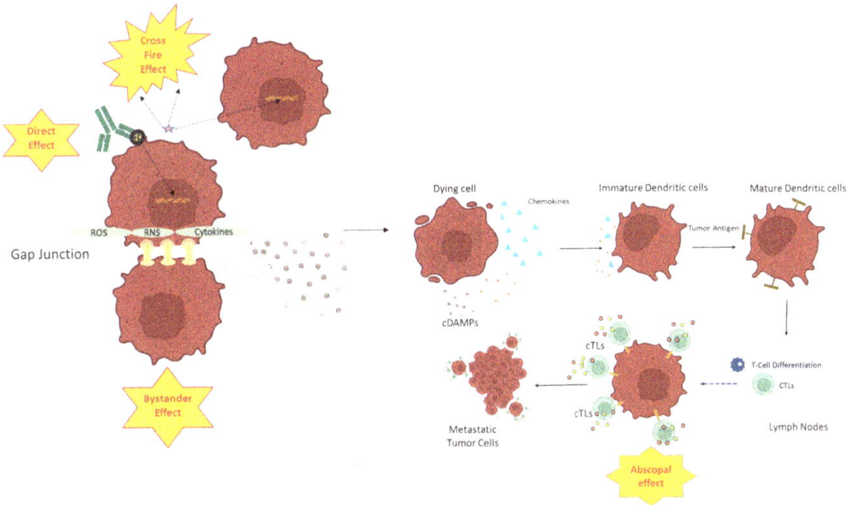

Fig. 2 Tumor cell killing mechanisms of ionizing radiations. (**a**) Radiolabeled antibodies bind to the antigen present on the tumor cells and induce direct damage to cellular DNA or other macromolecules (direct hit). The surrounded tumor cells also receive radiation doses through the cross-fire effect. (**b**) Bystander effect: the damage induced to the tumor cells leads to the generation of cytokines, ROS, RNS, and other death effect molecules into the tumor microenvironment, which, when bound to nearby tumor cells, induce cell death. (**c**) Abscopal effect: the damaged tumor cells release cDAMPs, chemokines, and other tumor antigens into the tumor microenvironment, which attracts macrophages and immature dendritic cells to the tumor site. cDAMPs modulate the ultimate function of dendritic cells, which then engulf tumor antigens and then cross present these antigens to boost T cell differentiation to CTLs. CTLs then migrate to distant metastatic tumor sites and kill malignant cells

tumor. This is the reason for using β emitting radionuclides for treating larger size tumors. Additionally, the use of β-emitting radionuclides also solves the problem of heterogenous antibody distribution [51]. However, in case of metastatic lesions or where the tumor size is small, β radiations increase unwanted radiation dose to the normal tissue. In such cases, α emitting radioisotopes are considered ideal as they have a higher-LET which means that they will deliver large amount of energy in a very small area [52]. The α emitting radioisotopes also perform better in treating hypoxic tumors. Hypoxic tumors are more resistant to the radiation-induced cell death [53]. This effect is quite evident in lower LET radiations (β-radiations); however, this is not the case with high LET radiations (α-radiations) [54]. Low oxygen concentration reduces the possibility of initially produced free radicals to react with oxygen molecules and thus provide more window of radical recombination, ultimately decreasing the possibility of radiation-induced DNA damage [55]. However, in case of high LET radionuclides, the radiations are quite dense, and they produce large number of free radicals in a very small volume, which cause multiple damages in the DNA molecules. These multiple damages to the DNA molecules increase the possibility of double-strand breaks in DNA, which are difficult to repair [56]. This

Table 1 Some commonly used radionuclides for RIT

Radionuclide	Half-life	Type of emission	Average (α or β) energy (KeV)	γ energy (KeV)	Average tissue range (β)	Advantages
I-131	8.02 days	β & γ	192	364	0.4 mm	Theranostic
Lu-177	6.65 days	β & γ	134	208	0.2 mm	Theranostic
Y-90	2.7 days	β	935	–	3 mm	Therapeutic
Re-188	16.9 h	β & γ	784	155	3 mm	Theranostic
At-211	7.2 h	α	6790	–	(55–70) μm	Therapeutic
Bi-213	45.6 min	β, α, γ	8400 (α)	440	α (50–80) μm	Theranostic

to certain extent explains better performance of α radiations over β radiations. A large number of radionuclides are currently under investigation for RIT; however, I-131, Lu-177, Y-90, Re-188, At-211, or Bi-213 are most commonly used [57]. The detailed physical properties of these radionuclides are given in Table 1.

5 Challenges in Radioimmunotherapy

Radioimmunotherapy has been successfully applied for the treatment of hematological diseases, especially in patients with non-Hodgkin's B-cell lymphoma. Non-Hodgkin's B-cell lymphoma is a general category of lymphoma where abnormal growth of lymphocytes is noted. Individuals exposed to Epstein-Barr virus or human T-cell lymphotropic virus who have immunodeficiencies or autoimmune diseases are at greater risk of developing this disease [58]. The promising results of RIT in non-Hodgkin's lymphoma have led to the commercial availability of two drugs, 131I-tositumomab (Bexxar) and 90Y-ibritumomab tiuxetan (Y-90 Zevalin), for the treatment of relapsed or refractory low-grade or follicular B-cell non-Hodgkin's lymphoma (NHL). Both these drugs are murine antiCD20 mAbs that are conjugated to Y-90 and I-131 respectively. I-131 is a theranostic radionuclide (produces both β and γ radiations); therefore, in the case of ^{131}I-tositumomab dosimetry is used for deciding the therapeutic dose. The therapy is a two-step protocol in which the first step is to examine the distribution of radiolabeled antibodies in the body at 1 h, 2–4 days, and 6–7 days (using conventional nuclear medicine imaging). The therapeutic dose of I-131 tositumomab is established based on the data obtained from these studies. After 1 or 2 weeks of the first step, Bexxar is infused, followed by the administration of I-131 linked Bexxar. It is important to protect the thyroid gland with this medication, as free I-131 can accumulate in thyroid, leading to thyroid toxicity. Therefore, thyroid-protective medication (thyroid blockers) is started 24 h prior to the I-131 Bexxar administration until 2 weeks of the second step.

Y-90 Zevalin (Acrotech Biopharma LLC) is administered at a fixed dosage of 0.4 mCi/kg (14.8 MBq per kg) if the platelet count is at least 150,000/mm^3 and 0.3 mCi/kg (11.1 MBq per kg) for lower platelet counts. The therapy regime is similar to I-131 Bexxar except for the first step. Instead on day one, rituximab 250 mg/m^2 is administered through an intravenous route, and after a week again 250 mg/m^2 of rituximab is infused, followed by Y-90 Zevalin after 4 h.

Despite success in hematological diseases, RIT demonstrated limited success in treating solid tumors. There are several differences between hematological and solid tumors that might contribute to the poor response of solid tumors to RIT [59]. The accessibility of mAbs to the cancer cells is one of the major challenges to RIT in solid tumors [60]. Tumors have highly chaotic vascular blood supply, with certain areas where blood supply is minimal [61]. These regions are called hypoxic regions of tumors, which are most difficult to treat as they are insensitive to the ionizing radiation. The poor tumor vasculature further results in an inhomogeneous distribution of radioactivity within the tumors [62]. The therapeutic index of RIT for solid tumors is low and insufficient for treating solid tumors [63]. This is due to the tendency of antibodies to distribute in other organs like the liver and bone marrow, thus limiting the radiation doses that are required to achieve the therapeutic response [64]. There are different approaches that are employed to increase therapeutic response.

6 Improving RIT for Solid Tumors

One of the approaches applied to improve the therapeutic index is interferon therapy, where the expression of surface antigens presents on the cancer cells is increased, which results in higher uptake and binding of radioimmunoconjugate [65]. RIT can also be combined with agents, which can improve its therapeutic efficacy. The major targets for this type of approach are the factors that hinder antibody distribution within the tumor tissue or that affect the radiosensitivity of cancer cells. For example, RIT is often combined with chemotherapy drugs, where chemotherapy drug increases the sensitivity of tumor cells toward RIT [66, 67]. The potentiation of the therapeutic effect of RIT was observed when the RIT radio-drug, 90Y-MX-DPTA BrE3 mAb, was administered in combination with topoisomerase-I inhibitor drug, topotecan. The combination of topotecan and 90Y-MX-DPTA BrE3 mAb substantially reduced tumors in human mammary carcinoma xenografts and increased the survival of animals [68]. In yet another study, the improvement in the efficacy of alpha-emitting radioimmunoconjugate was demonstrated when used in combination with a chemotherapeutic agent, gemcitabine. The tumor-bearing mice were treated with gemcitabine and ^{212}Pb-trastuzumab, which resulted in the improved survival of animals [69]. The other radio-sensitizing drugs like parecoxib (cyclooxygenase-2 inhibitor) and paclitaxel, etc. are also tested in combination with RIT which showed improvement in treatment efficacy of RIT [70, 71].

In another approach known as pre-targeted radioimmunotherapy (PRIT), antibodies and radionuclides are not bound together; instead, they are administered separately. This approach overcomes the long circulation time of antibodies and unwanted exposure to normal tissues related to the drawbacks of RIT [72, 73]. The bispecific mAbs with two different binding sites are used in this approach. One of

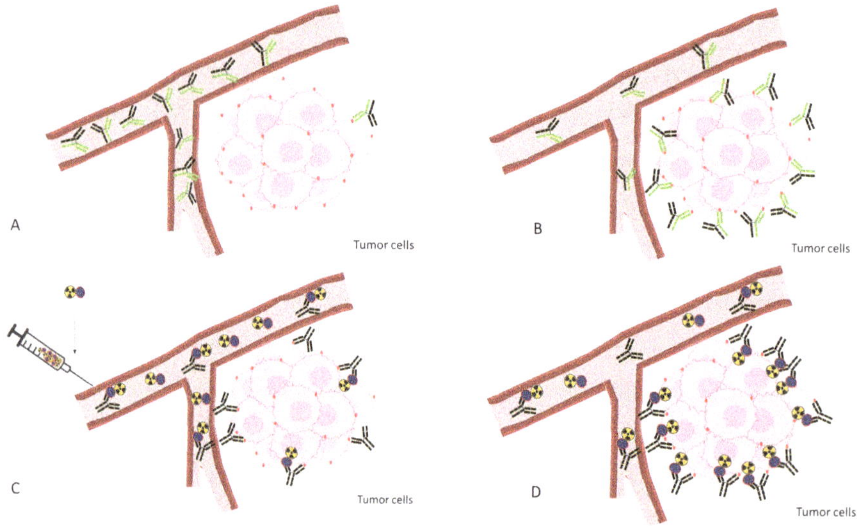

Fig. 3 General mechanistic plan of pretargeted radioimmunotherapy. (**a**) Bispecific mAb is injected first and allowed to be distributed within different body compartments. (**b**) Bispecific mAb is slowly accumulated within the tumor and gets attached to the surface antigen present on the tumor cells. The uptake within the tumor and simultaneous clearance from the remaining compartments of the body decrease background unbound antibodies. (**c**) After this, radioimmunoconjugate is injected, which gets distributed within the blood compartment. (**d**) radioimmunoconjugate diffuses into tumor tissue, where it binds to the second binding site of bispecific mAb. The remaining unbound radioimmunoconjugate is cleared quickly from the body through biological excretion processes

the binding sites attaches to the antigen present on the cancer cells, and the other site is specific to the radionuclide. This idea was first proposed by Reardan et al., in 1985, when they developed antibodies against metal chelates [74]. The schematic overview of PRIT is given in Fig. 3.

First, an antibody is administered the individual, which binds to the tumor antigen. Once antibody is sufficiently cleared from the other organs, then radionuclide is administered, which attaches to the antibody present on the surface of tumor cells. The unbound radiometal/radionuclide is cleared rapidly from the blood, thereby reducing unnecessary radiation exposure to hematological tissues. The effectiveness of this approach has been demonstrated in several clinical trials [75].

The tumors, which are accessible for direct injection, are better suited for compartmental RIT [76, 77]. In this approach, direct injection of radioimmunoconjugate into the tumor mass delivers a high radiation dose to the tumor without exposing other tissue to deleterious radiation dose. This approach is successfully applied to the tumors of the central nervous system, where radioimmunoconjugate is administered directly into the cerebrospinal fluid via intrathecal or intraventricular injection. For example, ^{131}I-3F8 and ^{131}I-omburtamab radiolabeled antibodies are administered directly into the central nervous system via intrathecal/intraOmmaya administration for the treatment of metastatic leptomeningeal tumors [78, 79].

Pandit-Taskar et al., 2019 demonstrated that ^{131}I-omburtamab when injected directly into the CSF compartment, resulted in high CSF to blood dose ratio. The group showed that a compartmental dose of 50–80 mCi can be administered without exceeding dose-limiting toxicity blood dose of 200 cGy [80].

7 Conclusion and Future Prospects

RIT delivers therapeutic radionuclides to malignant tissues with the help of mAbs. These radionuclides emit either α particles, β particles, or Auger electrons, which kill cancer cells. The efficacy of RIT is dependent on a number of factors, such as tumor type, size, target antigen, and the properties of mAbs. RIT is successfully applied to the treatment of hematological malignancies, where, in some cases, they proved better as compared to other modalities. However, the therapeutic index of RIT in solid tumors is still very low. Therefore, new modified methods of RIT are being evaluated with the aim of improving the therapeutic index and decreasing dose-related toxicities. Increasing radiosensitivity of tumors through chemotherapy is one of the approaches that shows better results. The other approaches, like PRIT, decrease radiotoxicity by separating tumor targeting from radioactivity. Future studies should investigate the combination of PRIT with tumor radio-sensitizing agents to improve therapeutic outcomes. Additionally, investigations should focus on the evaluation of radioimmunoconjugate, which can provide dosimetry data for patient-specific treatment.

Acknowledgments Authors gratefully acknowledge the help provided by Ms. Maya P. Shetty, Research scholar, Department of Nuclear Medicine, MCHP, Manipal Academy of Higher Education, Manipal, India in the preparation of the figures.

Conflict of Interest Authors declare no conflicts of interest.

References

1. Cancer today [Internet]. [cited 2023 Nov 30]. Available from: http://gco.iarc.fr/today/home
2. Hanahan D. Hallmarks of Cancer: New Dimensions. Cancer Discovery. 2022 Jan 12;12(1):31–46.
3. Hanahan D, Weinberg RA. Hallmarks of Cancer: The Next Generation. Cell. 2011 Mar 4;144(5):646–74.
4. Stewart BW. Mechanisms of carcinogenesis: from initiation and promotion to the hallmarks. In: Baan RA, Stewart BW, Straif K, editors. Tumour Site Concordance and Mechanisms of Carcinogenesis [Internet]. Lyon (FR): International Agency for Research on Cancer; 2019 [cited 2023 Nov 30]. (IARC Scientific Publications). Available from: http://www.ncbi.nlm.nih.gov/books/NBK570326/
5. Doe JE, Boobis AR, Dellarco V, Fenner-Crisp PA, Moretto A, Pastoor TP, et al. Chemical carcinogenicity revisited 2: Current knowledge of carcinogenesis shows that categorization

as a carcinogen or non-carcinogen is not scientifically credible. Regulatory Toxicology and Pharmacology. 2019 Apr 1;103:124–9.

6. Baba AI, Câtoi C. CARCINOGENESIS. In: Comparative Oncology [Internet]. The Publishing House of the Romanian Academy; 2007 [cited 2023 Nov 30]. Available from: https://www.ncbi.nlm.nih.gov/books/NBK9552/
7. Parija SC. Oncogenic Viruses. In: Parija SC, editor. Textbook of Microbiology and Immunology [Internet]. Singapore: Springer Nature; 2023 [cited 2023 Nov 30]. p. 911–20. Available from: https://doi.org/10.1007/978-981-19-3315-8_63
8. Chang Y, Moore PS, Weiss RA. Human oncogenic viruses: nature and discovery. Philosophical Transactions of the Royal Society B: Biological Sciences. 2017 Sep 11;372(1732):20160264.
9. Karabajakian A, Ray-Coquard I, Blay JY. Molecular Mechanisms of Kaposi Sarcoma Development. Cancers (Basel). 2022 Apr 7;14(8):1869.
10. Al-Khreisat MJ, Ismail NH, Tabnjh A, Hussain FA, Mohamed Yusoff AA, Johan MF, et al. Worldwide Prevalence of Epstein-Barr Virus in Patients with Burkitt Lymphoma: A Systematic Review and Meta-Analysis. Diagnostics (Basel). 2023 Jun 15;13(12):2068.
11. Tsao SW, Tsang CM, Lo KW. Epstein-Barr virus infection and nasopharyngeal carcinoma. Philos Trans R Soc Lond B Biol Sci. 2017 Oct 19;372(1732):20160270.
12. Han HH, Tian H, Zang Y, Sedgwick AC, Li J, Sessler JL, et al. Small-molecule fluorescence-based probes for interrogating major organ diseases. Chem Soc Rev. 2021 Aug 31;50(17):9391–429.
13. Assadi M, Gholamrezanezhad A. Radioimmunotherapy. In Nuclear Medicine and Immunology 2022 (pp. 281–295). Springer, Cham.
14. Duan S, Buxton IL. Evolution of Medical Approaches and Prominent Therapies in Breast Cancer. Cancers. 2022 May 16;14(10):2450.
15. Etrych T, Braunova A, Zogala D, Lambert L, Renesova N, Klener P. Targeted Drug Delivery and Theranostic Strategies in Malignant Lymphomas. Cancers. 2022 Jan 26;14(3):626.
16. Augustine R, Al Mamun A, Hasan A, Salam SA, Chandrasekaran R, Ahmed R, Thakor AS. Imaging cancer cells with nanostructures: Prospects of nanotechnology driven non-invasive cancer diagnosis. Advances in Colloid and Interface Science. 2021 Aug 1;294:102457.
17. Bodei L, Herrmann K, Schöder H, Scott AM, Lewis JS. Radiotheranostics in oncology: current challenges and emerging opportunities. Nature Reviews Clinical Oncology. 2022 Jun 20:1–7.
18. Bolcaen J, Kleynhans J, Nair S, Verhoeven J, Goethals I, Sathekge M, Vandevoorde C, Ebenhan T. A perspective on the radiopharmaceutical requirements for imaging and therapy of glioblastoma. Theranostics. 2021;11(16):7911.
19. Nguyen TH, Havari E, McLaren R, Zhang M, Jiang Y, Madden SL, Roberts B, Kaplan J, Shankara S. Alemtuzumab induction of intracellular signaling and apoptosis in malignant B lymphocytes. Leukemia & lymphoma. 2012 Apr 1;53(4):699–709.
20. Nigro CL, Macagno M, Sangiolo D, Bertolaccini L, Aglietta M, Merlano MC. NK-mediated antibody-dependent cell-mediated cytotoxicity in solid tumors: biological evidence and clinical perspectives. Annals of translational medicine. 2019 Mar;7(5).
21. Cao X, Chen J, Li B, Dang J, Zhang W, Zhong X, Wang C, Raoof M, Sun Z, Yu J, Fakih MG. Promoting antibody-dependent cellular phagocytosis for effective macrophage-based cancer immunotherapy. Science advances. 2022 Mar 18;8(11):eabl9171.
22. Kolev M, Markiewski MM. Targeting complement-mediated immunoregulation for cancer immunotherapy. In Seminars in immunology 2018 Jun 1 (Vol. 37, pp. 85–97). Academic Press.
23. Harsini S, Alavi A, Rezaei N. Introduction on Nuclear Medicine and Immunology. In Nuclear Medicine and Immunology 2022 (pp. 1–13). Springer, Cham.
24. Pei P, Liu T, Shen W, Liu Z, Yang K. Biomaterial-mediated internal radioisotope therapy. Materials Horizons. 2021;8(5):1348–66.
25. Herrero Álvarez N, Bauer D, Hernández-Gil J, Lewis JS. Recent advances in radiometals for combined imaging and therapy in cancer. ChemMedChem. 2021 Oct 6;16(19):2909–41.

26. Aarts F, Bleichrodt RP, Oyen WJ, Boerman OC. Intracavitary radioimmunotherapy to treat solid tumors. Cancer biotherapy & radiopharmaceuticals. 2008 Feb 1;23(1):92–107.
27. Franchi, S., Di Marco, V., & Tosato, M. (2022). Bismuth chelation for targeted alpha therapy: Current state of the art. *Nuclear Medicine and Biology*.
28. Vivier D, Sharma SK, Zeglis BM. Understanding the in vivo fate of radioimmunoconjugates for nuclear imaging. Journal of Labelled Compounds and Radiopharmaceuticals. 2018 Jul;61(9):672–92.
29. Yang H, Tong Z, Sun S, Mao Z. Enhancement of tumour penetration by nanomedicines through strategies based on transport processes and barriers. Journal of Controlled Release. 2020 Dec 10;328:28–44.
30. Dogra P, Butner JD, Nizzero S, Ruiz Ramírez J, Noureddine A, Peláez MJ, Elganainy D, Yang Z, Le AD, Goel S, Leong HS. Image-guided mathematical modeling for pharmacological evaluation of nanomaterials and monoclonal antibodies. Wiley Interdisciplinary Reviews: Nanomedicine and Nanobiotechnology. 2020 Sep;12(5):e1628.
31. Verhoeven M, Seimbille Y, Dalm SU. Therapeutic applications of pretargeting. Pharmaceutics. 2019 Sep 1;11(9):434.
32. Bates A, Power CA. David vs. Goliath: the structure, function, and clinical prospects of antibody fragments. Antibodies. 2019 Apr 9;8(2):28.
33. Goel A, Baranowska-Kortylewicz J, Hinrichs SH, Wisecarver J, Pavlinkova G, Augustine S, Colcher D, Booth BJ, Batra SK. 99mTc-labeled divalent and tetravalent CC49 single-chain Fv's: novel imaging agents for rapid in vivo localization of human colon carcinoma. Journal of Nuclear Medicine. 2001 Oct 1;42(10):1519–27.
34. Wittel UA, Jain M, Goel A, Chauhan SC, Colcher D, Batra SK. The in vivo characteristics of genetically engineered divalent and tetravalent single-chain antibody constructs. Nuclear medicine and biology. 2005 Feb 1;32(2):157–64.
35. Berndorff D, Borkowski S, Sieger S, Rother A, Friebe M, Viti F, Hilger CS, Cyr JE, Dinkelborg LM. Radioimmunotherapy of solid tumors by targeting extra domain B fibronectin: identification of the best-suited radioimmunoconjugate. Clinical Cancer Research. 2005 Oct 1;11(19):7053s–63s.
36. Polito L, Calafato G, Bortolotti M, Chiarelli Olivari C, Maiello S, Bolognesi A. Antibody Conjugates for Sarcoma Therapy: How Far along Are We?. Biomedicines. 2021 Aug 8;9(8):978.
37. Dondelinger M, Filée P, Sauvage E, Quinting B, Muyldermans S, Galleni M, Vandevenne MS. Understanding the significance and implications of antibody numbering and antigen-binding surface/residue definition. Frontiers in immunology. 2018 Oct 16;9:2278.
38. Xenaki KT, Oliveira S, van Bergen En Henegouwen PM. Antibody or antibody fragments: implications for molecular imaging and targeted therapy of solid tumors. Frontiers in immunology. 2017 Oct 12;8:1287.
39. Khera E, Thurber GM. Pharmacokinetic and immunological considerations for expanding the therapeutic window of next-generation antibody–drug conjugates. BioDrugs. 2018 Oct;32(5):465–80.
40. Rudnick, S. I., & Adams, G. P. (2009). Affinity and avidity in antibody-based tumor targeting. *Cancer Biotherapy and Radiopharmaceuticals*, *24*(2), 155–161.
41. Drago, J. Z., Modi, S., & Chandarlapaty, S. (2021). Unlocking the potential of antibody–drug conjugates for cancer therapy. *Nature Reviews Clinical Oncology*, *18*(6), 327–344.
42. Adams GP, Schier R, McCall AM, Simmons HH, Horak EM, Alpaugh RK, Marks JD, Weiner LM. High affinity restricts the localization and tumor penetration of single-chain fv antibody molecules. Cancer research. 2001 Jun 15;61(12):4750–5.
43. Arms L, Robson AL, Woldu A, Martin A, Palmer W, Flynn J, Hua S. Considerations for using optical clearing techniques for 3D imaging of nanoparticle biodistribution. International journal of pharmaceutics. 2020 Oct 15;588:119739.

44. Knapp FF, Dash A. Introduction: Radiopharmaceuticals play an important role in both diagnostic and therapeutic nuclear medicine. In Radiopharmaceuticals for Therapy 2016 (pp. 3–23). Springer, New Delhi.
45. Payolla FB, Massabni AC, Orvig C. Radiopharmaceuticals for diagnosis in nuclear medicine: a short review. Eclética Química. 2019;44(3):11–9.
46. Keisari Y, Kelson I. The potentiation of anti-tumor immunity by tumor abolition with alpha particles, protons, or carbon ion radiation and its enforcement by combination with immunoadjuvants or inhibitors of immune suppressor cells and checkpoint molecules. Cells. 2021 Jan 25;10(2):228.
47. Lepareur N, Lacœuille F, Bouvry C, Hindré F, Garcion E, Chérel M, Noiret N, Garin E, Knapp Jr FR. Rhenium-188 labeled radiopharmaceuticals: current clinical applications in oncology and promising perspectives. Frontiers in Medicine. 2019 Jun 14;6:132.
48. Bailly C, Bodet-Milin C, Guerard F, Chouin N, Gaschet J, Cherel M, Davodeau F, Faivre-Chauvet A, Kraeber-Bodéré F, Bourgeois M. Radioimmunotherapy of lymphomas. In Nuclear Medicine Therapy 2019 (pp. 113–121). Springer, Cham.
49. Dadachova E, Casadevall A. Treatment of infection with radiolabeled antibodies. QJ Nucl Med Mol Imaging. 2006 Sep 1;50(3):193–204.
50. Aghevlian S, Boyle AJ, Reilly RM. Radioimmunotherapy of cancer with high linear energy transfer (LET) radiation delivered by radionuclides emitting α-particles or Auger electrons. Advanced drug delivery reviews. 2017 Jan 15;109:102–18.
51. Sgouros G, Bodei L, McDevitt MR, Nedrow JR. Radiopharmaceutical therapy in cancer: clinical advances and challenges. Nature Reviews Drug Discovery. 2020 Sep;19(9):589–608.
52. Naskar N, Lahiri S. Theranostic terbium radioisotopes: challenges in production for clinical application. Frontiers in Medicine. 2021 May 31;8:675014.
53. Chédeville AL, Madureira PA. The role of hypoxia in glioblastoma radiotherapy resistance. Cancers. 2021 Feb 1;13(3):542.
54. Valable S, Gérault AN, Lambert G, Leblond MM, Anfray C, Toutain J, Bordji K, Petit E, Bernaudin M, Pérès EA. Impact of hypoxia on carbon ion therapy in glioblastoma cells: Modulation by let and hypoxia-dependent genes. Cancers. 2020 Jul 23;12(8):2019.
55. Wardman P. Radiotherapy using high-intensity pulsed radiation beams (FLASH): a radiation-chemical perspective. Radiation research. 2020 Dec 1;194(6):607–17.
56. Reisz JA, Bansal N, Qian J, Zhao W, Furdui CM. Effects of ionizing radiation on biological molecules – mechanisms of damage and emerging methods of detection. Antioxidants & redox signaling. 2014 Jul 10;21(2):260–92.
57. Martins CD, Kramer-Marek G, Oyen WJ. Radioimmunotherapy for delivery of cytotoxic radioisotopes: current status and challenges. Expert opinion on drug delivery. 2018 Feb 1;15(2):185–96.
58. Bartholomä MD. Radioimmunotherapy of solid tumors: Approaches on the verge of clinical application. Journal of Labelled Compounds and Radiopharmaceuticals. 2018 Jul;61(9):715–26.
59. Shan L, Liu Y, Wang P. Recombinant immunotoxin therapy of solid tumors: challenges and strategies. Journal of basic and clinical medicine. 2013 Jan 1;2(2):1.
60. He B, Ganss R. Modulation of the vascular-immune environment in metastatic cancer. Cancers. 2021 Feb 15;13(4):810.
61. Brighi C, Puttick S, Rose S, Whittaker AK. The potential for remodelling the tumour vasculature in glioblastoma. Advanced Drug Delivery Reviews. 2018 Nov 1;136:49–61.
62. Tafreshi NK, Doligalski ML, Tichacek CJ, Pandya DN, Budzevich MM, El-Haddad G, Khushalani NI, Moros EG, McLaughlin ML, Wadas TJ, Morse DL. Development of targeted alpha particle therapy for solid tumors. Molecules. 2019 Nov 26;24(23):4314.
63. Wahl RL, Sgouros G, Iravani A, Jacene H, Pryma D, Saboury B, Capala J, Graves SA. Normal-tissue tolerance to radiopharmaceutical therapies, the knowns and the unknowns. Journal of Nuclear Medicine. 2021 Dec 1;62(Supplement 3):23S–35S.

64. Cao X, Liang Y, Hu Z, Li H, Yang J, Hsu EJ, Zhu J, Zhou J, Fu YX. Next generation of tumor-activating type I IFN enhances anti-tumor immune responses to overcome therapy resistance. Nature communications. 2021 Oct 7;12(1):1-1.
65. Al-Ejeh F, Shi W, Miranda M, Simpson PT, Vargas AC, Song S, Wiegmans AP, Swarbrick A, Welm AL, Brown MP, Chenevix-Trench G. Treatment of triple-negative breast cancer using anti-EGFR–directed radioimmunotherapy combined with radiosensitizing chemotherapy and PARP inhibitor. Journal of Nuclear Medicine. 2013 Jun 1;54(6):913–21.
66. Wu L, de Perrot M. Radio-immunotherapy and chemo-immunotherapy as a novel treatment paradigm in malignant pleural mesothelioma. Translational Lung Cancer Research. 2017 Jun;6(3):325.
67. Ng B, Kramer E, Liebes L, Wasserheit C, Hochster H, Blank E, Ceriani R, Furmanski P. Radiosensitization of tumor-targeted radioimmunotherapy with prolonged topotecan infusion in human breast cancer xenografts. Cancer research. 2001 Apr 1;61(7):2996–3001.
68. Milenic DE, Garmestani K, Brady ED, Albert PS, Abdulla A, Flynn J, Brechbiel MW. Potentiation of high-LET radiation by gemcitabine: targeting HER2 with trastuzumab to treat disseminated peritoneal disease. Clinical cancer research. 2007 Mar 15;13(6):1926–35.
69. Koppe MJ, Oyen WJ, Bleichrodt RP, Hendriks T, Verhofstad AA, Goldenberg DM, Boerman OC. Combination therapy using the cyclooxygenase-2 inhibitor Parecoxib and radioimmunotherapy in nude mice with small peritoneal metastases of colonic origin. Cancer Immunology, Immunotherapy. 2006 Jan;55(1):47–55.
70. Jang BS, Lee SM, Kim HS, Shin IS, Razjouyan F, Wang S, Yao Z, Pastan I, Dreher MR, Paik CH. Combined-modality radioimmunotherapy: synergistic effect of paclitaxel and additive effect of bevacizumab. Nuclear medicine and biology. 2012 May 1;39(4):472–83.
71. Chandler CS, Bell MM, Chung SK, Veach DR, Fung EK, Punzalan B, Burnes Vargas D, Patel M, Xu H, Guo HF, Santich BH. Intraperitoneal Pretargeted Radioimmunotherapy for Colorectal Peritoneal CarcinomatosisIntraperitoneal PRIT for Peritoneal Carcinomatosis. Molecular Cancer Therapeutics. 2022 Jan 1;21(1):125–37.
72. Keinänen O, Fung K, Brennan JM, Zia N, Harris M, van Dam E, Biggin C, Hedt A, Stoner J, Donnelly PS, Lewis JS. Harnessing 64Cu/67Cu for a theranostic approach to pretargeted radioimmunotherapy. Proceedings of the National Academy of Sciences. 2020 Nov 10;117(45):28316–27.
73. Reardan DT, Meares CF, Goodwin DA, McTigue M, David GS, Stone MR, Leung JP, Bartholomew RM, Frincke JM. Antibodies against metal chelates. Nature. 1985 Jul;316(6025):265–8.
74. Jallinoja VI, Houghton JL. Current Landscape in Clinical Pretargeted Radioimmunoimaging and Therapy. Journal of Nuclear Medicine. 2021 Sep 1;62(9):1200–6.
75. Kasten BB, Ferrone S, Zinn KR, Buchsbaum DJ. B7-H3-targeted radioimmunotherapy of human cancer. Current medicinal chemistry. 2020 Jul 1;27(24):4016–38.
76. Modak S, Zanzonico P, Grkovski M, Slotkin EK, Carrasquillo JA, Lyashchenko SK, Lewis JS, Cheung IY, Heaton T, LaQuaglia MP, Cheung NK. B7H3-directed intraperitoneal radioimmunotherapy with radioiodinated omburtamab for desmoplastic small round cell tumor and other peritoneal tumors: results of a phase I study. Journal of Clinical Oncology. 2020 Dec 12;38(36):4283.
77. Kramer K, Humm JL, Souweidane MM, Zanzonico PB, Dunkel IJ, Gerald WL, Khakoo Y, Yeh SD, Yeung HW, Finn RD, Wolden SL. Phase I study of targeted radioimmunotherapy for leptomeningeal cancers using intra-Ommaya 131-I-3F8. Journal of Clinical Oncology. 2007 Dec 1;25(34):5465–70.
78. Kramer K, Pandit-Taskar N, Humm JL, Zanzonico PB, Haque S, Dunkel IJ, Wolden SL, Donzelli M, Goldman DA, Lewis JS, Lyashchenko SK. A phase II study of radioimmunotherapy with intraventricular 131I-3F8 for medulloblastoma. Pediatric blood & cancer. 2018 Jan;65(1):e26754.
79. Yerrabelli RS, He P, Fung EK, Kramer K, Zanzonico PB, Humm JL, Guo H, Pandit-Taskar N, Larson SM, Cheung NK. IntraOmmaya compartmental radioimmunotherapy using

131I-omburtamab – pharmacokinetic modeling to optimize therapeutic index. European journal of nuclear medicine and molecular imaging. 2021 Apr;48(4):1166–77.
80. Pandit-Taskar N, Zanzonico PB, Kramer K, Grkovski M, Fung EK, Shi W, Zhang Z, Lyashchenko SK, Fung AM, Pentlow KS, Carrasquillo JA. Biodistribution and dosimetry of intraventricularly administered 124I-omburtamab in patients with metastatic leptomeningeal tumors. Journal of Nuclear Medicine. 2019 Dec 1;60(12):1794–801.

Index

C. Somboonwit et al. (eds.), *Global Virology IV: Viral Disease Diagnosis and Treatment Delivery in the 21st Century*,
https://doi.org/10.1007/978-3-031-57369-9

GPSR Compliance
The European Union's (EU) General Product Safety Regulation (GPSR) is a set of rules that requires consumer products to be safe and our obligations to ensure this.

If you have any concerns about our products, you can contact us on

ProductSafety@springernature.com

In case Publisher is established outside the EU, the EU authorized representative is:

Springer Nature Customer Service Center GmbH
Europaplatz 3
69115 Heidelberg, Germany

www.ingramcontent.com/pod-product-compliance
Lightning Source LLC
Chambersburg PA
CBHW072003110925
32461CB00003B/32
9783031573682